Joint/Interagency SMARTBOOK

thelightningpress.com

joint strategic & operational PLANNING

Planning for Planners

The Lightning Press
Santacroce

The Lightning Press

2227 Arrowhead Blvd.
Lakeland, FL 33813
24-hour Voicemail/Fax/Order: 1-800-997-8827
E-mail: SMARTbooks@TheLightningPress.com

www.TheLightningPress.com

Joint/Interagency SMARTbook 1: Joint Strategic & Operational Planning

Planning for Planners

ISBN: 978-1-935886-58-7

Notice of Liability

The information in this SMARTbook and quick reference guide is distributed on an "As Is" basis, without warranty. While every precaution has been taken to ensure the reliability and accuracy of all data and contents, neither the author nor The Lightning Press shall have any liability to any person or entity with respect to liability, loss, or damage caused directly or indirectly by the contents of this book. If there is a discrepancy, refer to the source document.

All references used to compile the SMARTbooks are in the public domain and are available to the general public through official public websites and designated as approved for public release with unlimited distribution. The SMARTbooks do not contain ITAR-controlled technical data, classified, or other sensitive material restricted from public release. SMARTbooks are reference books that address general military principles, fundamentals and concepts rather than technical data or equipment operating procedures.

"The views presented in this publication are those of the author and do not necessarily represent the views of the Department of Defense or its components."

Printed and bound in the United States of America.

Note to Readers

The dawn of the 21st Century presents a global environment characterized by regional instability, failed states, increased weapons proliferation, global terrorism and unconventional threats to United States citizens, interests and territories. If we are to be successful as a nation, we must embrace the realities of this environment and operate with clarity from within. It is this setting that mandates a flexible, adaptive approach to planning and an ever-greater cooperation between all the elements of national power, supported by and coordinated with, that of our allies and various intergovernmental, nongovernmental and regional security organizations. It is within this chaotic environment that planners will craft their trade.

Joint/Interagency SMARTbook 1: Joint Strategic & Operational Planning (Planning for Planners) was developed to assist planners at all levels in understanding how to plan within this environment utilizing the Joint Operational Planning Process; an orderly, logical, analytical progression enabling planners to sequentially follow it to a rational conclusion. By utilizing this planning process, which is conceptually easy-to-understand and applicable in all environments, any plan can come to life.

This new revision of *Planning for Planners* incorporates the latest thinking on Adaptive Planning and Execution (APEX), Global Force Management (GFM), Campaign Planning and Assessment Fundamentals. *Planning for Planners* has been utilized since 2007 by war colleges, joint staffs, Services, combatant commands and allies as a step-by-step guide to understanding the complex world of global planning and force management. Paramount to planning is flexibility. The ultimate aspiration of this book is to help develop flexible planners who can cope with the inevitable changes that occur during the planning process.

Military Reference and National Power SMARTbooks

In addition to *The Joint Forces Operations & Doctrine SMARTbook*, The Lightning Press offers three specific Joint/Interagency SMARTbooks in it's National Power series, plus more than a dozen related and supporting reference titles:

SMARTbooks - DIME is our DOMAIN!

SMARTbooks: Reference Essentials for the Instruments of National Power (D-I-M-E: Diplomatic, Informational, Military, Economic)! Recognized as a "whole of government" doctrinal reference standard by military, national security and government professionals around the world, SMARTbooks comprise a comprehensive professional library designed with all levels of Soldiers, Sailors, Airmen, Marines and Civilians in mind.

SMARTbooks can be used as quick reference guides during actual operations, as study guides at education and professional development courses, and as lesson plans and checklists in support of training. Visit **www.TheLightningPress.com**!

Introduction: Planning for Planners

Today's global environment is characterized by regional instability, failed states, increased weapons proliferation, global terrorism, and unconventional threats to United States citizens, interests, and territories. If we are to be successful as a nation, we must embrace the reality of this environment and operate from within with clarity. It is this environment that mandates a flexible, adaptive approach to planning and an ever greater cooperation between all the elements of national power supported by and coordinated with that of our allies and various intergovernmental, nongovernmental, and regional security organizations. It is for this reason that *Planning for Planners* was developed.

The criteria for deciding to employ United States military forces exemplify the dynamic link among the people, the government, and the military. The people of the United States do not take the commitment of their armed forces lightly. They charge the government to commit forces only after due consideration of the range of options and likely outcomes. Moreover, the people expect the military to accomplish its missions in compliance with national values. The American people expect decisive victory and abhor unnecessary casualties. They prefer quick resolution of conflicts and reserve the right to reconsider their support should any of these conditions not be met. They demand timely and accurate information on the conduct of military operations.

> *"One should know one's enemies, their alliances, their resources and nature of their country, in order to plan a campaign. One should know what to expect of one's friends, what resources one has, and foresee the future effects to determine what one has to fear or hope from political maneuvers."*
>
> *Frederick the Great*
> *Instructions for His Generals, 1747*

The responsibility for the conduct and use of United States military forces is derived from the people and loaned to the government. The Department of Defense commits forces only after appropriate direction from the President and in support of national strategy. The national strategy of the United States dictates where, when, and with what means the armed forces will conduct military campaigns and operations. The necessity to plan and conduct joint and combined operations across the operational continuum dictates a comprehensive understanding of the military strategy of the United States, and proficiency in current Service and joint *doctrine*.

Never static, always dynamic, *doctrine* is firmly rooted in the realities of current capabilities. At the same time, it reaches out with a measure of confidence to the future. Doctrine captures the lessons of past wars, reflects the nature of war, conflict and crisis in its own time, and anticipates the intellectual and technological developments that will ensure victory now and in the future.

Doctrine derives from a variety of sources that profoundly affect its development: strategy, history, technology, the nature of the threats the nation and its armed forces face, inter-service relationships, and political decisions that allocate resources and designate roles and missions. Doctrine seeks to meet the challenges facing the armed forces by providing the guidance to deal with the range of threats to which its elements may be exposed. It reflects the strategic context in which armed forces will operate, sets a marker for the incorporation of developing technologies, and optimizes the use of all available resources. It also incorporates the lessons learned from the many missions, operations and campaigns of the United States.

Scenario

You're new on a Joint Operations staff for a geographic combatant command. It's 0200 and the phone rings. The Chief of Staff is on the phone and relays the following Warning Order to you:

> *"A magnitude 7.6 earthquake has struck parts of Pakistan and India. By far the biggest in its magnitude and scale that we've seen in current history. Pakistan has an estimated 79,000+ injured, 73,000+ dead with possibly 3 million displaced and homeless. Have an initial mission analysis brief for the Combatant Commander by 0700…….. GO!"*

You suddenly feel the full affect of the proverbial "planning fire hose."

What do you do?

Where do you begin?

#1- Take a breath.

#2- Pick up your well-worn and dove-tailed "Planner's SMARTbook" and get to work.

#3- Delegate!

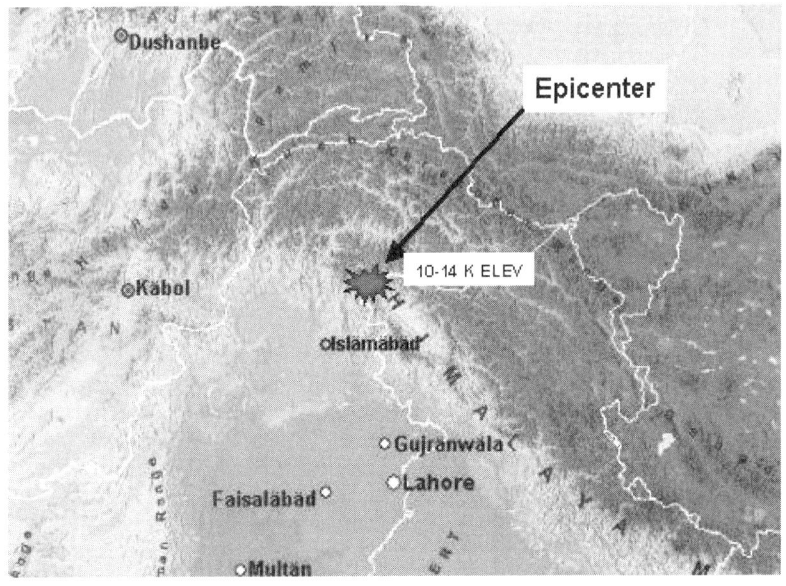

Continued on next page

Introduction (Cont.)

Doctrinal principles set forth in planning are developed and written as the starting point for any variation or deviation from the planning process. One must understand doctrine prior to digressing from it. As noted by Dr. Douglas V. Johnson II at the Strategic Studies Institute in his article *Doctrine that Works*, "doctrine should set forth principles and precious little more."[1] With that thought in mind *Planning for Planners* was designed to promulgate information from several source documents and utilize best practices to fill in where the principles of Joint doctrine departs.

Planning for Planners presents the planning process as described by doctrine, best practices and *common sense* while focusing on the concepts of operational planning and the often misunderstood world of global force management. It is presented in a logical flow which will enable planners to sequentially follow the process to a logical conclusion.

Baron von Steuben's 1779 *Regulations for the Order and Discipline of the Troops of the United States* was not penned in a setting of well-ordered formations and well-disciplined troops but, at a time of turmoil during a winter at Valley Forge. Baron von Steuben's doctrine, maybe our first written doctrine, set forth principles and created a discipline that went on to defeat the greatest army on the face of the earth. This doctrine, written over 200 years ago, and followed by others, has led to a highly professional armed force that generations later stands foremost in the world. Doctrine reflects the collective wisdom of our armed forces against the background of history and it reflects the lessons learned from recent experiences and the setting of today's strategic and technological realities. It considers the nature of today's threats and tomorrows challenges.[2]

Joint/Interagency SMARTbook 1 will assist planners at all levels with these challenges. It will furnish the planner with an understanding of doctrine and the intricate world of global planning. The ultimate aspiration being to develop planners who can cope with the inevitable change that occurs during the planning process.

Joint Operation Planning

Joint operation planning is the overarching process that guides us in developing plans for the employment of forces and capabilities within the context of national strategic objectives and national military strategy to shape events, meet contingencies, and respond to unforeseen crises. *Planning for Planners* focuses on this joint operation planning and the global force management processes.

The Joint Operation Planning Process is an orderly, analytical planning process consisting of a set of logical steps to analyze a mission, develop and compare potential courses of action, select the best course of action, and produce a plan or order. This planning process underpins planning at all levels and for missions across the full range of contingencies. It applies to all planners and helps them organize their planning activities, share a common understanding of the method, purpose and end state and to develop effective plans and executable orders.

Planning provides an awareness and opportunity to study potential future events amongst multiple alternatives in a controlled environment. By planning we can evaluate complex systems and environments allowing us to break these down into small, manageable segments for analysis, assisting directly in the increased probability of success. In this way, deliberately planning for contingencies allows us to manage identified

Continued from previous page

Continued from previous page

[1] *Dr. Douglas V. Johnson II, Strategic Studies Institute, Doctrine That Works, www.StrategicStudiesInstitute.army.mil/pdffiles/pub724.pdf.*

[2] *FM 100-5 Operations, Headquarters, Department of the Army.*

risks and influence the operational environment in which we have chosen to interact, in a deliberate way. The plans generated in this process represent actions to be taken if an identified risk occurs or a trigger event has presented itself.

The variance in any plan is the constant change in the operational environment (system). Whether a crisis or contingency scenario, we plan in a chaotic environment. In the time it takes us to plan, the likelihood that the operational environment has changed is a certain, whether by action or inaction, affecting the plan (i.e., assumptions change or are not validated, leaders change, the operational environment fluctuates, apportionment tables are poor assumptions, disputed borders fluctuate, weather changes the rules, plans change at contact, enemy gets a vote, etc.).

Variables are hard to predict because each environment and situation have their own unique challenges which can certainly affect an orderly plan. Given the size and scope of an operational environment a plan can only anticipate, or forecast, for a short duration without being updated. This is known as the plans horizon. In a fluid crisis situation the plans horizon may be very short and contain greater risk, causing the planner to constantly re-evaluate and update the plan. Inversely, for a contingency plan, the plans horizon may be relatively static with less risk allowing time for greater analysis. The number of variables within the operational environment and the interactions between those variables and known components of the operational environment increases exponentially with the number of variables, thus potentially allowing for many new and sometimes subtle planning changes to emerge.

As an example of a plans horizon, or stability, lets look at an environment that constantly influences us; the weather:

A forecaster endeavors to anticipate the path of a tropical cyclone and utilizes historical models and probabilities to predict the tropical cyclones path and warn residents. When a low pressure area first forms and the storm begins to take shape along the equator, forecasters are working within a complex environment with constant and multiple variables (i.e., winds, temperatures, currents, pressures, etc.) and few facts (i.e., exact location at this moment, jet stream location, etc.). As variables amplify and the storm begins to move the storm's horizon shifts yet again, and the forecaster updates the assessment. Over days of surveillance, gathering information, updating, and studying the variables, the actual track of the storm begins to emerge and the storms horizon becomes more durable and predictable. The forecaster continuously narrows the storm's estimated track, eventually forecasting with some certainty the tropical cyclones land fall.

Planners employ the same technique by utilizing current knowledge of the operational environment to anticipate events, calculate what those may be by means of an in-depth analysis, update, and plan accordingly. But always remember, plans are orderly; probabilities and variables are not. Just as a tropical storm has a self-organizing phase within its environment, so must the planner.

So the challenge is how to plan within an environment with continuously changing and emerging variables. The planner must understand that every plan is unique and never as perfect as you want it, there are too many variables. But with constant awareness each iteration of the plan will improve the prospect of success as the variables become known and are planned for.

Simplicity should be the aspiration for every plan. Prepare clear, uncomplicated plans and concise orders to ensure a through understanding. A plan need not be more complicated than the underlying principles which generate it.

About the Author

Michael A. Santacroce has 35 years of joint and interagency experience working within the Department of Defense as a Joint Staff, Combatant Command and Service Planner. As Faculty and Chair for the Joint Advanced Warfighting School, Campaign Planning and Operational Art, Mike taught advanced planning to leaders from all branches of the government and our allies. His current SMARTbook, *Planning for Planners*, walks the prospective or advanced planner through joint strategic and operational planning as well as the complex world of global force management.

During his Marine Corps career Mike served in a multiple of demanding leadership, senior staff, strategic and operational planning positions. As a Marine aviator he flew the AV-8B Harrier Jump Jet and participated in operations globally. He commanded a Marine Harrier Squadron (VMA-214 Blacksheep) and later led a Marine Air Group (Forward) for combat operations in Iraq. A seasoned military professional and teacher, Mike has a unique understanding of operations and planning at all levels. Mike retired with more than 30 years of military service.

> *"The inspiration of a noble cause involving human interests wide and far, enables men to do things they did not dream themselves capable of before, and which they were not capable of alone."*
>
> *Joshua L. Chamberlain, October 3, 1889. Monument dedication ceremony, Gettysburg, Pa.*

Today's preparation determines tomorrow's achievements. Dedicated to all planners; may this work assist you in some small way.

> *Planning for Planners is reviewed continually and updated as required. Point of contact is the author, Col (Ret) Mike Santacroce, USMC, at mike.santacroce@thelightningpress.com*

Table of Contents

0

Chap 17 – COA Analysis and Wargaming......17-1

Chap 18 – COA Comparison......18-1

Chap 19 – COA Selection & Approval......19-1

Appendixes

Strategic Organization 1

"Our Nation's cause has always been larger than our Nation's defense. We fight, as we always fight, for a just peace - a peace that favors liberty. We will defend the peace against the threats from terrorists and tyrants. We will preserve the peace by building good relations among the great powers. And we will extend the peace by encouraging free and open societies on every continent."

President Bush, West Point, New York, June 1, 2002

1. Background

a. Civilian Control of the Military. Since the founding of the nation, civilian control of the military has been an absolute and unquestioned principle. The Constitution incorporates this principle by giving both the President and Congress the power to ensure civilian supremacy. The *Constitution* establishes the President as the Commander-in-Chief, but gives the Congress the power "to declare war," to "raise and support Armies – provide and maintain a Navy – (and) to make rules for the government and regulation of the land and naval forces."

b. Joint Organization before 1900. As established by the Constitution, coordination between the War Department and Navy Department was effected by the President as the Commander in Chief. Army and naval forces functioned autonomously with the President as their only common superior. Despite Service autonomy, early American history reflects the importance of joint operations. Admiral MacDonough's naval operations on Lake Champlain were a vital factor in the ground campaigns of the War of 1812. The joint teamwork displayed by General Grant and Admiral Porter in the Vicksburg Campaign of 1863 stands as a fine early example of joint military planning and execution. However, instances of confusion, poor inter-Service cooperation and lack of coordinated, joint military action had a negative impact on operations in the Cuban campaign of the Spanish-American War (1898). By the turn of the century, advances in technology and the growing international involvement of the United States required greater cooperation between the military departments.

c. Joint History through World War I. As a result of the unimpressive joint military operations in the Spanish-American War, in 1903 the Secretary of War and the Secretary of the Navy created the Joint Army and Navy Board charged to address "all matters calling for cooperation of the two Services." The Joint Army and Navy Board was to be a continuing body that could plan for joint operations and resolve problems of common concern to the two Services. Unfortunately, the Joint Board accomplished little, because it could not direct implementation of concepts or enforce decisions, being limited to commenting on problems submitted to it by the secretaries of the two military departments. It was described as "a planning and deliberative body rather than a center of executive authority." As a result, it had little or no impact on the conduct of joint operations during the First World War. Even as late as World War I, questions of seniority and command relationships between the Chief of Staff of the Army and American Expeditionary Forces in Europe were just being resolved.

d. Joint History through World War II. After World War I, the two Service secretaries agreed to reestablish and revitalize the Joint Board. Membership was expanded to six: the chiefs of the two Services, their deputies, the Chief of War Plans Division for the Army and Director of Plans Division for the Navy. More importantly, a working staff (named the Joint Planning Committee) made up of members of the plans divisions of both Service staffs was authorized. The new Joint Board could initiate recommendations on its own. Unfortunately, the 1919 board was given no more legal authority or responsibility than its 1903 predecessor; and, although its 1935 publication, *Joint Action Board of the Army and Navy*, gave some guidance for the unified operations of World War II, the board itself was not influential in the war. The board was officially disbanded in 1947.

2. The Security Envronment

Today's security environment is not unlike those of historic times. The commanders during those eras considered the enemy extremely complex and fluid with continually changing coalitions, alliances, partnerships, and new threats constantly appearing and disappearing. Today, with the national and transnational threats we face, our political and military leaders conduct operations in an ever-more complex, interconnected, and increasingly global operational environment. This increase in the scope of the operational environment may not necessarily result from actions by the confronted adversary alone, but is likely to result from other adversaries exploiting opportunities as a consequence of an overextended or distracted United States (U.S.) or coalition. These adversaries encompass a variety of actors from transnational organizations to states or even ad hoc state coalitions and individuals.

To prepare the United States for today's threats and contingencies we have, over time, established a system of checks and balances to include numerous governmental organizations that are involved in the implementation of United States security policy. However, constitutionally, the ultimate authority and responsibility for the national defense rests with the President.

> *"As in a building, which, however fair and beautiful, the superstructure is radically marred and imperfect if the foundation be insecure-so, if the strategy be wrong, the skill of the general on the battlefield, the valor of the soldier, the brilliancy of victory, however otherwise decisive, fail of their effect."*
>
> *-A.T. Mahan*

Joint Operations, Unified Action, & the Range of Military Operations (ROMO)

Ref: JP 3-0, Joint Operations (Aug '11), chap. 1.

Services may accomplish tasks and missions in support of Department of Defense (DOD) objectives. However, the DOD primarily employs two or more services in a single operation, particularly in combat, through joint operations. The general term, joint operations, describes military actions conducted by joint forces or by Service forces employed under command relationships. A joint force is one composed of significant elements, assigned or attached, of two or more military departments operating under a single joint force commander. Joint operations exploit the advantages of interdependent Service capabilities through unified action, and joint planning integrates military power with other instruments of national power to achieve a desired military end state.

Unified Action

Whereas the term joint operations focuses on the integrated actions of the Armed Forces of the United States in a unified effort, the term unified action has a broader connotation. JFCs are challenged to achieve and maintain operational coherence given the requirement to operate in conjunction with interorganizational partners. CCDRs play a pivotal role in unifying joint force actions, since all of the elements and actions that comprise unified action normally are present at the CCDR's level. However, subordinate JFCs also integrate and synchronize their operations directly with the operations of other military forces and the activities of nonmilitary organizations in the operational area to promote unified action. Unified action is a comprehensive approach that synchronizes, coordinates, and when appropriate, integrates military operations with the activities of other governmental and nongovernmental organizations to achieve unity of effort.

The Range of Military Operations (ROMO)

The range of military operations is a fundamental construct that provides context. Military operations vary in scope, purpose, and conflict intensity across a range that extends from military engagement, security cooperation, and deterrence activities to crisis response and limited contingency operations and, if necessary, to major operations and campaigns. Use of joint capabilities in military engagement, security cooperation, and deterrence activities helps shape the operational environment and keep the day-to-day tensions between nations or groups below the threshold of armed conflict while maintaining US global influence.

- **Military Engagement, Security Cooperation, and Deterrence.** These ongoing activities establish, shape, maintain, and refine relations with other nations and domestic civil authorities (e.g., state governors or local law enforcement). The general strategic and operational objective is to protect US interests at home and abroad.
- **Crisis Response & Limited Contingency Operations.** A crisis response or limited contingency operation can be a single small-scale, limited-duration operation or a significant part of a major operation of extended duration involving combat.
- **Major Operations and Campaigns.** When required to achieve national strategic objectives or protect national interests, the US national leadership may decide to conduct a major operation or campaign normally involving large-scale combat.

Refer to The Joint Forces Operations & Doctrine SMARTbook (Guide to Joint, Multinational & Interagency Operations) for complete discussion of joint operations, unified action, and the range of military operations. Additional topics include joint doctrine fundamentals, joint operation planning, joint logistics, joint task forces, information operations, multinational operations, and IGO/NGO coordination.

3. National Strategic Direction

The common thread that integrates and synchronizes the activities of the Joint Staff (JS), combatant commands (CCMDs), Services, and Combat Support Agencies (CSAs) is strategic direction. As an overarching term, strategic direction encompasses the processes and products by which the President, Secretary of Defense (SECDEF), and Chairman of the Joint Chiefs of Staff (CJCS) provide strategic guidance. Combatant Commanders (CCDRs), once provided the direction and guidance each prepares strategy and campaign plans in the context of national security and foreign policy goals. These strategic guidance documents are the principle source for DOD global campaign plans, theater strategies, CCMD campaign plans, operation plans, contingency plans, base plans, and commanders estimates.

Figure I-1. National Strategic Direction (JP5-0).

a. The President provides strategic guidance through the National Security Strategy (NSS), Presidential Policy Document (PDD), and other strategic documents in conjunction with additional guidance from other members of the National Security Council (NSC)[4] (Figure I-1).

b. The President and SECDEF, through the CJCS, direct the national effort that supports combatant and subordinate commanders. The principal forum for deliberation of national security policy issues requiring Presidential Decisions (PDs) that will directly affect the CCDRs actions is the NSC. Knowledge of the history and relationships between elements of the national security structure is essential to understanding the role of JS organizations.

[4] *Joint Pub 5-0, Joint Operations Planning.*

c. <u>The National Security Council System (NSC).</u> Department of Defense (DOD) participation in the interagency process is grounded within the Constitution and established by law in the National Security Act of 1947 (NSA 47).

(1) The NSC is a product of NSA 47. NSA 47 codified and refined the interagency process used during World War II, modeled in part on Franklin D. Roosevelt's 1919 proposal for a "Joint Plan-Making Body" to deal with the overlapping authorities of the Departments of State, War, and Navy. Because of the diverse interests of individual agencies, previous attempts at interagency coordination failed due to lack of national-level perspectives, a staff for continuity, and adequate appreciation for the need of an institutionalized coordination process. Evolving from the World War II experience (during which the Secretary of State was not invited to War Council meetings), the first State-War-Navy Coordinating Committee was formed in 1945.

(a) From the earliest days of this nation, the President has had the primary responsibility for national security stemming from his constitutional powers both as Commander-in-Chief of the Armed Forces and his authority to make treaties and appoint cabinet members and ambassadors. The intent of NSA 47 was to assist the President with respect to the integration of domestic, foreign, and military policies relating to national security. Most current United States Government (USG) interagency actions flow from these beginnings.

(b) Within the constitutional and statutory system, interagency actions at the national level may be based on both personality and process, consisting of persuasion, negotiation, and consensus building, as well as adherence to bureaucratic procedure.

(2) The NSC is the principal forum for deliberation of national security policy issues requiring Presidential decision. The NSC advises and assists the President in integrating all aspects of national security policy — domestic, foreign, military, intelligence, and economic (in conjunction with the National Economic Council).

(a) Together with supporting interagency working groups (some permanent and others ad hoc), high-level steering groups, executive committees, and task forces, the National Security Council System (NSCS) provides the foundation for interagency coordination in the development and implementation of national security policy. The NSC develops policy options, considers implications, coordinates operational problems that require interdepartmental consideration, develops recommendations for the President, and monitors policy implementation. The national security staff is the President's principal staff for national security issues and also serves as the President's principal arm for coordinating these policies among various government agencies. NSC documents are established to inform USG departments and agencies of Presidential actions (JP 3-08, *Interorganizational Coordination During Joint Operations*).

(b) Each administration typically adopts different names for its NSC documents. For example, The Reagan Administration used the terms NSDD (National Security Decision Directive) and NSSD (National Security Study Directive). The George H. W. Bush Administration used NSR (National Security Review) and NSD (National Security Directive), while the Clinton Administration used the terms PDD (Presidential Decision Directive) and PRD (Presidential Review Directive). The George W. Bush administration used NSPD (National Security Presidential Directive) and Homeland Security Presidential Directive (HSPD). The Obama Administration uses the terms PPD and PSD (Presidential Study Directive).

(3) <u>National Security Council Membership.</u> The President chairs the NSC. As prescribed in PDD, the NSC shall have as its regular attendees (both statutory and non-statutory) the President, the Vice President, the Secretary of State, the Secretary of the Treasury, the Secretary of Defense, the Secretary of Homeland Security, the Secretary of Energy, and the Assistant to the President for National Security Affairs. The Director of Central Intelligence and the CJCS, as statutory advisors to the NSC, shall also attend NSC meetings. The Chief of Staff to the President and the Assistant to the President for Economic Policy are invited to attend any NSC meeting. The Counsel to the President shall be consulted regarding the agenda of NSC meetings, and shall attend any meeting when, in consultation with the Assistant to the President for National Security Affairs, he deems it appropriate.

(4) NSC Organization (Figure I-2). The members of the NSC constitute the President's personal and principal staff for national security issues. The council tracks and directs the development, execution, and implementation of national security policies for the President, but does not normally implement policy. Rather, it takes a central coordinating or monitoring role in the development of policy and options, depending on the desires of the President and the National Security Advisor. There are three levels of formal interagency committees for coordinating and making decisions on national security issues. The advisory bodies include:

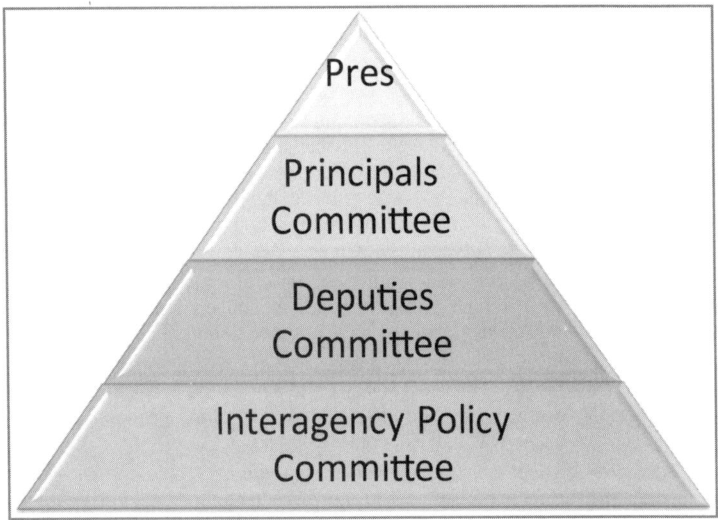

Figure I-2. National policy-making process is built on consensus.

(a) The NSC Principals Committee (NSC/PC) is the senior Cabinet-level interagency forum for consideration of policy issues affecting national security. The Principals Committee meets at the call of and is chaired by the National Security Advisor.

(b) The NSC/Deputies Committee (NSC/DC) is the senior sub-Cabinet-level (deputy secretary-level) interagency forum for consideration of policy issues affecting national security. The NSC/DC prescribes and reviews the work of the NSC Interagency Policy Committee (NSC/IPC). The NSC/DC ensures that NSC/IPC issues have been properly analyzed and prepared for decision. The NSC/DC shall focus significant attention on policy implementation. Periodic reviews of the Administration's major foreign policy initiatives shall be scheduled to ensure that they are being implemented in a timely and effective manner. Such reviews should periodically consider whether existing policy directives should be revamped or rescinded. Finally, the NSC/DC shall be responsible for day-to-day crisis management, reporting to the NSC. Any NSC principal or deputy, as well as the National Security Advisor, may request a meeting of the NSC/DC in its crisis management capacity. The Deputies Committee meets at the call of and is chaired by the Deputy National Security Advisor.

(c) NSC/IPCs are the main day-to-day action committees for interagency coordination of national security policy. NSC/IPCs manage the development and implementation of national security policies by multiple agencies of the USG, provide policy analysis for consideration by the more senior committees of the NSCS, and ensure timely responses to decisions made by the President. The NSC/IPCs shall be established at the direction of the NSC/DC, and be chaired by the NSC (or NEC, as appropriate); at its discretion, the NSC/DC may add co-chairs to any NSC/IPC if desirable. The NSC/IPCs shall convene on a regular basis to review and coordinate the implementation of Presidential decisions in their policy areas. Strict guidelines shall be established governing the operation of the IPCs, including participants, decision-making paths, and time frame.

1 Six NSC/IPCs are established for the following regions: Europe and Eurasia, Western Hemisphere, East Asia, South Asia, Near East and North Africa, and Africa.

2 Each of the NSC/IPCs shall be chaired by an official of Under Secretary or Assistant Secretary rank to be designated by the Secretary of State. The oversight of ongoing operations assigned by the Deputies Committee is performed by the appropriate NSC/IPCs, which may create subordinate working groups. Each NSC/IPC includes representatives from the executive departments, and offices and agencies represented in the NSC/DC. Additional NSC/IPCs may be established as appropriate by the President or the National Security Advisor.

3 Functional NSC/IPCs are established for specific purposes as issues or crises arise and for developing long-term strategies. There are multiple functional NSC/IPCs each chaired by a person of Under Secretary or Assistant Secretary rank designated by the indicated authority. [5]

(d) During a rapidly developing crisis, the President may request the National Security Advisor to convene the NSC. The NSC reviews the situation, determines a preliminary COA, and tasks the Principals and Deputies Committees.

(e) Under more routine conditions, concerns focus on broader aspects of national policy and long-term strategy perspectives. National Security Presidential Directives outline specific national interests, overall national policy objectives, and tasks for the appropriate components of the executive branch.

(5) DOD Role in the National Security Council System.

(a) Key DOD players in the NSCS come from within the Office of the Secretary of Defense (OSD) and the JS. The SECDEF is a regular member of the NSC and the NSC/PC. The Deputy Secretary of Defense is a member of the NSC/DC. In addition to membership, an Under Secretary of Defense may chair a NSC/IPC.

(b) *The NSCS is the channel for the CJCS to discharge substantial statutory responsibilities as the principal military advisor to the President, the SECDEF, and the NSC.* The CJCS regularly attends NSC meetings and provides advice and views in this capacity. The other members of the Joint Chiefs of Staff (JCS) may submit advice or an opinion in disagreement with that of the CJCS, or advice or an opinion in addition to the advice provided by the CJCS.

(c) The Military Departments which implement, but do not participate directly in national security policy-making activities of the interagency process, are represented by the CJCS.

(d) Of note and worth mentioning here are the geographic boundary differences between the DOS Bureaus and the DOD geographic commands. It's important we recognize these seams and boundary differences to ensure smooth coordination between these two interagency partners, see Chapter 3.

(6) The Joint Staff Role in the National Security Council System.

(a) The JS provides operational input and staff support through the CJCS (or designee) for policy decisions made by the Office of the Secretary of Defense (OSD). It coordinates with the CCMDs, Services, and other agencies and prepares appropriate directives, such as warning, alert, and execute orders, for SECDEF approval. This preparation includes definition of command and interagency relationships.

(b) When CCMDs require interagency coordination, the JS, in concert with the OSD, routinely accomplishes that coordination.

(c) Within the JS, the offices of the CJCS, Secretary of the JS, Operations (J3), Logistics (J4), Plans and Policy (J5), and Operational Plans and Joint Force Development Directorates are focal points for NSC-related actions. The J3 provides advice on execution

[5] *National Security Presidential Directive 1 (NSPD-1)*

of military operations, the J4 assesses logistic implications of contemplated operations, and the J5 often serves to focus DOD on a particular NSC matter for policy and planning purposes and as the Joint Staff touch point for Promote Cooperation (PC), as OSD/Joint Staff program to coordinate plans with other agencies. Each of the JS directorates coordinates with the Military Departments to solicit Service input in the planning process. The SECDEF may also designate one of the Services as the executive agent for direction and coordination of DOD activities in support of specific mission areas.

(7) The CCDRs' Role in the National Security Council System. Although CCDRs sometimes participate directly in the interagency process by directly communicating with committees and groups of the NSC system and by working to integrate the military with diplomatic, economic, and informational instruments of national power, the normal conduit for information between the President, SECDEF, NSC, and a CCMD is the CJCS. CCDRs may communicate with the Deputies Committee during development of the POLMIL plan with the Joint Staff in a coordinating role.

4. National Strategic Guidance

Joint plans and orders are developed with the strategic and military objectives and end states in mind. The commander and planners derive their understanding of those end states from strategic guidance.

a. The National Security Strategy (NSS). The NSS is a comprehensive report required annually by Title 50, USC, Section 404a. It is prepared by the executive branch of the government for Congress and outlines the major national security concerns of the US and how the administration plans to address them using all instruments of national power. The document is purposely general in content, and its implementation relies on elaborating guidance provided in supporting documents (such as the National Defense Strategy (NDS), the Guidance for Employment of the Force (GEF), and the National Military Strategy (NMS).

b. National Defense Strategy (NDS). The NDS addresses how the Armed Forces of the United States will fight and win America's wars and describes how DOD will support the objectives outlined in the NSS. It also provides a framework for other DOD strategic guidance, specifically on deliberate planning, force development, and intelligence.

c. Quadrennial Defense Review (QDR). The QDR articulates a national defense strategy consistent with the most recent NSS by defining force structure, modernization plans, and a budget plan allowing the military to successfully execute the full range of missions within that strategy.

d. Unified Command Plan (UCP). The UCP establishes CCMD missions and responsibilities; addresses assignment of forces; delineates geographic areas of responsibility for geographic combatant commanders (GCCs); and specifies responsibilities for functional combatant commanders (FCCs).

e. Guidance for Employment of the Force[6]. The GEF provides two-year direction to CCMDs for operational planning, force management, security cooperation, and posture planning. The GEF is the method through which OSD translates strategic priorities set in the NSS, NDS, and QDR into implementable direction for operational activities. In 2008 a transition in national guidance began with a new strategic guidance hierarchy that includes the GEF and Guidance for Development of the Force (GDF), with an aligned JSCP. The GEF fulfills the SECDEF's responsibility under Title 10 USC to provide the CJCS with "written policy guidance for the preparation and review of contingency plans, every two years or more frequently as needed." The GEF issues both the President's guidance for contingency planning, and conveys the SECDEF's guidance for near-term (2-5 years), steady state activities and defense posture. It translates the national security objectives and high-level

[6]*Guidance for Employment of the Force (GEF)*

GEF: Providing the "What" (policy guidance for planning).

- *Strategic endstates (theater or functional) for campaign planning.*
- *Prioritized deliberate planning scenarios and endstates.*
- *Global posture and global force management guidance.*
- *Relative security cooperation and global force management priorities.*
- *Strategic assumptions.*
- *Provides overarching DOD and USG nuclear policy.*

JSCP: Implementing the "What" (provides plan guidance).

- *Formally tasks campaign, deliberate, and posture planning requirements to CCMDs.*
- *Detailed planning tasks and considerations for campaign, deliberate, and posture plans.*
- *Specifies type of plan required for deliberate plans.*
- *Planning assumptions.*
- *Provides detailed guidance to apportion forces.*
- *Nuclear guidance remains in the "Nuclear Supplement to the JSCP."*

strategy found in documents like the NSS and defense strategy articulated in the Quadrennial Defense Review (QDR) and the Defense Planning Guidance (DPG) into DOD priorities and comprehensive planning direction to guide DOD components in the employment of forces. It does this through a series of prioritized objectives, and by explicitly listing the DOD's top-priority contingency planning efforts. The GEF consolidates and integrates DOD planning guidance related to operations and other military activities into a single, overarching guidance document which takes into account lessons from Operation Iraqi Freedom (OIF), Operation Enduring Freedom (OEF) and other key-named operations around the world. The GEF is produced in conjunction with the CJCS JSCP, which provides additional planning guidance and direction for the development of contingency plans, and with Department of State (DOS) input.

(1) The key concept among these lessons – captured in the QDR – is the idea that the DOD requires a framework for integrating efforts to shape the strategic environment towards deterring major conflicts, precluding major instability from arising, enhancing the governance or military capacity of partner countries, or preparing for catastrophic events.

(a) Planning and resultant activities or operations should aim to defuse strategic problems before they become crises and resolve crises before they reach a critical stage requiring large-scale military operations. Steady-state activities should support these ends and, at the same time, set the conditions for success should military operations become necessary. The campaign plan construct within the GEF is designed to do this. The GEF-tasked campaign plans will be the central mechanism for integrating DOD steady-state activities, including shaping and deterrence actions associated with contingency plans.

A statement of U.S. national defense strategy is not required as a separate document. The only statutory requirement is the mandate that the QDR "delineate a national defense strategy" and that the report on the QDR include "a comprehensive discussion of the national defense strategy of the United States." (CRS,QDR 2010: Overview and Implications for National Security Planning. Stephen Daggett, May 2010).

(b) Should shaping or deterrent measures fail, contingency plans must provide the President, SECDEF and CCDR's multiple military options for managing crises or conflicts and ending them on terms favorable to the United States.

(c) Under this concept, contingency plans become "branches" to the campaign plan. Contingency plans are built to account for the possibility that steady-state shaping measures, security cooperation activities, and operations could fail to prevent aggression, preclude large-scale instability in a key state or region, or mitigate the effects of a major disaster.

(d) Stability operations will likely be a significant component of many contingencies. Planning should ensure in such cases that the command is prepared to conduct and integrate stability operations in the earliest phases of the operation and throughout its duration. A plan's concept of operations should be informed by the longer-term need to restore the countries involved to functioning and responsible members of international society. Well-designed stability operations will play a crucial role in achieving this end.

(2) The GEF considers the range of national and defense priorities and establishes a set of prioritized objectives to focus DOD near-term activity and planning. Objectives are prioritized in three levels in order to establish an integrated set of global priorities to guide overall resource expenditure. CCMDs will use this prioritization to guide the order in which they use limited resources, however CCMDs are not limited to pursuing only the objectives listed within the GEF, but should seek guidance before establishing additional objectives.

(3) Resource Informed Planning. Resource informed planning requires the development of plans based on the near-term availability and readiness of the force, the capacity and capability of the logistics and transportation systems, preferred munitions availability, and the level of contract support required to offset the known shortfall in military capability. The GEF provides direction and planning factors for the resources available to support achieving steady state campaign objectives and responding to emerging threats over the next 2 to 5 year time horizon. These resources include allocation of forces, security cooperation, contingency, and operations and maintenance funding. The DOD is directed to emphasize a more supply based decision process for force management to better balance readiness, operational, and contingency planning demands.

(4) One of the most important features of the GEF is that it complements the security goals outlined in the DOS Joint Strategic Plan (JSP). Through campaign and contingency planning CCMD operations and activities align with national security objectives and complement the DOS's country-specific Mission Strategic Plans (MSPs). It is critically important that CCMD words and actions complement each other in shaping perceptions to support U.S. policy goals.

f. National Military Strategy (NMS). The NMS, derived from the NSS and NDS, prioritizes and focuses the efforts of the Armed Forces of the United States while conveying the CJCS's advice with regard to the security environment and the necessary military actions to protect vital US interests. The NMS defines the national military objectives (i.e., ends), how to accomplish these objectives (i.e., ways), and addresses the military capabilities required to execute the strategy (i.e., means).

g. Joint Strategic Capabilities Plan (JSCP). The JSCP is the primary vehicle through which the CJCS exercises responsibility for directing the preparation of joint plans. The JSCP provides military strategic and operational guidance to CCDRs, Service Chiefs, Combat Support Agencies (CSAs), and applicable defense agencies for preparation of campaign plans and contingency plans based on current military capabilities. It serves as the link between strategic guidance provided in the GEF and the joint operation planning activities and products that accomplish that guidance.

h. Global Force Management Implementation Guidance (GFMIG). The GFMIG integrates complementary assignment, apportionment, and allocation information into a single GFM document. GFM aligns force assignment, apportionment, and allocation methodologies in support of the NDS, joint force availability requirements, and joint force assessments. It provides comprehensive insights into the global availability of US military resources and provides senior decision makers a process to quickly and accurately assess the impact and risk of proposed changes in forces assignment, apportionment, and allocation.

5. The Joint Strategic Planning System (JSPS)

The JSPS establishes the administrative framework for the CJCS to advise the SECDEF, President and to provide strategic direction to the CCDR's. The CJCS is charged by the National Security Act of 1947 with preparing strategic plans and providing for the strategic direction of the Armed Forces. The JSPS is the primary means by which the CJCS carries out statutory responsibilities assigned in Titles 6, 10, 22 and 50 of the United States Code (USC). The primary roles are:

- Conduct independent assessments.
- Provide independent advice to the President, Secretary of Defense, National Security Council (NSC), and Homeland Security Council (HSC).
- Assist the President and SECDEF in providing unified strategic direction to the Armed Forces (includes execution).

a. The JSPS provides formal structure to the CJCSs statutory responsibilities and considers the strategic environment and the alignment of ends, ways, means, risk, and risk mitigation over time to provide the best possible assessments, advice, and direction of the Armed Forces in support of senior leaders and processes at the national and Office of the Secretary of Defense (OSD) level. JSPS constitutes a continuing process in which each document, program, or plan is an outgrowth of preceding cycles and of documents formulated earlier and in which development proceeds concurrently.[7]

b. Joint strategic planning begins the process which creates the forces whose capabilities form the basis for theater operation plans. It culminates with planning guidance for the CCDRs to develop campaign and contingency plans. The JSPS provides the framework for strategic planning and direction of the armed forces as displayed in Figure I-3. The CJCS:

- Requires development of and reviews strategic plans.
- Prepares and reviews contingency plans. Advises the President and SECDEF on requirements, programs, and budgets.
- Provides net assessments on the capabilities of the Armed Forces of the United States and its allies relative to potential adversaries.

c. JSPS is a flexible and interactive system intended to provide supporting military advice to the DOD Planning, Programming, Budgeting, and Execution System (PPBES) and strategic direction for use in APEX.[8] Through the JSPS, the JCS and the CCDRs:

- Review the national security environment and U.S. national security objectives and evaluate the threat.
- Assess current strategy and existing or proposed programs and budgets.
- Propose military strategy, programs, and forces necessary to achieve those national security objectives in a resource-limited environment consistent with policies and priorities established by the President and SECDEF.

d. Although all JSPS documents are prepared in consultation with other members of the JCS and the CCDRs, the final approval authority for all JSPS documents is the CJCS.

[7] *CJCSI 3100.1, Joint Strategic Planning System*

[8] *APEX is replacing JOPES series publications, the next revision of the CJCSM 3122 series will be the APEX 3130 series.*

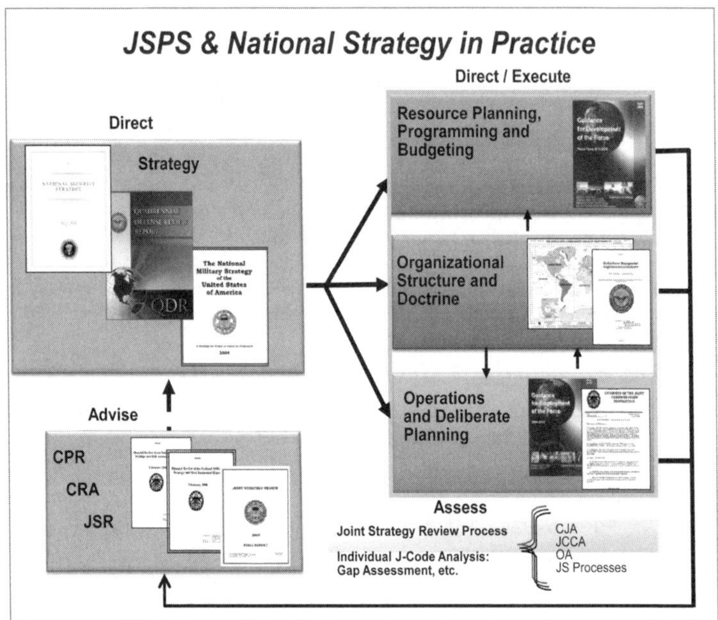

Figure I-3. JSPS & National Strategy in Practice

Most JSPS documents are published biennially; however, all documents are subject to annual review and may be changed as required. The products of JSPS that gives direction to strategic and operational planning are the NMS and JSCP.

e. The NMS and JSCP provide the strategic guidance and direction for joint strategic planning to the CCDR and for the other categories of military planning. CCDRs prepare strategic estimates, strategies, and plans to accomplish their assigned missions based on strategic direction and guidance from the President, SECDEF, and CJCS. CCDR's and their subordinate Joint Force Commander's (JFC's) primarily accomplish theater strategic and operational level planning. It is at this level where campaigns and major operations are planned, conducted and sustained to accomplish strategic objectives within their operational areas. Activities at this level link tactics and strategy by establishing operational objectives required to accomplish strategic objectives, sequencing events to achieve the operational objectives, initiating actions, and applying resources to bring about and sustain these events.

f. The JSPS is linked to the broader Strategic planning process and strategic documents (Figure I-4).

g. After providing direction to the Joint Force, the CJCS implements and monitors that direction. During execution, the JSPS helps the CJCS execute near-, mid-, and far-term processes to employ, manage and develop the Joint Force.

6. Global Context

Strategic guidance can at times be overwhelming. There are currently over ten National Strategies and constantly revised Regional Strategies/Plans that require our attention (Figure I-5). In 1999 the United States had one National Strategy with ten supporting strategies from the OSD and JS. The year 2008 found us with twelve National Strategies and sixteen OSD and JS supporting strategies for a total of twenty-eight strategies. Today, we top sixteen National Strategies and counting (Figure I-5 and I-6, page 1-14).

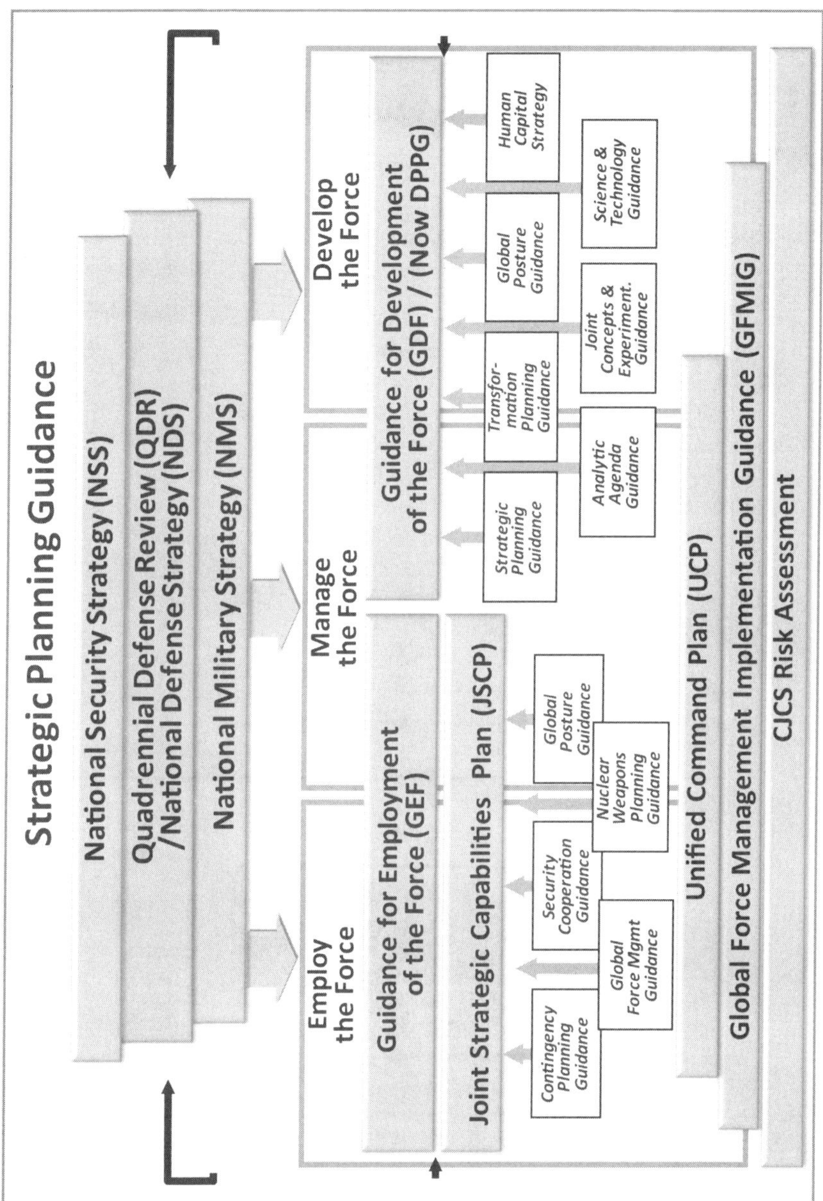

Figure I-4. Strategic Planning Process and Documents

7. Summary

Joint plans and orders are developed with the strategic and military objectives and end states in mind. The commander and planners derive their understanding of those end states from strategic guidance. Strategic guidance comes in many forms and provides the purpose and focus of joint operation planning. Joint operation planners must know where to look for the guidance to ensure that plans are consistent with national priorities and are directed toward achieving national security goals and objectives.

Current National Strategies & Proliferation of National Guidance since 1999

Current National Strategies (2002 to Present)

- National Security Strategy – May 2010
- National Strategy for Combating Terrorism – September 2006
- National Strategy for Public Diplomacy and Strategic Communications – May 2007
- National Counterintelligence Strategy – 2008
- National Intelligence Strategy – August 2009
- National Strategy to Combat WMD – December 2002
- National Strategy to Combat Terrorist Travel – February 006
- National Strategy to Secure Cyberspace – February 2003
- National Strategy for Homeland Security – October 2007
- National Strategy for Maritime Security – September 2005
- National Strategy for Information Sharing – October 2007
- National Strategy for Victory in Iraq – November 2005
- National Strategy for Pandemic Influenza - November 2005
- National Strategy to Secure Cyberspace - February 2003
- National Strategy for Countering Biological Threats - November 2009
- National Security Space Strategy - January 2011

Figure I-5. National Strategic Guidance from 2002 to present

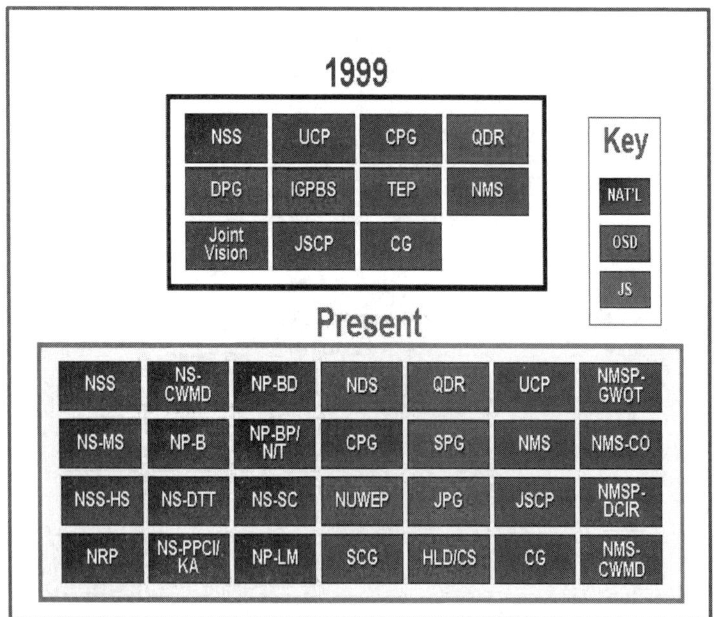

Figure I-6. National Guidance Proliferation from 1999 to present

Planning 2

I. PLANNING AND PLANS

> *Plan: (noun)*
>
> *1. A method or scheme for achieving or doing something. An aim: goal. (Webster's II)*
>
> *2. A scheme or method of acting, doing, proceeding, making, etc., developed in advance: battle plans. (Dictonary.com)*

1. Planning

Planning is the process of thinking about and organizing the activities required to achieve a desired goal (forethought). It is an anticipatory decision making process that helps in coping with complexities and combines forecasting of developments with the preparation of scenarios and how to react to them. Planning is conducted for different planning horizons, from long-range to short-range. Depending on the echelon and circumstances, units may plan in years, months, or weeks, or in days, hours, and minutes.

> *Planning*
>
> *The process by which commanders and the staff translate the commander's visualization into a specific course of action for preparation and execution, focusing on the expected results. (FM 3-0)*

The defining challenges to effective planning are uncertainty and time. Uncertainty increases with the length of the planning horizon and the rate of change in an operational environment. A tension exists between the desire to plan far into the future to facilitate preparation and the fact that the farther into the future the commander plans, the less certain the plan will remain relevant. Given the uncertain nature of the operational environment, the object of planning is not to eliminate uncertainty, but to develop a framework for action in the midst of such uncertainty.[1]

a. Planning provides an informed forecast of how future events may unfold. It entails identifying and evaluating potential decisions and actions in advance to include thinking through consequences of certain actions. Planning involves thinking about ways to influence the future as well as how to respond to potential events. Put simply, planning is thinking critically and creatively about what to do and how to do it, while anticipating changes along the way.

b. Planning keeps us oriented on future objectives despite the requirements of current operations. By anticipating events beforehand, planning helps the commander seize, retain, or exploit the initiative. As a result, the force anticipates events and acts purposefully and effectively before the adversary can act or before situations deteriorate. In addition, planning helps anticipate favorable turns of events that could be exploited during shaping operations.

[1] *ADP 5-0, The Operations Process*

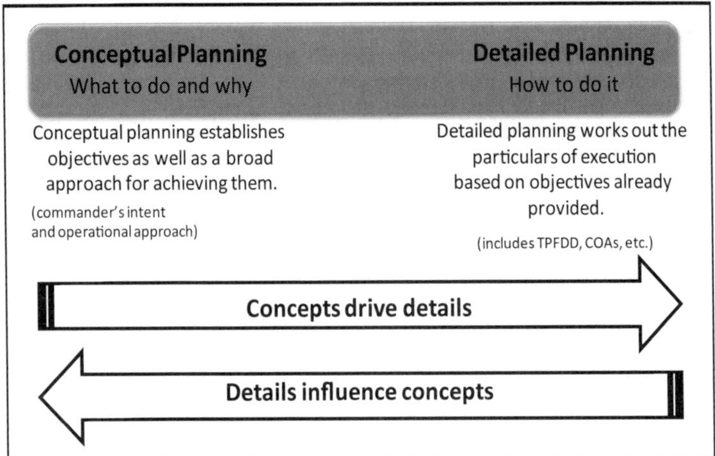
Conceptual Planning
What to do and why
Detailed Planning
How to do it
Conceptual planning establishes objectives as well as a broad approach for achieving them.
(commander's intent and operational approach)
Detailed planning works out the particulars of execution based on objectives already provided.
(includes TPFDD, COAs, etc.)
Concepts drive details
Details influence concepts

Figure II-1. Integrated Planning

2. Defining Challenges

a. Planning is also the art and science of understanding a situation, envisioning a desired future, and laying out an operational approach to achieve that future. Planning is both a continuous and a cyclical activity of the operations process which translates strategic guidance and direction. Based on this understanding and operational approach, planning continues with the development of a fully synchronized campaign plan, operation plan or order that arranges potential actions in time, space, and purpose to guide the force during execution.[2]

b. While planning may start an iteration of the operations process, planning does not stop with production of a plan or an order. During preparation and execution, the plan is continuously refined as assessments and situational understanding improves. Supporting commands, subordinates and others provide feedback as to what is working, what is not working, and how the force can do things better.

c. Planning may be based on defined tasks identified in the GEF and the JSCP, or it may be based on the need for a military response to an unforeseen current event, emergency, or time-sensitive crisis. The value of following the well-established Joint Operation Planning Process (JOPP) has been reinforced through operational and exercise experiences. Key to the process is the detailed analysis necessary to produce the requisite plans and orders that will direct subordinates. In addition to the required analysis, planners must strive to ensure the generated solution does not further exacerbate the problem or limit future options.

3. Integrated Planning

Planning consists of two separate, but closely related, components: a conceptual component and a detailed component (Figure II-1). Conceptual planning involves understanding the operational environment and the problem, determining the operation's end state, and visualizing an operational approach. Conceptual planning generally corresponds to operational art and is the focus of the commander with staff support. Detailed planning translates the broad operational approach into a complete and practical plan. Generally, detailed planning is associated with the science of operations including the synchronization of the forces in time, space, and purpose. Detailed planning works out the scheduling, coordination, or

[2] *ADP 5-0, The Operations Process*

technical problems involved with moving, sustaining, and synchronizing the actions of force as a whole toward a common goal. Effective planning requires the integration of both the conceptual and detailed components of planning.[3]

4. Operational Art And Planning

Conceptual planning is directly associated with operational art—the cognitive approach by commanders and staffs—supported by their skill, knowledge, experience, creativity, and judgment—to develop strategies, campaigns, and operations to organize and employ military forces by integrating ends, ways, and means.[4] Operational art is a thought process that guides conceptual and detailed planning to produce executable plans and orders.

a. In applying operational art, commanders and their staffs use a set of intellectual tools to help them communicate a common vision of the operational environment as well as visualizing and describing the operational approach. Collectively, this set of tools is known as the elements of operational art (Chapter VI). These tools help commanders understand, visualize, and describe combinations of combat power and help them formulate their intent and guidance. Commanders selectively use these tools in any operation. However, their application is broadest in the context of long-term operations.

b. The elements of operational art support the commander in identifying objectives that link tactical missions to the desired end state. They help refine and focus the operational approach that forms the basis for developing a detailed plan or order. During execution, commanders and staffs consider the elements of operational art as they assess the situation. They adjust current and future operations and plans as the operation unfolds.[5]

5. Understand and Develop Solutions to Problems

A *problem* is an issue or obstacle that makes it difficult to achieve a desired goal or objective. In a broad sense, a problem exists when an individual becomes aware of a significant difference between what actually is and what is desired. In the context of operations, an operational problem is the issue or set of issues that impede commanders from achieving their desired end state or objectives.[6]

To *understand* something is to grasp its nature and significance. Understanding includes establishing context—the set of circumstances that surround a particular event or situation. Throughout the operations process, commanders develop and improve their understanding of their operational environment and the problem. An operational environment is a composite of the conditions, circumstances, and influences that affect the employment of capabilities and bear on the decisions of the commander. Both conceptual and detailed planning assist commanders in developing their initial understanding of the operational environment and the problem. Based on personal observations and inputs from others (to include running estimates), commanders improve their understanding and modify their visualization throughout the conduct of operations.

a. Throughout operations, commanders face various problems, often requiring unique and creative solutions. Planning helps commanders and staffs understand problems and develop solutions. Not all problems require the same level of planning. For simple problems, commanders often identify them and quickly decide on a solution, sometimes on the spot. Planning is critical, however, when a problem is actually a set of interrelated issues, and the solution to each affects the others. For unfamiliar situations, planning offers ways to deal with the complete set of problems as a whole. In general, the more complex a situation is, the more important and involved the planning effort becomes.

[3] *ADRP 5-0, The Operations Process*

[4] *Joint Pub 3-0, Doctrine for Joint Operations*

[5] *ADRP 5-0, The Operations Process*

[6] *Ibid*

b. Just as planning is only part of the operations process, planning is only part of problem solving. In addition to planning, problem solving includes implementing the planned solution (execution), learning from implementation of the solution (assessment), and modifying or developing a new solution as required. The object of problem solving is not just to solve near-term problems, but to do so in a way that forms the basis for long-term success.

6. Plans

A product of planning is a plan or order—a directive for future action. Commanders issue plans and orders to subordinates to communicate their understanding of the situation and their visualization of an operation. A plan is a continuous, evolving framework of anticipated actions that maximizes opportunities. It guides subordinates as they progress through each phase of the operation. Any plan is a framework from which to adapt, not a script to be followed to the letter. The measure of a good plan is not whether execution transpires as planned, but whether the plan facilitates effective action in the face of unforeseen events. Good plans and orders foster initiative. Plans and orders come in many forms and vary in scope, complexity, and length of time addressed. Generally, a plan is developed well in advance of execution and is not executed until directed. A plan becomes an order when directed for execution based on a specific time or an event. Some planning results in written orders complete with attachments. Other planning produces brief fragmentary orders issued verbally and followed in writing.a. Plans are developed utilizing the planning steps developed for the JOPP. JOPP is an orderly, analytical process, which consists of a set of logical steps to examine a mission, analyze, and compare alternate courses of action (COAs); select the best COA; and produce a plan or order. The JOPP process is utilized for "deliberate" planning for a contingency plan or crisis action planning in response to a real world crisis. The greatest difference between deliberate and crisis action planning is the time allotted to conduct each planning step. We will discuss this in greater detail in the following paragraphs.

b. Although the four planning functions of strategic guidance, concept development, plan development, and plan assessment are generally sequential, they often run simultaneously in the effort to accelerate the overall planning process. SecDef or the CCDR may direct the planning staff to refine or adapt a plan by reentering the planning process at any of the earlier functions. The time spent accomplishing each activity and function depends on the nature of the problem.

(1) In time-sensitive cases, activities and functions may be accomplished simultaneously and compressed so that all decisions are reached in open forum and orders are combined and initially may be issued orally.

(2) A crisis could be so time critical, or a single COA so obvious, that the firstwritten directive might be a DEPORD or an EXORD.

c. Plans encompasses four levels of planning detail, with an associated planning product for each level. See following page for each level.

II. THE CONTINGENCY PLAN AND DELIBERATE PLANNING

1. Contingency Plans

Under the GEF, campaign plans provide the vehicle for linking steady-state shaping activities to current operations and contingencies. Contingency plans under this concept become "branch" plans to the overarching theater campaign plan. A contingency is a situation that likely would involve military forces in response to natural and man-made disasters, terrorists, subversives, military operations by foreign powers, or other situations as directed by the President or SECDEF. Contingency plans are built to account for the possibility that

Planning: Levels of Planning Detail

Level 1 Planning Detail — CDR's Estimate/Concept of Operations/Course of Action (COA)

This level of planning involves the least amount of detail, and focuses on producing a developed COA. The product for this level can be a COA briefing, command directive, CDR's estimate, or a memorandum. The CDR's estimate provides the SECDEF with military COAs to meet a potential contingency. The estimate reflects the supported CDR's analysis of the various COAs available to accomplish an assigned mission and contains a recommended COA.

Level 2 Planning Detail — Base Plan

A base plan describes the CONOPS, major forces, concepts of support, and anticipated timelines for completing the mission. It normally does not include annexes or Time Phased Force Deployment Data (TPFDD).

Level 3 Planning Detail — Concept Plan (CONPLAN)

A CONPLAN is an operation plan in an abbreviated format that may require considerable expansion or alteration to convert it into an OPLAN or operations order (OPORD). It includes a base plan with selected annexes (A, B, C, D, J, K, S, V, Y and Z) required by the JFC and a supported CDR's estimate of the plan's feasibility. It may also produce a transportation feasible TPFDD if applicable.

Level 4 Planning Detail — Operation Plan (OPLAN)

An OPLAN is a complete and detailed joint plan containing a full description of the concept of operations (CONOPS), all annexes applicable to the plan, and a TPFDD. It identifies the specific forces, functional support, and resources required to execute the plan and provides closure estimates for their flow into the theater. OPLANs can be quickly developed into an OPORD. An OPLAN is normally prepared when:

(a) The contingency is critical to national security and requires detailed prior planning.

(b) The magnitude or timing of the contingency requires detailed planning.

(c) Detailed planning is required to support multinational planning.

(d) The feasibility of the plan's CONOPS cannot be determined without detailed planning.

(e) Detailed planning is necessary to determine force deployment, employment, and sustainment requirements, determine available resources to fill identified requirements, and validate shortfalls.

> *"In preparing for battle i have always found that plans are useless, but planning is indispensable."*
>
> *General Dwight D. Eisenhower*

Figure II-2. Joint Operation Planning

steady-state shaping measures, security cooperation activities, and operations could fail to prevent aggression, preclude large-scale instability in a key state or region, or mitigate the effects of a major disaster. Contingency plans address scenarios that put one or more U.S. strategic objectives or endstates in jeopardy and leave the United States no other recourse than to address the problem at hand through military operations. Military operations can be in response to many scenarios, including armed aggression, regional instability, a humanitarian crisis, or a natural disaster. Contingency plans should provide a range of military options coordinated with total U.S. Government response.

2. Deliberate Planning

Deliberate planning is the deliberate process in which the JOPP is utilized to develop responses to contingencies. Deliberate planning is planning that occurs in non-crisis situations based on hypothetical contingencies. Following the guidance provided by the JSCP, CCDRs prepare, submit, and continuously refine their plans and continue until the requirement no longer exists. Planning guidance is provided in the GEF, JSCP, Strategic Guidance Statements (SGS) and through SECDEF and CCDR IPRs, which exist to stimulate a synchronized dialogue between the supported CCDR, the SECDEF and other appropriate senior leaders.

a. Deliberate planning is an iterative process and is adaptive to situational changes within the operational and planning environments. The process allows for changes in plan priorities, changes to the review and approval process of either a single plan or a category of plans, and contains the flexibility to adjust the specified development time line for producing and refining plans. Deliberate planning also facilitates the transition to crisis planning by having a library of information available (intelligence, logistics, forces, etc) for disparate regions of the globe.

b. The JSCP links the JSPS to joint operation planning, identifies broad scenarios for plan development, specifies the type of joint plan required, and provides additional planning guidance as necessary. A CCDR may also initiate a contingency plan by preparing plans not specifically assigned but considered necessary to discharge command responsibilities. If a contingency develops that warrants deliberate planning but was not anticipated in the GEF or JSCP, the SECDEF, through the CJCS, tasks the appropriate supported CCDR and applicable supporting CCDRs, Services, and CSAs to begin deliberate planning in response to the new contingency. The primary mechanism for tasking a contingency plan outside of the GEF/JSCP cycle will be through SGS from the SECDEF and endorsed by message from the CJCS to the CCDRs.

c. Plans are produced and updated periodically to ensure relevancy. Deliberate planning most often addresses military options requiring combat operations; however, plans must account for other types of joint operations across the range of military operations. For example, operations during Phase IV (Stabilize) of a campaign, and most stability operations, are very complex and require extensive planning and coordination with non-DOD organizations with the military in support of other agencies. Deliberate planning occurs in prescribed cycles in accordance with formally established procedures that complement and support other DOD planning cycles. In coordination with the JPEC, the JS develops and issues a planning schedule that coordinates plan development activities and establishes submission dates for joint plans.

d. Plan Management. The CJCS reviews deliberate planning accomplished for contingency plans specified in the JSCP, combined military plans, military plans of international treaty organizations, and as otherwise specifically directed by the SECDEF.

(1) The JS Director for Operational Plans and Joint Force Development (JS J7) is responsible for the plan management processes for all contingency plans, to include plans maintained by the JS Director of Operations (DJ-3) that are not in execution. DJ-3 is responsible for managing the process of developing operations plans in a crisis action environment, overseeing the execution of operations, and maintaining subject matter experts (SME) on all J3 developed plans.

(2) The J-5/Joint Operational War Plans Division (JOWPD) serves as the office of primary responsibility (OPR) within the JS for all contingency plan matters, to include bilateral military plans and military plans of international treaty organizations not specifically designated otherwise. This consists of both the management of contingency plans and the plan review process, including but not limited to review of the TPFDD, final plan, and facilitation of contingency plan IPRs with the SECDEF.

(3) JOWPD is the primary liaison for the CCDR with both the Office of the Chairman of the Joint Chiefs of Staff (OCJCS) and the Office of the Under Secretary of Defense for Policy (OUSD(P)) for development of deliberate plans and the plan review process.

(4) To achieve rapid planning with greater efficiency, this process features early planning guidance with frequent dialogue in the form of IPRs between the JS and planners and the commanders and DOD senior leaders. This dialogue promotes an understanding of, and agreement on, the mission, planning assumptions, threat definitions, interagency, and allied planning cooperation, risks, courses of action, and other key factors (see Chapter VIII, IPR's).

3. Deliberate Planning and the GFM Process

(See Chapter 7, GFM, for greater details).

a. GFM Process during Deliberate Planning. The apportionment tables maintained in Section IV of the GFMIG provides the number of forces reasonably expected to be available for planning. Theses tables should be used as a beginning assumption in planning.

b. Force Identification Planning. The purpose of force identification planning is to identify all forces needed to accomplish the supported commander's CONOPS and phase the forces into the theater. It consists of determining the force requirements by operation phase, mission, mission priority, mission sequence, and operations area (OA). It includes major force phasing; integration planning; force list structure development; followed by force list development. Force identification planning is the responsibility of the CCDR supported by the Service component commanders, Services, JS, and JS Joint Force Coordinator/Joint Force Providers (JS JFC/JFPs). The primary objectives of force identification planning are to apply the right force to the mission while ensuring force visibility, force mobility, and adaptability. Force identification planning begins during CONOPS development. During the planning stage, these forces are planning assumptions called preferred forces. The supported commander determines force requirements, identifies and resolves or reports shortfalls, develops a TPFDD and letter of instruction (LOI), and designs force modules to align and time-phase the forces in accordance with (IAW) the CONOPS. Major combat forces are selected from those apportioned for planning and included in the supported commander's CONOPS by operation phase, mission, and mission priority. The supported CCDR Service components then collaboratively make tentative assessments of the combat support (CS) and combat service support (CSS) required IAW the CONOPS.

a. For some high priority plans, as the plan is either approved or nearing approval, the CJCS may direct the JS JFC/JFPs to contingency source a plan to support CJCS and/or SECDEF's strategic risk assessments. Contingency sourcing allows greater fidelity in force planning. Contingency sourcing is based on planning assumptions and contingency sourcing guidance. To the degree that the contingency sourcing planning assumptions and sourcing guidance mirrors the conditions at execution, contingency sourcing may validate the supported CCDR preferred forces (Chapter 7).

b. If the plan is directed to be executed, the force requirements are forwarded to the JS for validation and tasked to the JS JFC/JFPs to provide sourcing recommendations to the SECDEF. When the SECDEF accepts the sourcing recommendation, the FP is ordered to deploy the force to meet the requested force requirement. After the actual forces are sourced, the CCDR refines the force plan to ensure it supports the CONOPS, provides force visibility, and enables flexibility.

> *"Everybody has a plan, until they get punched in the face."*
> *Mike Tyson*

III. CRISIS AND CRISIS ACTION PLANNING

1. Crisis

A crisis is an incident or situation involving a threat to the United States, its territories, citizens, military forces, possessions, or vital interest. It typically develops rapidly and creates a condition of such diplomatic, economic, political, or military importance that the President or SECDEF considers a commitment of U.S. military forces and resources to achieve national objectives. It may occur with little or no warning (e.g., Indonesian Tsunami or Haitian Earthquake) and it is fast-breaking and requires accelerated decision making and timelines. Sometimes a single crisis may spawn another crisis elsewhere.

2. Crisis Planning Relationship to Deliberate Planning

Crisis Action planning (CAP) is accomplished utilizing JOPP and can respond to crises spanning the full range of military operations. Strategic planning supports CAP by anticipating potential contingencies and developing plans that facilitate execution planning during crises. As discussed earlier, deliberate planning prepares for a hypothetical military contingency based on the best available intelligence, while using forces and resources projected to be available for the period during which the plan will be effective. It relies heavily on assumptions regarding the political and military circumstances that will exist when the plan is implemented. Even though every crisis situation cannot be anticipated, the distributed collaborative environment, detailed analysis, and coordination which occur during deliberate planning may facilitate effective decision-making, execution, and redeployment planning as a crisis unfolds. During CAP, assumptions and projections made during deliberate planning are replaced with facts and actual conditions. Figure II-3 compares deliberate planning and CAP with time, environment, forces, etc.

a. CAP encompasses the activities associated with the time-sensitive development of OPORDs for the deployment, employment, and sustainment of assigned, attached, and allocated forces and resources in response to an actual situation that may result in actual military operations. While deliberate planning normally is conducted in anticipation of future events, CAP is based on circumstances that exist at the time planning occurs. There are always situations arising in the present that might require U.S. military response. Such situations may approximate those previously planned for in deliberate planning, though it is unlikely they would be identical, and sometimes they will be completely unanticipated. The time available to plan responses to such real-time events is short. In as little as a few days, CCDRs and staffs must develop and approve a feasible COA, publish the plan or order, prepare forces, ensure sufficient communications systems support, and arrange sustainment for the employment of U.S. military forces.

b. In a crisis, situational awareness is continuously fed by the latest intelligence and operations reports. An adequate and feasible military response in a crisis demands flexible procedures that consider time available, rapid and effective communications, and relevant previous planning products whenever possible.

c. In a crisis or time-sensitive situation, the CCDR uses CAP to adjust previous deliberate planning and converts these plans to executable OPORDs or develops OPORDs from scratch when no useful contingency plans exist. To maintain plan viability it is imperative that all Steps of the JOPP are conducted and thought through, although some may be done sequentially. Time-sensitivities are associated with CAP and the JOPP may be abbreviated for time.

d. CAP activities are similar to deliberate planning activities, but CAP is based on dynamic, real-world conditions vice assumptions. CAP procedures provide for the rapid and effective exchange of information and analysis, the timely preparation of military COAs for consideration by the President or SECDEF, and the prompt transmission of their decisions to the JPEC. CAP activities may be performed sequentially or in parallel, with supporting and subordinate plans or OPORDs being developed concurrently. The exact flow of the procedures is largely determined by the time available to complete the planning and by the significance of the crisis. Capabilities such as collaboration and decision-support tools will increase the ability of the planning process to adapt quickly to changing situations and improve the transition from deliberate planning to CAP. The following paragraphs summarize the activities and interaction that occur during CAP.

(1) When the President, SECDEF, or CJCS decides to develop military options, the CJCS issues a Planning Directive to the JPEC initiating the development of COAs and requesting that the supported CCDR submit a CDR's Estimate of the situation with a recommended COA to resolve the situation. Normally, the directive will be a Warning Order (WARNORD), but a Planning Order (PLANORD) or Alert order (ALERTORD) may be used if the nature and timing of the crisis warrant accelerated planning. In a quickly evolving crisis, the initial WARNORD may be communicated vocally with a follow-on record copy to ensure that the JPEC is kept informed. If the directive contains force deployment preparation or deployment orders, SECDEF approval is required.

(2) The WARNORD describes the situation, establishes command relationships, and identifies the mission and any planning constraints. It may identify forces and strategic mobility resources, or it may request that the supported CCDR develop these factors. It may establish tentative dates and times to commence mobilization, deployment, employment, or it may solicit the recommendations of the supported CCDR regarding these dates and times. If the President, SECDEF, or CJCS directs development of a specific COA,

the WARNORD will describe the COA and request the supported CCDR's assessment. A WARNORD sample can be found in CJCSM 3122.02.

(3) In response to the WARNORD, the supported CCDR, in collaboration with subordinate and supporting CCDRs and the rest of the JPEC, reviews existing joint plans for applicability and develops, analyzes, and compares COAs. Based on the supported CCDR's guidance, supporting CCDRs begin their planning activities.

(4) Although an existing plan almost never completely aligns with an emerging crisis, it can be used to facilitate rapid COA development. An existing plan can be modified to fit the specific situation. An existing CONPLAN can be fully developed beyond the stage of an approved CONOPS. The Time Phased Force Deployment Lists (TPFDL) related to specific plans are stored in the JOPES database and available to the JPEC for review.

(5) The CJCS, in consultation with other members of the JCS and CCDRs, reviews and evaluates the supported CCDR's estimate and provides recommendations and advice to the President and SECDEF for COA selection. The supported CCDR's COAs may be refined or revised, or new COAs may have to be developed to accommodate a changing situation. The President or SECDEF selects a COA and directs that detailed planning be initiated.

(6) On receiving the decision of the President or SECDEF, the CJCS issues an alert order (ALERTORD) to the JPEC to announce the decision. The SECDEF approves the ALERTORD. The order is a record communication that the President or SECDEF has approved the detailed development of a military plan to help resolve the crisis. The contents of an ALERTORD may vary, and sections may be deleted if the information has already been published, but it should always describe the selected COA in sufficient detail to allow the supported CCDR, in collaboration with other members of the JPEC, to conduct the detailed planning required to deploy, employ, and sustain forces. However, the ALERTORD does not authorize execution of the approved COA.

(7) The supported CCDR develops the OPORD and supporting TPFDD using an approved COA. Understandably, the speed of completion is greatly affected by the amount of prior planning and the planning time available. The supported CCDR and subordinate describe the CONOPS in OPORD format. They update and adjust planning accomplished during COA development for any new force and sustainment requirements and source forces and lift resources. All members of the JPEC identify and resolve shortfalls and limitations.

(8) The supported CCDR submits the completed OPORD for approval to the SECDEF or President via the CJCS. After an OPORD is approved, the President or SECDEF may decide to begin deployment in anticipation of executing the operation or as a show of resolve, execute the operation, place planning on hold, or cancel planning pending resolution by some other means. Detailed planning may transition to execution as directed or become realigned with continuous situational awareness, which may prompt planning product adjustments and/or updates.

(9) In CAP, plan development continues after the President decides to execute the OPORD or to return to the pre-crisis situation. When the crisis does not lead to execution, the CJCS provides guidance regarding continued planning under either crisis- action or deliberate planning procedures.

e. CAP provides the CJCS and CCDRs with a process for getting vital decision-making information up the chain of command to the President and SECDEF. CAP facilitates information-sharing among the members of the JPEC and the integration of military advice from the CJCS in the analysis of military options. Additionally, CAP allows the President and SECDEF to communicate their decisions rapidly and accurately through the CJCS to the CCDRs, subordinate and supporting CCDRs, the Services, and CSAs to initiate detailed military planning, change deployment posture of the identified force, and execute military options. It also outlines the mechanisms for monitoring the execution of the operation.

Deliberate Planning and Crisis Action Planning Comparison

Ref: JP 5-0, Joint Operation Planning (Aug '11), fig. II-8, p. II-30.

	Deliberate Planning	Crisis Action Planning
Time available	As defined in authoritative directives (normally 6+ months)	Situation dependent (hours, days, up to 12 months)
Environment	Distributed, collaborative planning	Distributed, collaborative planning and execution
JPEC involvement	Full JPEC participation (Note: JPEC participation may be limited for security reasons.)	Full JPEC participation (Note: JPEC participation may be limited for security reasons.)
APEX operational activities	Situational awareness Planning	Situational awareness Planning Execution
APEX functions	Stategic guidance Concept development Plan development Plan assessment	Strategic guidance Concept development Plan development Plan assessment
Document assigning planning task	CJCS issues: 1. JSCP 2. Planning directive 3. WARNORD (for short suspense planning)	CJCS issues: 1. WARNORD 2. PLANORD 3. SecDef-approved ALERTORD
Forces for planning	Apportioned in JSCP	Allocated in WARNORD, PLANORD, or ALERTORD
Planning guidance	CJCS issues JSCP or WARNORD CCDR issues PLANDIR and TPFDD LOI	CJCS issues WARNORD, PLANORD, or ALERTORD CCDR issues WARNORD, PLANORD, or ALERTORD and TPFDD LOI to subordinates, supporting commands, and supporting agencies
COA selection	CCDR selects COA and submits strategic concept to CJCS for review and SecDef approval	CCDR develops commander's estimate with recommended COA
CONOPS approval	SecDef approves CSC, disapproves or approves for further planning	President/SecDef approve COA, disapproves or approves further planning
Final planning product	Campaign Plan Level 1–4 contingency plan	OPORD
Final planning product approval	CCDR submits final plan to CJCS for review and SecDef for approval	CCDR submits final plan to President/SecDef for approval
Execution document	Not applicable	CJCS issues SecDef-approved EXORD CCDR issues EXORD

Legend

ALERTORD	alert order	JSCP	Joint Strategic Capabilities Plan
APEX	Adaptive Planning and Execution	LOI	letter of instruction
CCDR	combatant commander	PLANDIR	planning directive
CJCS	Chairman of the Joint Chiefs of Staff	PLANORD	planning order
COA	course of action	OPORD	operations order
CONOPS	concept of operations	SecDef	Secretary of Defense
CSC	commanders' strategic concept	TPFDD	time-phased force and deployment data
EXORD	execution order		
JPEC	joint planning and execution community	WARNORD	warning order

Refer to The Joint Forces Operations & Doctrine SMARTbook (Guide to Joint, Multinational & Interagency Operations) for complete discussion of deliberate and crisis action planning. Additional topics include joint doctrine fundamentals, joint operations and unified action, joint operation planning, joint logistics, joint task forces, information operations, multinational operations, and IGO/NGO coordination.

3. GFM Process during CAP

During CAP, *preferred force identification* is used the same as it was during deliberate planning. Contingency sourcing is rarely used during CAP due to the time constraints involved, but if time allows, the option exists for the CJCS to direct JS Joint Force Coordinator/Joint Force Providers (JS JFC/JFPs) to contingency source a CAP.

(1) In deliberate planning and CAP the difference in force planning is the level of detail done with the force requirements for the plan. For deliberate planning, the number of planning assumptions prevents generating the detailed force requirements needed by the JS JFC/JFPs to begin execution sourcing. During CAP, a known event has occurred and there are fewer assumptions. The focus of CAP is usually on transitioning to execution quickly. The detailed information requirements specified to support the execution sourcing process, either emergent or annual, preclude completion until most assumptions are validated.

(2) During CAP, the CCDR will utilize assigned and already allocated forces to respond to the situation. If the CCDR identifies additional forces are required to respond, then a force request is submitted. Once this request is endorsed by the CCDR, that force request is considered a CCDR requirement to execute the operation as planned. The force request is sent from the CCDR to the SECDEF via the JS. The vehicle for the force request is a message called a Request for Forces (RFF) to the SECDEF and JS info the JS JFC/JFPs, FPs, OSD, and all other CCDRs as specified in CJCSM 3130.06, GFM Allocation Policies and Procedures. Each individual force requested is serialized with a Force Tracking Number (FTN). An RFF message may contain one or more FTNs. To request Joint Individual Augmentations (JIAs) for a Joint Task Force Headquarters, the message is called a Joint Manning Document (JMD) Emergent JIA Request. The initial force or JIA request to perform a mission is an emergent request (see the GFMIG, Section III).

(3) If the plan is directed to be executed, the force requirements are forwarded to the JS for validation and tasked to the JS JFC/JFPs to provide sourcing recommendations to the SECDEF. When the SECDEF accepts the sourcing recommendation, the FP is ordered to deploy the force to meet the requested force requirement via a deployment order.

4. Abbreviated Procedures

The activities in the preceding discussion have been described sequentially. During a crisis, they may be conducted concurrently or even eliminated, depending on prevailing conditions. In some situations, no formal WARNORD is issued, and the first record communication that the JPEC receives is the PLANORD or ALERTORD containing the COA to be used for plan development. It is also possible that the President or SECDEF may decide to commit forces shortly after an event occurs, thereby significantly compressing planning activities. No specific length of time can be associated with any particular planning activity. Severe time constraints may require crisis participants to pass information verbally, including the decision to commit forces. Verbal force requests and verbal orders to deploy should be followed up with hard copy and electronic means as soon as practicable (see VOCO in Chapter 7, GFM).

The primary goal of planning is not the development of elaborate plans that inevitably must be changed; a more enduring goal is the development of planners who can cope with the inevitable change.

Campaign Planning

1. Campaign Plans

Campaign plans are intended to focus and direct steady-state activities that can prevent or mitigate conflict and set the conditions necessary for successful execution of contingency plans. In linking steady-state objectives with resources and activities, campaign plans enable resource-informed planning and permit prioritization across DOD. The UCP, GEF and JSCP are the core strategic guidance directives for campaign planning. The GEF and the JSCP directs that campaign plans focus on theater strategy implementation through the command's steady-state, foundational activities, which include ongoing operations, security cooperation, and other shaping, preventative, or deterrence actions.

a. Campaign planning operationalizes a CCMD's theater and functional strategy by comprehensively and coherently integrating all of its directed steady-state (actual) and contingency (potential) operations and activities. The CCMD's strategy and resultant campaign should be designed to achieve the prioritized theater or functional objectives provided in the GEF and serve as the integrating framework that informs and synchronizes all subordinate and supporting planning and operations.[1]

b. The format of a campaign plan is the APEX basic operation plan format tailored by the CCMD with supporting, posture, and country plans organized and formatted as annexes and appendices of the overarching campaign plan.

2. Functional Campaign Planning

The GEF gives specific functional planning guidance to geographic combatant commanders (GCC's) and functional combatant commanders (FCC's) for functional campaign-related contingency planning efforts. Functional planning guidance addresses security challenges that are often global in nature or affect more than one GCC. Though functional planning guidance often leads to planning by FCCs, GCCs must ensure their theater campaign plans (TCPs) support achievement of strategic end states and objectives contained in the functional planning guidance in the GEF, including the guidance for cyberspace, space, countering WMD, ballistic missile defense, pandemic influenza and infectious diseases, and countering terrorism.

a. The JSCP tasks CCDRs to develop functional campaign plans (FCPs) when achieving strategic objectives requires joint operations and activities conducted in multiple area of responsibilities (AORs). FCPs establish the strategic and operational framework within which subordinate campaign plans are developed. The FCP's framework also facilitates coordinating and synchronizing the many interdependent, cross-AOR missions such as security cooperation, intelligence collection, and coalition support.

b. The UCP assigns select CCMDs a "synchronizing planning" role for several mission areas, a "global synchronizer." A global synchronizer is the CCMD responsible for the alignment of specified planning and related activities of other CCMD's, Services, Defense agencies and activities, and, as directed, other USG agencies to facilitate coordinated and decentralized execution across geographic or other boundaries. An example is United States Northern Command's (USNORTHCOM's) FCP for Pandemic Influenza (PI). This plan provides the strategic framework that guides the development of the GCCs PI plans.

[1] *CJCSM 3130.01 series*

These regional plans, synchronized with both the TCP and USNORTHCOM's FCP, direct the execution of operations and activities in each GCC's AOR.

(1) Unless directed by the SECDEF, the global synchronizer's role is not to execute specific plans, but to align and harmonize plans and recommend sequencing of actions to achieve the GEF stated strategic endstates and objectives.

3. Theater Campaign Planning

The GEF also gives regional planning guidance for theater campaign and related contingency planning efforts. Theater campaign planning operationalizes the CCMD's strategy by integrating all of its directed steady-state (actual) and contingency (potential) operations and activities. The resultant theater campaign serves as the integrating framework that informs and synchronizes all regional planning and operations. The theater campaign's operational approach and design provides the structure and linkage to integrate all subordinates and supporting US military regional and functional operations, activities, events, and investments in support of the theater strategy. Strategy implementation, and the broader theater engagement, ebbs and flows within a dynamic operational environment as US military operations, activities, and investments are executed in concert with other USG activities, create effects, and those effects are assessed by the CCMD against tangible goals linked to desired objectives. Contingency plans are developed for anticipated (potential) threats and catastrophic events as branch plans under the theater campaign. As events or circumstances dictate, the CCDRs may direct through an appropriate order (e.g., PLANORD, WARNORD, ALERTORD) a given contingency plan be placed under enhanced plan management because political circumstances or indications and warning have changed its priority, visibility, likelihood, or risk to national security. Issuance of an EXORD signals the transition of that contingency plan to an OPORD for implementation as a branch plan under the theater campaign.

a. GCCs prepare TCPs to achieve GEF-directed objectives for their UCP-assigned AORs. The GCCs are responsible for integrating the planning of designated missions assigned to FCCs into their theater campaigns. FCCs prepare FCPs to achieve GEF-directed objectives for their UCP-assigned missions and responsibilities. FCCs are responsible as directed for synchronizing planning across CCMDs, Services, and Defense Agencies for designated missions. CCDRs document the full scope of their theater or functional campaigns in the set of plans that includes the overarching theater or functional campaign plan, and all of its subordinate, supporting, posture, country (for GCCs), operation (for operations currently in execution), and contingency plans.[2]

b. CCDRs use their campaign plans to articulate resource requirements in a comprehensive manner vice an incremental basis and also to provide a vehicle for conducting a comprehensive assessment of how the CCMD activities are contributing to the achievement of IMOs and strategic objectives (see Chapter 21 and 22).

c. Each CCDR's TCP could include the following elements:

4. Key Stakeholders

Theater campaign planning is a multi-dimensional team activity with many stakeholders both inside and outside the DOD. Each of these stakeholders plays an important role in the design and outcome of the theater campaign:

- OSD
- Joint Staff
- Military Departments
- CSAs

[2] *CJCSM 3130.01 series*

- CCMDs
- Service Component Commands
- DOS
- US Ambassadors and Country Teams
- Senior Defense Official and Country Team Security Cooperation Offices
- USAID
- Others; DOJ (Drug Enforcement Administration, FBI), DHS Immigration and Customs Enforcement, Customs and Border Protection, US Coast Guard)

5. Theater Campaign Strategic Guidance

Key applicable strategic guidance for TCPs originates both external to DOD, primarily from DOS, and internal to DOD.

a. DOS Strategic Guidance:

- QDDR
- Integrated Country Strategy (ICS)
- USAID Country Development Cooperation Strategy (CDCS)
- Region Security Initiatives (RSI) Plans.

b. DOD Strategic Guidance.

- NMS/GEF/DPG/JSCP/GFMIG

6. Nested Planning

The nesting relationship of TCPs, FCPs, subordinate campaign plans, and contingency plans is portrayed in (Figure III-1). GCCs must ensure their TCP and contingency plans support the achievement of strategic objectives. Similarly, FCCs must ensure their functional plans support the achievement of strategic objectives. Subordinate campaign plans are nested under the TCP and synchronized with the FCPs. Contingency plans are branches to TCPs and/or subordinate campaign plans. The relationship of TCPs, FCPs, subordinate campaign plans, and contingency plans is unique to each CCDR.

Security Environment:	Campaign Plans will include:	Campaign Plans:
Transnational terrorism	Security Cooperation	Annex-Posture Plans
Spread of WMD	Information operations	Branch/Contingency Plans
Regional Instability	Intelligence	Annex-Security Cooperation Plan**
Increasing powerful states	Strategic Communications*	
Competition for natural Resources	Interagency Cooperation*	*must be addressed at a minimum
Natural disasters and pandemics	Alliance/Partners/Coalition contributions	**Campaign plans address all instruments of national power
Cyber & Space vulnerability and competition	Stability Operations	

Table III-1 TCP requirements

Strategic guidance requires all regional and contingency planning to nest under a theater campaign (Figure III-2). Theater campaign planning is mostly art; requiring the judicious application of operational design to arrive at a concept for theater engagement that provides the conceptual linkage of GEF-directed objectives, through resource informed ways, and means, to regional operational actions. Theater campaign planning directs and justifies the resources required by the CCDR to accomplish assigned theater objectives.[3]

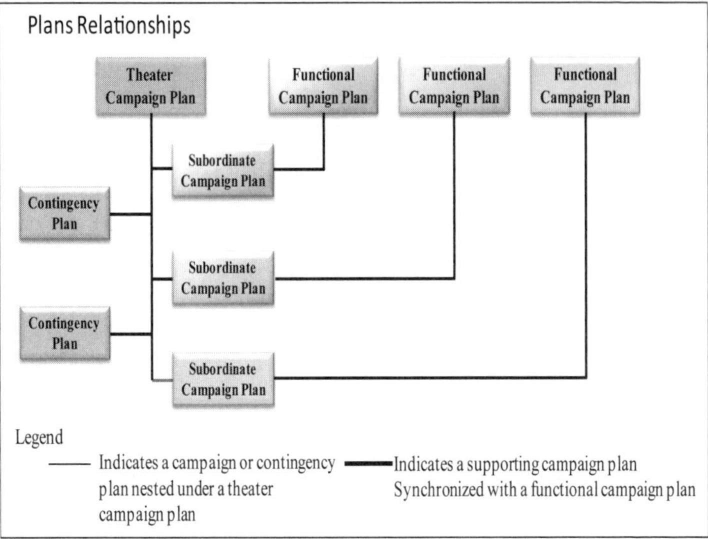

Figure III-1. Plans Relationship (CJCSM 3141.01 series)

Theater campaign planning is inherently intergovernmental. It is informed by the strategic planning of other USG agencies, in particular the DOS and USAID. The intent is for the theater campaign design to complement and support the DOS's broader foreign policy objectives and, to the extent possible, not undermine or work at cross purposes to the goal and activities of other USG agencies in the region. Theater campaign planning is also heavily informed by detailed country planning. Country-level plans help quantify and justify aggregate theater military resource requirements. In peacetime, regional military actions occur in a world where the US Ambassadors' objectives have primacy. Therefore, regional US military operations, activities, events, and investments are prioritized, aligned, and/or integrated with US developmental and diplomatic actions at the country level to achieve unity of effort and husband scarce resources. In the end, theater campaign planners seek to synchronize and nest the planned operations, activities, events, and investments across posture planning, country planning, security cooperation planning, contingency planning, Phase 0 (Shape Phase) integration, strategic communication planning, interagency planning, and multi-national planning in TCPs to promote overall regional unity of effort.

a. Posture Plans. GCCs prepare Theater Posture Plans (TPPs) which outline their posture strategy, link national and theater objectives with the means to achieve them, and identify posture requirements and initiatives to meet TCP objectives. The TPP is the single source document used to advocate for changes to posture and to support resource decisions. It describes the forces, footprint, and agreements present in a theater. FCCs also prepare posture plans to enable their assigned missions and support GCCs.

(1) Posture Planning. TPPs describe the forces, footprint, and agreements present in a theater. They propose a set of posture initiatives and other posture changes, along with the corresponding cost data, necessary to support the DODs activities as described in the commands TCP. TPPs also must account for the desires of the FCCs, other GCCs, and Service's, then balance these possibly competing desires as much as possible.

[3]CJCSM 3130A, APEX Overview and Framework

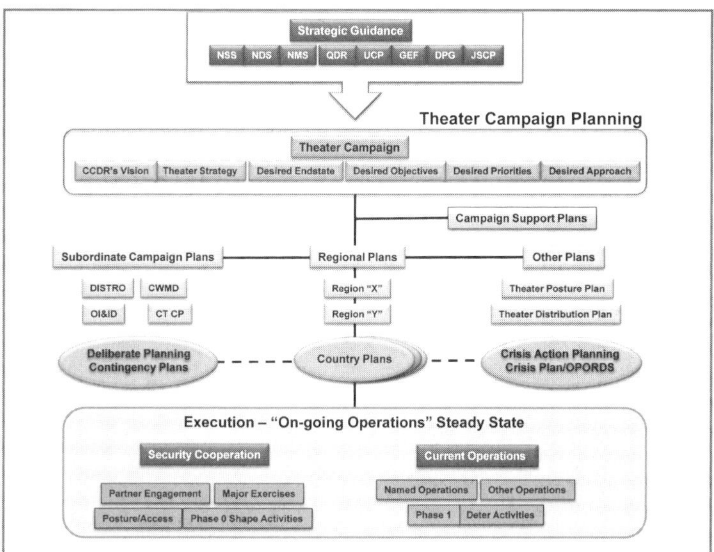

Figure III-2. Nested Plans in Generic TCP Framework (CJCG 3130)

Theater planners must ensure theater objectives that run counter to global and regional objectives are properly aligned and prioritized to ensure that those objectives with the highest priority are elevated and the risk associated with the theater objectives that are counter are well understood. Also, planners must understand that TPPs are integrally linked to the Services ability to resource them both from a fiscal and a force requirements perspective.

b. Country Plans. GCCs prepare selected country plans in collaboration with respective security cooperation organizations. Country plans are nested within TCPs by designing country objectives to support TCP IMOs, link to GEF-directed objectives, and compliment the TCP's activities. A country plan describes how the CCMD, working with the US country team, will engage with the partner country to achieve both US and partner country security objectives, and the role the partner has agreed to play or is expected to play in the theater campaign; and, it must be consistent with the bilateral security agreements that govern the US-partner country relationship.

(1) Country Planning. Usually developed in parallel with TCPs which allows appropriate nesting of country plans within TCPs and allows country planners to inform and be informed by the higher level discussions between CCMD and relevant steady-state actors. Country planners ultimately make a determination on "what the US wants," and "does not want," each country to do in support of US strategy. These country-level assessments inform the development of a country engagement concept (for country-level plans) and the larger theater campaign concept of engagement. Country planning begins with an assessment of the partner country's security environment. In addition to understanding geopolitical trends or conditions that influence a partner nation, planners also assess significant internal and external threats to both the partner nation and neighboring countries in the region. Theater planners evaluate the capabilities and resources of partner nations and their ability and willingness to participate with US forces in coordinated operations, activities, and events. Theater planners also identify key security-related opportunities in context with the goals and activities of other USG agencies in the region. Planners must also review the DOS and USAID strategies to gain appreciation of diplomatic and developmental objectives and goals for a particular country.

c. Security Cooperation Plans. Security cooperation encompasses activities undertaken by the DOD to assist and enable international partners to apply capability and capacity; provide US access to territory, information, and resources; and/or take a political action in support of US goals. It includes, but not limited to, DOD interactions with foreign defense and security establishments and DOD-administered security assistance programs. Security cooperation activities, formerly outlined in Theater Engagement Plans, are now in CCMDs campaign, posture, and country plans and the Services, Defense agencies, and National Guard Bureaus (NGB) campaign support plans (CSPs). Examples of security assistance programs are the Foreign Military Sales Program, Economic Support Fund, and the International Military Education and Training Program to name a few.

(1) Security Cooperation Planning. Security cooperation planning consists of a focused program of bilateral and multilateral defense activities conducted with foreign countries to serve mutual security interests and build defense partnerships. Security cooperation efforts also should be aligned with and support strategic communication themes, messages, and actions. The SECDEF identifies security cooperation objectives, assesses the effectiveness of security cooperation activities, and revises goals when required to ensure continued support for United States interests abroad. Although they can shift over time, examples of typical security cooperation objectives include: creating favorable military regional balances of power, advancing mutual defense or security arrangements, building allied and friendly military capabilities for self-defense and multinational operations, and preventing conflict and crisis. The GEF gives a framework to CCDRs for integrating efforts to shape the strategic environment and it gives them theater or functional strategic objectives appropriately prioritized for each CCMD.

d. Contingency Plans. Contingency plans are branches of theater and functional campaign plans that are deliberately planned for designated threats, catastrophic events, and contingent missions without a crisis at-hand. Contingency plans are nested with theater and functional campaign plans by designing the campaigns activities to contribute to preventing, preparing for, and mitigating the contingencies. Planners utilize the JOPP to plan contingency operations.

(1) Deliberately Planning for a Contingency. Deliberate planning is conducted to respond to defined planning tasks identified in the GEF and the JSCP or as directed by the CCDR. It is based on derived assumptions and forces apportioned from Section IV of the GFMIG. Contingency plans are built to account for the possibility the TCP steady-state activities could fail to prevent aggression, preclude large-scale instability in a key state or region, or mitigate the effects of a major disaster. Under the GEFs campaign planning direction, contingency plans are conceptually considered branches of the overarching TCPs. Deliberate planning for a contingency informs the theater campaign and prepares the CCMD for transition to crisis planning when necessary. Planners developing contingency plans identify shaping, condition setting, prevention, or preparation for entry task requirements to theater campaign planners that can be accomplished within the scope of the TCP's steady-state activities. Contingency planners may also identify shaping, condition setting, prevention, or preparation for entry task requirements specific to their plan that would only be implemented in the event of an emerging crisis (see Chapter 2).

e. Subordinate and Supporting Plans. CCMDs prepare suburbanite campaign plans for regions or functions as the CCDR considers necessary to carry out the missions assigned to the command and as directed in the GEF. Component commands prepare supporting plans at the discretion of the component commander or CCDR, or as directed in the GEF. Supporting plans may include organize, train and equip responsibilities such as exercises, readiness, interoperability, and capabilities development.

f. Interagency Coordination. The quality of DOD theater campaign planning improves with early and regular participation of other US departments and agencies. Interagency coordination is the interaction that occurs among US agencies, including DOD, for the purpose of accomplishing theater objectives. In general, security cooperation will align with

Regional Priorities/Geographic Commands

The DOS has six Bureaus covering regional priorities.[4]

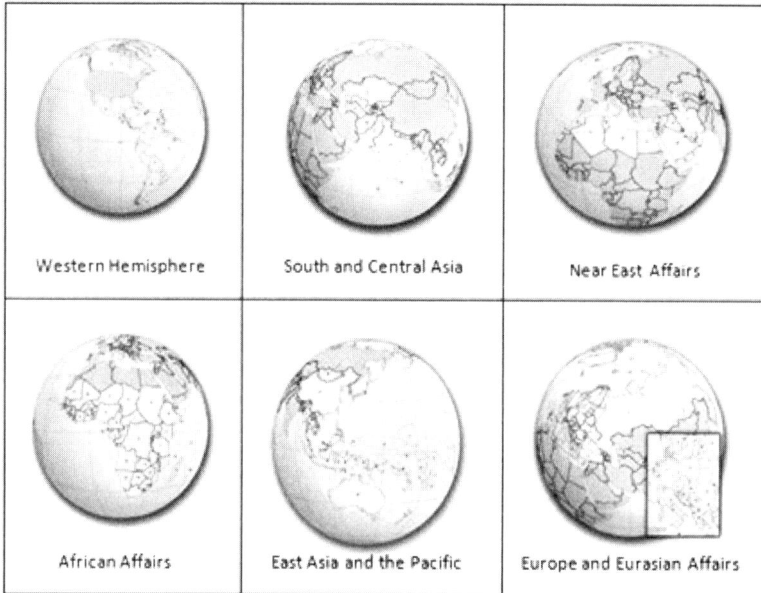

Figure III-3. Department of State Six Regional Bureaus. [4] Department of State/US Agency for International Development , FY 2007-2012 Revised Strategic Plan.

The DoD has six geographic commands:

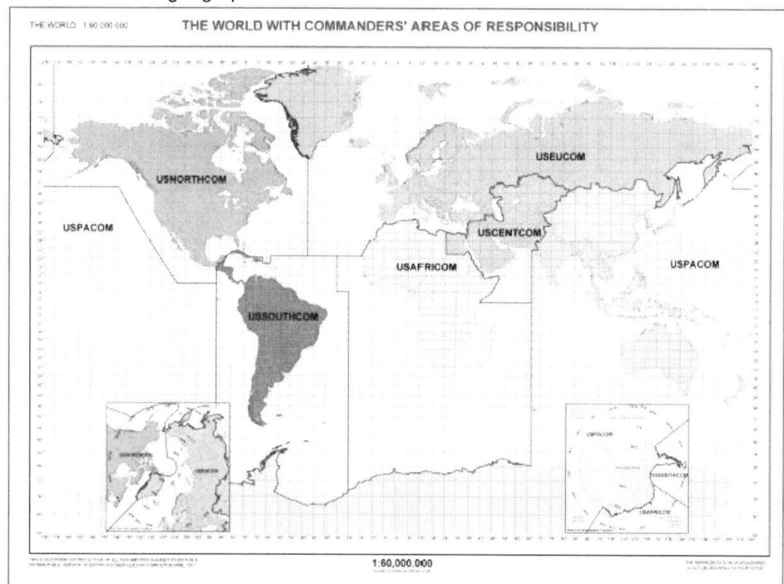

Figure III-4. DOD Geographic Commands (UCP).

broader USG policy. Interagency coordination forges the vital link between the US military and the other instruments of national power.

(1) Interagency Planning. To accomplish required coordination, CCMD planners interact with non-DOD agencies and organizations to ensure mutual understanding of the capabilities, limitations, and consequences of military and nonmilitary actions as well as the understanding of regional objectives. Formally, CCDRs will work through the DOD/ Joint Staff program known as Promote Cooperation (PC). PC is the forum where CCDRs coordinate their plans with other agencies in the national capital region. PC generates collaborative development of DOD plans with civilian agencies and non-DOD entities. PC events provide CCDRs with a means of directly engaging USG departments and agencies to better inform plan development and identify intergovernmental policy issues to advance plan development.

(a) Of note and worth mentioning here are the geographic boundary differences between the DOS Bureaus and the DOD geographic commands (Figures III-3 through III-5). It's important we recognize these seams and boundary differences to ensure smooth coordination between these two interagency partners.

(b) When overlaid with each other we see the potential coordination challenges that face both the DOS and DOD when working across boundaries. Close coordination is required between DOS Bureaus and DOD geographic CCMD's to ensure national security issues and priorities are addressed. (Fig III-5)

Figure III-5. U.S. Department of State Regional Bureaus Overlaid on CCDR's AOR's

g. Multinational Plans. Theater campaign planning considers the capabilities and activities of allies and partners including regional security organizations and other multinational organizations to complement US efforts to achieve regional objectives. Multinational integration in theater campaign planning is accomplished in national and international channels. Collective security goals, strategies, and combined plans are developed in accordance with individual treaty or alliance procedures. Host nation support and contingency mutual support agreements are usually developed through national planning channels.

(1) Multinational Planning. Joint operation planning is integrated with alliance or coalition planning at the theater and operational level by the commander of US national forces dedicated to the alliance or coalition military organization. Contingency plans identify assumed contributions and requested support, where possible, to comply with strategic guidance to incorporate international coordination into DOD planning. Theater planners also consult allies and partners on tailored regional security architectures.

7. Strategy

Strategy is a broad statement of the GCCs long-term vision for the AOR and the FCC's long-term vision for the global employment of functional capabilities. Strategy includes a description of the factors of the environment key to the achievement of the CCMD's GEF-directed objectives, the CCDR's approach to applying military power in concert with other elements of national power in pursuit of the objectives, the resources needed to affect the approach, and the risks inherent in implementation. Developing strategy requires understanding the complexity of the environment, translating national-level objectives into desired conditions, and building flexible, adaptable approaches that enable military means to work with other elements of national power to achieve those desired objectives. CCDRs publish a theater strategy to serve as the framework for the TCP, which ties ends, ways, means and risk together over time, and provides an action plan for the strategy.

> *Strategy evolves over time in a continuous, iterative process; there is no static, single, or final strategy or plan. Commanders and strategists should never assume the plans they create will remain static or be executed as conceived, but should create strategy with the assumption that strategy will need to evolve.*

a. Strategic Estimate. The CCDR and staff, with input from subordinate and supporting commands and agencies, prepare a strategic estimate by analyzing and describing the political, military and economic factors, and the threats and opportunities that facilitate or hinder achievement of the objective over the timeframe of the strategy. The CCMD's input to the Chairman's Comprehensive Joint assessment (CJA) is produced annually and informs the strategic estimate and its periodic updates.

b. Policy-Strategy Dynamic. Strategy is always subordinate to policy. However, there is a two-way dependent relationship between policy and strategy. CCDRs bridge the inevitable friction that policy and politics create when developing the theater strategy. Military strategy must be clear, achievable, and flexible to react to changing policy. Policy may evolve as the theater strategy is implemented in a dynamic operational environment. Also, policy may change in reaction to unanticipated opportunities or in reaction to unanticipated challenges. The CCDR's role is to keep national policy makers informed about changes in the theater's operational environment that effect such policy decisions and provide advice on the potential outcomes of proposed policy changes.

c. Theater Strategy Components. The theater strategy consists of a description of key factors about the environment that provide context for the strategy and affect the achievement of the desired objectives in the theater, a description of the desired strategic objectives (ends), a strategic approach to apply military power in concert with the other elements of national power over time to achieve the desired objectives (ways), a description of resources needed to source the operational approach (means), and a description of the risks in implementing the strategy.

(1) Environment. Planners describe the current environment of the theater, as well as the desired future environment that meets national and regional policy objectives. This provides context for the strategy. While strategy is subordinate to policy, so is it subordi-

nate to the environment – as the environment changes, so must the strategy. The CCDR and staff conduct a theater strategic estimate which describes the broad strategic factors that influence the theater strategic environment. This continually updated estimate helps to determine missions, objectives, and potential activities required in the TCP.

(2) Ends. The ends for both the strategy and campaign are the GEF-directed objectives. CCDRs use the Secretary's prioritization to guide the order in which they employ limited resources, accepting risk on lower priority objectives before accepting risk on higher priority objectives.

(3) Ways. The strategic approach describes the ways that the CCDR will employ the command's total joint force along with other elements of national power to advance toward its objective. Although military operations, activities, and investments may achieve some objectives without the involvement of non-DOD agencies, the commander's strategic approach should be complementary with partner agencies' national security and foreign policy efforts.

(4) Means. The strategy's means are the resources and authorities required to conduct the strategic approach. If there is a reasonable expectation that required means will not become available, then the CCMD must develop an alternative approach within the means that are available or can reasonably be expected to become available. The CCDR takes unresolved issues of means to the DODs senior leaders to identify shortfalls.

(5) Risk. CCDRs assess how strongly U.S. interests are help within their AOR, how those interests can be threatened, and their ability to execute assigned missions to protect them. This is documented in the CCDR's strategic estimate and in the annual Chairman's Comprehensive Joint Assessment (CJA). CCDRs and DODs senior leaders work together to reach a common understanding of campaign risk, decide what risk is acceptable, and minimize the effects of accepted risk by establishing appropriate risk controls.[5]

Campaign Risk *is the strategic risk assessed by the CCMD at theater level combined with the military risk.*

Military Risk *is the risk to mission assessed by the CCMD combined with the risk to the force assessed by the Services.*

8. Theater Campaign Design

Typically, complex tasks or problems are better understood applying operational art and design techniques to assist understanding and visualization. Operational art and design helps planners bound the theater engagement "problem" in such a way that it can be solved by the CCMD. Three interrelated activities collectively provide understanding and visualization of the theater campaign's purpose. These activities include framing the operational environment, framing the problem to be solved, and developing an appropriate operational approach to solve the problem. For greater discussion on these design activities see Chapter VI, Operational Art, Design, and the Joint Operations Cycle.

[5]*CJCSM 3130.01 series*

Joint Operation Planning

Joint operation planning is the problem-solving piece of the "design." It is procedural, follows the steps of the JOPP and produces the requisite plans and orders to direct action. While not prescriptive, it provides a common framework for joint planning and provides interagency and multinational partners an outline for how the U.S. joint forces plan and where to provide their inputs as stakeholders. Joint Operations Planning consists of the following:

- Joint Operation Planning
- Stability Operations
- GFM and Force Projection Planning
- Operation Phasing
- Joint Operation Planning Organization and Responsibility
- In-Progress Reviews

1. Joint Operation Planning

Joint operation planning is the overarching process that guides CCDR's in developing plans for the employment of military power within the context of national strategic objectives and national military strategy to shape events, meet contingencies, and respond to unforeseen crises. *Planners Smart Book* focuses on the planning process and those aspects of GFM which fall into the steps of joint planning.

a. Planning is triggered when directed by strategic guidance or when the continuous monitoring of global events indicates the need to prepare military options. It is a collaborative process that can be iterative and/or parallel to provide actionable direction to CCDRs and their staffs across multiple echelons of command.

b. Joint operation planning includes all activities that must be accomplished to plan for an anticipated operation — the force planning, mobilization, deployment, distribution, employment, sustainment, redeployment and demobilization of forces. Planners recommend and CCDRs define criteria for the termination of joint operations and link these criteria to the transition to stabilization and achievement of the objectives and endstate.

2. Stability Operations

Stability operations are core U.S. military missions that the DOD shall be prepared to conduct and support. Stability operations shall be given priority comparable to combat operations and be explicitly addressed and integrated across all DOD activities including doctrine, organizations, training, education, exercises, materiel, leadership, personnel, facilities, and planning.[1]

a. Per DOD Instruction (DODI) 3000.05, Stability Operations, all military plans shall address stability operation requirements throughout all phases of an operation or plan as appropriate. Stability operations dimensions of military plans shall be:

(1) Conduct stability operations activities throughout all phases of conflict and across the range of military operations, including in combat and non-combat environments. The magnitude of stability operations missions may range from small-scale, short-duration to large-scale, long-duration.

[1] *DODI 3000.05, Stability Operations*

(2) Support stability operations activities led by other USG departments or agencies (hereafter referred to collectively as "USG agencies"), foreign governments and security forces, international governmental organizations, or when otherwise directed.

(3) Lead stability operations includes activities to establish civil security and civil control, restore essential services, repair and protect critical infrastructure, and deliver humanitarian assistance until such time as it is feasible to transition lead responsibility to other USG agencies, foreign governments and security forces, or international governmental organizations. In such circumstances, the DOD will operate within USG and, as appropriate, international structures for managing civil-military operations, and will seek to enable the deployment and utilization of the appropriate civilian capabilities.

3. GFM and Force Projection Planning

GFM is an integral building block to Joint Planning and is conducted simultaneously. At any given time there could be multiple requirements to employ military forces. Each operation could have a different strategic priority, and could be of a different size and scope. To effectively support multiple requirements, and apply the right level of priority and resources to each, requires effective GFM. The national importance of these missions is reflected in the elevated movement priorities that can be invoked by the President or SECDEF. See Chapter VII, Global Force Management, for greater detail.[2]

a. Background. GFM has transformed the former reactive force management process into a near real-time, proactive process. Historically, and prior to GFM, the DOD conducted strategic force management through a decentralized, ad hoc process that framed decision opportunities for the SECDEF. For Operation Enduring Freedom (OEF) and Operation Iraqi Freedom (OIF), the SECDEF made crisis force management decisions in response to CCDR's request for forces (RFF) or capabilities. To support these decisions, the CJCS hosted ad hoc "wargames" to identify forces to support those OEF/OIF requests and determine risk mitigation options. This process was cumbersome and time intensive.[3]

(1) GFM today enables the SECDEF to make proactive, risk-informed force management decisions by integrating and aligning the three processes of force assignment, apportionment, and allocation in support of the NDS, joint force availability requirements, and joint force assessments. This process facilitates alignment of operational forces against known allocation requirements in advance of planning and deployment preparation timelines.

(2) The end result is a timely allocation of forces/capabilities necessary to execute CCMD missions (including Theater Security Cooperation (TSC) tasks), timely alignment of forces against future requirements, and informed SECDEF decisions on the risk associated with allocation decisions while eliminating ad hoc assessments. The Director JS J3 (DJ-3) has been designated the Joint Force Coordinator (JFC)[4] for identifying and recommending sourcing solutions for conventional forces, in coordination with the Military Departments and other CCMDs, and as the conventional Joint Force Provider (JFP). U.S. Special Opera-

[2]*CJCSM 3130.06, GFM Policies and Procedures*

[3]*Global Force Management Implementation Guidance.*

[4]*CJCS, through the Director, J-3 (DJ-3), will serves as the Joint Force Coordinator (JFC) responsible for providing recommended sourcing solutions for all validated force and JIA requirements. In support of the DJ-3 the Joint Staff Deputy Director for Regional Operations and Force Management (J-35) assumes the responsibilities of the JFC. As such the JFC will coordinate with the Joint Staff J-3, Secretaries of the Military Departments, CCDRs, JFPs, JFM, and DoD Agencies. The Joint Force Coordinator (JFC) is referred to in current DoD GFM guidance and policy as the JFC. However, for clarity in this document, and distinction between the Joint Force Commander (JFC) and Joint Force Coordinator (JFC), the Joint Force Coordinator will be referred to as the Joint Staff Joint Force Coordinator (JS JFC).*

tions Command (USSOCOM) is the JFP for special operations forces and U.S. Transportation Command (USTRANSCOM) is the JFP for mobility forces. U.S. Strategic Command (USSTRATCOM) is the Joint Force Manager (JFM) for ISR and Missile Defense.

b. Baseline Documents. The UCP, GFMIG to include - "Forces For Unified Commands Memorandum" (Forces For) (within Section II of GFMIG), and GFM Allocation Business Rules (within Annex C of the GFMIG), the CJCSI 3110.01, Joint Strategic Capabilities Plan, CJCSM 3130.06, GFM Allocation Policies and Procedures, and JP-1, Doctrine for the Armed Forces of the United States, are the baseline documents that establish the policy and procedures in support of GFM. GFM includes: (1) direction from the SECDEF as to assignment of forces to CCMDs, (2) the forces/capabilities allocation process that provides access to all available forces – including military, DOD Combat Support Agencies (CSAs), and other federal agency resources (by request) — to support CCMDs for both steady state and rotational requirements and requests for capabilities or forces in response to crises or emergent contingencies, (3) include apportionment guidance provided in the GFMIG, and (4) inform joint force, structure, and capability assessment process.

c. The Assignment, Allocation, and Apportionment Relationship.[5] GFM guides the global sourcing processes of CCMD force requirements. It provides the JS and force providers (FPs) a decision framework for making force assignment and allocation recommendations to the SECDEF and apportionment recommendations to the CJCS. GFM is used to align U.S. forces in support of the NDS. It provides comprehensive insights into the global availability of U.S. military forces/capabilities and provides senior decision-makers a process to quickly and accurately assess the impact and risk of proposed changes in forces/capability assignment, apportionment, and allocation. Authorities that govern the three processes are listed in Chapter VII, GFM Overview.

d. GFM Process.[6] The GFM process begins and ends with the SECDEF. GFM aligns force apportionment, assignment, and allocation methodologies in support of the NDS and joint forces availability requirements. It provides comprehensive insights into the global availability of U.S. military forces/capabilities for plans and operations. It also provides senior decision-makers a process to assess quickly and accurately the impact and risk of proposed changes in forces/capability assignment, apportionment, and allocation.

(1) Force Visibility. Force visibility provides the current and accurate status of forces at the strategic and operational level; their current mission; future missions; location; mission priority and readiness status. Force visibility provides information on the location, operational tempo, assets and sustainment requirements of a force as part of an overall capability for a CCDR. Force visibility integrates operations and logistics information and facilitates GFM, and enhances the capability of the entire JPEC to adapt rapidly to unforeseen events, to respond and ensure capability delivery. Force visibility enhances situational awareness and is required to support force sourcing, allocation, assignment of forces; force position; sustainment forecasting and delivery; and forecasting for future force requirements. Force visibility is achieved through effective force and phase planning for contingencies and crises; detailed deployment planning; and sound reporting procedures.[7]

[5]GFMIG, also refer to Chapter XIV, Key-Step 7, and Chapter VII, GFM

[6]For additional information, see Guidance for Employment of the Force (GEF), Global Force Management Implementation Guidance (GFMIG), and Chairman of the Joint Chiefs of Staff Manual (CJCSM) 3122.01Series.

[7]JP 3-35, Deployment and Redeployment Operations.

4. Full Range of Joint Operations Activities

Joint operations planning encompasses the full range of activities required to conduct joint operations. Besides force sourcing these activities include the mobilization, deployment, employment, sustainment, redeployment, and demobilization of forces.

a. Mobilization.[8] Mobilization is the process of assembling and organizing national resources to support national objectives in time of war or other emergencies. Mobilization includes assembling and organizing personnel and materiel for active duty military forces, activating the RC (including federalizing the National Guard), extending terms of service, surging and mobilizing the industrial base and training bases, and bringing the Armed Forces of the United States to a state of readiness for war or other national emergency. There are two processes implied in this description; (1) the military mobilization process by which the nation's Armed Forces are brought to an increased state of readiness and (2) the national mobilization process of mobilizing the interdependent resource areas (see JP 4-05, Chapter IV, "Resource Areas") to meet non-defense needs as well as sustaining the Armed Forces across the range of military operations.

(1) From a national strategic perspective, the importance of a responsive mobilization capability to our national security is implicit in the President's NSS and its derivatives, the NDS, and the NMS.

(2) Mobilization Planning and Operation Plans. The GEF, GDF, and DOD Master Mobilization Plan provide SECDEF guidance for mobilization planning in support of joint operations. There are five mobilization tenets that describe the characteristics of successful mobilization and provide the foundation for mobilization doctrine, these are: objective, timeliness, unity of effort, flexibility, and sustainability.

(a) Mobilization planning complements and supports joint operation planning. It is accomplished primarily by the Services and their major subordinate commands based on SECDEF guidance. It requires development of supporting plans by other federal agencies. Just as the Services mobilize their reserve organizations and individuals to augment military capability, supporting federal agencies must oversee mobilization of the support base required to sustain the mobilized force.

(b) Mobilization plans reflect requirements for force expansion with reserve component (RC) units and Joint Individual Augmentees (JIAs) and for expansion of the Continental United States (CONUS) base to sustain the mobilized force for as long as necessary to achieve military and national security objectives. Mobilization plans explain how force and resource expansion is to be accomplished.

(c) Mobilization is a complex, time-sensitive process with many participants and activities. Mobilization plans must be carefully integrated among participants and the resource areas. Mobilization execution must be sequenced and carefully synchronized to ensure that resources are available to the supported and supporting commanders when needed. The CJCS, supported by the JS, integrates mobilization planning and monitors the status and progress of mobilization execution. Figure IV-1 shows levels of mobilization and levels of military commitment. CJCS advises SECDEF on establishing priorities; allocating resource shortages among claimants; and redirecting execution activities, when necessary, to eliminate bottlenecks and overcome unforeseen problems.

(3) CJCSI 3110.13, Mobilization Guidance for the Joint Strategic Capabilities Plan, guides the Military Departments and CCDRs in preparing mobilization plans that support the Operation Plan (OPLAN) developed in the deliberate planning process. The planning guidance is focused on the areas of manpower and industrial mobilization. Manpower mobilization requirements derived for each plan establish the level of mobilization assumed for each plan and drive the determination of mobilization requirements in the resource areas. The industrial mobilization guidance requires the Military Departments to conduct industrial preparedness planning and to maintain a production base that will support planning requirements.

[8] JP 4-05, Joint Mobilization Planning, provides fundamental principles and guidance for the planning and conduct of joint military mobilization and demobilization.

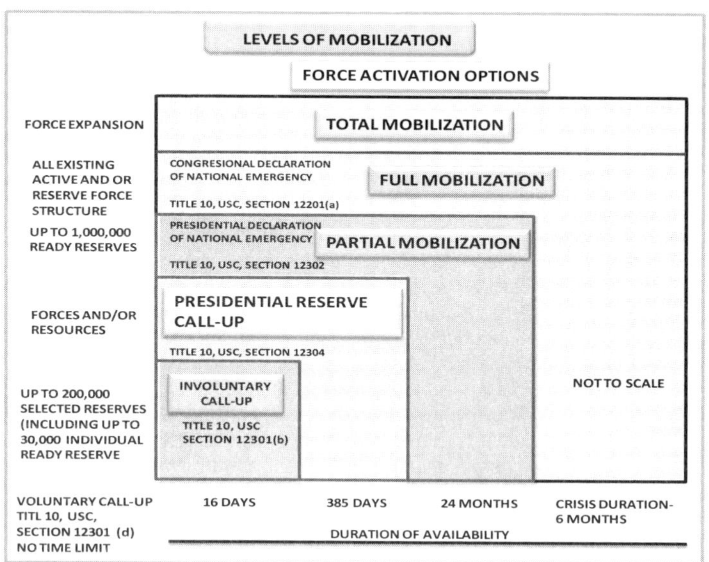

Figure IV-1. Levels of Mobilization (JP 3-35)

(4) During the planning process, the Military Departments furnish mobilization-related information to the CCDRs, who incorporate it into the OPLANs under development or revision.

(5) Manpower mobilization information furnished by the Military Departments for inclusion in OPLANs provides the foundation for detailed planning in the other resource areas. This information comprises the number of AC and RC personnel required by Service and skill for each option included in the OPLAN.

b Deployment. Deployment encompasses the movement of forces and their sustainment resources from their original locations to a specific destination to conduct joint operations. It specifically includes movement of forces and their requisite sustaining resources within the United States, within theaters, and between theaters. Deployment operations encompass four major nodes for the distribution process: (1) point of origin, (2) port of embarkation (POE), (3) port of debarkation (POD), and (4) destination; and three segments: (1) point of origin to POE, (2) POE to POD, and (3) POD to destination. GCCs are responsible for coordinating with the USTRANSCOM and supporting CCDRs to provide an integrated transportation system from origin to destination during deployment operations.

(1) Supported CCDRs are responsible for deployment operations planned and executed during joint force missions in their AORs. Supported CCDRs have four major responsibilities relative to deployment operations: (1) build and validate movement requirements based on the CONOPS; (2) determine pre-deployment standards; (3) balance and regulate the transportation flow; (4) and manage effectively. The primary task for supporting CCDRs is to ensure that the supported CCDR receives the timely and complete support needed to accomplish the mission. Supporting CCDRs have five major deployment responsibilities: source, prepare, and verify forces; ensure units retain their visibility and mobility; ensure units report movement requirements rapidly and accurately; regulate the flow; and coordinate effectively. Normally, several functional CCMDs are involved in every phase of a joint operation. Three functional CCMDs that could be involved in deployment of the joint force are USSOCOM, USSTRATCOM, and USTRANSCOM.

(2) Deployment planning is based primarily on mission requirements and the time available to accomplish the mission. During deployment operations (deployment, Joint

Reception, Staging, Onward Movement, and Integration (JRSOI), and redeployment), supported CCDRs are responsible for building and validating requirements, determining pre-deployment standards, and balancing, regulating, and effectively managing the transportation flow.

(3) Supporting CCMDs and agencies source requirements not available to the supported CCDR and are responsible for: verifying supporting unit movement data; regulating the support deployment flow; and coordinating effectively during deployment operations.

(4) Regardless of whether deliberate planning or crisis action planning (CAP) is used, joint planning determines the requirements for joint force employment to achieve the military objectives. Once the supported CCDR's strategic concept is approved by the CJCS, it becomes the CONOPs upon which further planning is developed. Planning is based on CCDR(s) and Service(s) guidance and joint doctrine. The supporting and subordinate CDRs use the supported CCDR's CONOPS and the apportioned or allocated combat forces as the basis to determine necessary support, including forces and sustaining supplies for the operation (mission analysis). The supported CCDR's staff organization is established and command relationships are formulated to assist the CDR in determining priorities and assigning tasks for conducting deployment operations. Supported CCDRs may task assigned Service components with the majority of responsibility for deployment operations based upon various factors (e.g., dominant user, most capable Service). Each supporting or subordinate CDR who is assigned a task in the CCDR's strategic concept prepares a supporting plan. The CCDR consolidates these plans to build a recommended phasing of forces and support, and performs a transportation analysis of the entire movement from the POE to the final destination. In essence, the supported CCDR uses the information to validate the adequacy of the theater and determine whether the infrastructure is satisfactory for employment of assets, forces, facilities, and supporting systems. Joint Intelligence Preparation of the Operational Environment (JIPOE) provides the framework for determining methods of accomplishing the assigned tasks. Following these actions, the supported CCDR, with USTRANSCOM support, hosts the TPFDD refinement conference. (JP 3-35, Joint Deployment and Redeployment Operations, discusses joint deployment planning in greater detail.)

c. Employment. Employment encompasses the use of military forces and capabilities within an operational area (OA). Employment planning provides the foundation for, determines the scope of, and is limited by mobilization, deployment, and sustainment planning. Employment is primarily the responsibility of the supported CCDRs and their subordinate and supporting CDRs. JP 5-0, Joint Operation Planning, JP 3-0, Joint Operations, this publication, and numerous other publications in the Joint Doctrine system discuss joint employment planning in greater detail.

d. Sustainment. Sustainment is the provision of logistics and personnel services required to maintain and prolong operations until successful mission accomplishment. The focus of sustainment in joint operations is to provide the CCDR with the means to enable freedom of action and endurance and extend operational reach. Effective sustainment determines the depth to which the joint force can conduct decisive operations, allowing the CCDR to seize, retain and exploit the initiative. Sustainment is primarily the responsibility of the supported CCDRs and their Service component CDRs in close cooperation with the Services, combat support agencies, and supporting commands.

e. Redeployment. Redeployment is the transfer of deployed forces and accompanying materiel from one operational area to support another JFC's operational requirements within a new operational area, or to home/de-mobilization station as a result of end-of-mission or rotation.

(1) Redeployment Planning and Execution. Similar to deployment operations, redeployment planning decisions are based on the operational environment in the OA at the time of redeployment. The redeployment process consists of four phases; redeployment planning, pre-redeployment activities, movement and JRSOI. The supported CCDR is

responsible for redeployment planning. This planning should be considered at the outset of an operation and continually be refined as the operation matures. The individual activities within each phase of redeployment are similar to those described in the deployment process, however significant differences exist during the JRSOI phase. These differences are apparent when the force is redeploying to a new OA or redeploying back to home or de-mobilization station. These distinctions and the command relationships during redeployment are discussed in detail in JP 3-35.

(2) Redeployment operations are the sum of activities required to plan, prepare, and move forces and accompanying materiel from origin to destinations within a new operational area or to home station to achieve the operational status required to execute its next mission or demobilize (JP 3-35 discusses joint deployment planning in greater detail).

f. Demobilization. Demobilization is the process of transitioning from a conflict or wartime military establishment and defense-based civilian economy to a peacetime configuration while maintaining national security and economic vitality. Implied in this description are two types of activities: those associated with reducing the percentage of the nation's production capacity devoted to the Armed Forces and defense industry, and those undertaken to maintain national security and economic vitality. These tasks, which historically compete for resources, can make the management of demobilization even more complex and challenging than mobilization. It involves more than releasing personnel from active duty, deactivating units, and reorganizing the RC. Although these activities drive the process, capability or capacity in the other resource areas must be reduced and reorganized at the same time. As in mobilization, activities in each resource area during demobilization will affect each of the others. For this reason, close coordination among resource area proponents is just as important during demobilization as it is during mobilization.

(1) From a national perspective, the results of a successful demobilization process should put the United States in a position to respond to future challenges to our national security. Policies should be established to regulate the pace of demobilization and retain the military capability required to ensure post-conflict national security commitments. Force structure changes are not inherent to the demobilization process. Industrial base and other civil sector resources mobilized during the conflict will be released to fuel the post-conflict national economy.

(2) The scope of demobilization will vary according to the extent of the preceding mobilization. The scope of mobilization can range from a relatively brief use of a few volunteer reservists to a protracted force and resource expansion well beyond the original peacetime levels.

(3) From a joint military perspective, demobilization plans should reflect the post conflict missions of supported GCCs and be synchronized with plans for recovery, reconstitution, and redeployment operations. DOD policies for the release of reservists and RC units ordered to active duty should first reflect military requirements and then considerations of equity and fairness for military personnel and their families. The demobilization personnel management programs of the Military Departments will be challenged to facilitate the return of Service members and their families to civilian life and need to provide substantive transition assistance, such as screening for medical care requirements and potential long-term health care support, and assistance in availing themselves of statutory reemployment rights, as members reenter the civilian workforce. National Guard units, reserve units and members ordered to active duty to augment the AC will, consistent with operational requirements, receive priority for redeployment. They will be released from active duty as expeditiously as possible.

5. Global Demand

In order to distribute a limited number of forces among the competing CCDR demands, the JS JFC/JFPs, FPs, Service Components, JS, OSD and the SECDEF must understand the entire global demand on the force pool. By understanding global demand, the risks of allocating forces to a given operation can be better understood (for a full discussion on global demand see Chapter 7, GFM).

6. Sequencing Actions and Phasing

Part of the art of planning is determining the sequence of actions that best accomplishes the mission. The concept of operations describes in sequence the start of the operation to the projected status of the force at the operation's end, or endstate. If the situation dictates a significant change in mission, tasks, task organization, or priorities of support during the operation, the commander may phase the operation.[9] A phase is a planning and execution tool used to divide an operation in duration or activity.

a. Phasing. A phase is a definitive stage of an operation or campaign during which a large portion of the forces and capabilities are involved in similar or mutually supporting activities for a common purpose. Phasing, which can be used in any operation regardless of size, helps the commanders organize large operations by integrating and synchronizing subordinate operations. Phasing helps commanders and staffs visualize, design, and plan the entire operation or campaign and define requirements in terms of forces, resources, time, space, and purpose. It helps them systematically achieve military objectives that cannot be attained all at once by arranging smaller, related operations in a logical sequence. Phasing also helps commanders mitigate risk in the more dangerous or difficult portions of an operation.

b. Operation Phasing Model

(1) Phasing is critical to arranging all tasks of an operation that cannot be conducted simultaneously. It describes how the commander envisions the overall operation unfolding. It is the logical expression of the commander's visualization in time. Within a phase, a large portion of the force executes similar or mutually supporting activities. Achieving a specified condition or set of conditions typically marks the end of a phase.

(2) The phasing model (Figure IV-2) displays all six phases: shape, deter, seize the initiative, dominate, stabilize the environment, and enable civil authority. Each phase must be considered during operation planning and plan assessment. This construct is prescriptive in nature and is meant to provide planners a consistent template while not imparting additional constraints on the flexibility of CCDRs. CCDRs are not obligated to execute all phases, but are expected to demonstrate consideration of all phases during their planning. The six-phase model is not intended to be a universally prescriptive template for all conceivable joint operations and may be tailored to the character and duration of the operation to which it applies.

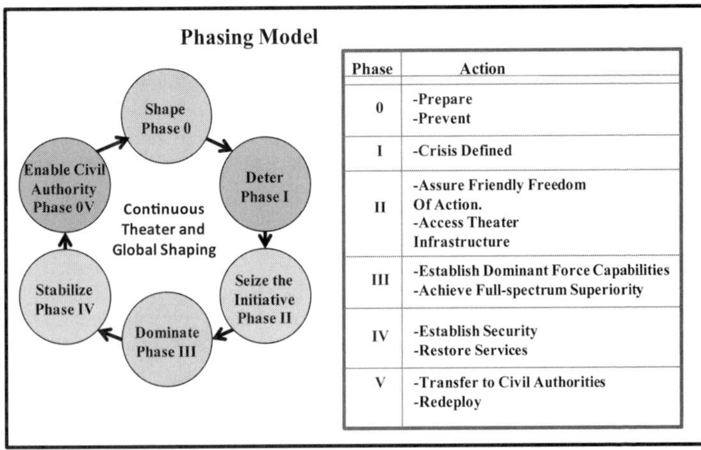

Phase	Action
0	-Prepare -Prevent
I	-Crisis Defined
II	-Assure Friendly Freedom Of Action. -Access Theater Infrastructure
III	-Establish Dominant Force Capabilities -Achieve Full-spectrum Superiority
IV	-Establish Security -Restore Services
V	-Transfer to Civil Authorities -Redeploy

Figure IV-2. Phasing Model

[9] ADRP 3-0

> ***Inducement***: *Increases the benefits of and/or reduces the cost of compliance (increasing overall utility of complying with our demands).*
>
> ***Persuasion***: *Alters the preferences against which the costs and benefits are evaluated (changing the decision context).*

c. A phase can be characterized by the "focus" that is placed on it. Phases are distinct in time, space, and/or purpose from one another, but must be planned in support of each other and should represent a natural progression and subdivision of the campaign or operation. Each phase should have a set of starting conditions (that define the start of the phase) and ending conditions (that define the end of the phase). The ending conditions of one phase are the starting conditions for the next phase. Phases are necessarily linked and gain significance in the larger context of the campaign.

The six phases are described as follows:

(1) Shape. Shaping Operations are focused on partners, potential partners and those that might impede our efforts or provide indirect support to adversaries. Shaping supports deterrence by showing resolve, strengthening partnership and fostering regional security. Insofar as the influencing of potential adversaries is concerned, shaping utilizes inducementand persuasion. Shaping activities set the foundations for operational access as well as develop the relationships and organizational precursors that enable effective partnerships in time of crisis.

(a) Participation in effective regional security frameworks with other instruments of national and multi-national power is critical. Pre-crisis shaping activities by their nature rely heavily on the non-military contributors to unified action; for example, the State Department as the lead agency for United States foreign policy leads the individual country teams, funds security assistance and is responsible for the integration of information as an instrument of national power. Also, the State Department's Office of the Coordinator for Reconstruction and Stabilization (S/CRS) has the mission to lead, coordinate and institutionalize U.S. Government civilian capacity to prevent or prepare for post-conflict situations, and to help stabilize and reconstruct societies in transition from conflict or civil strife.

(b) Ultimately, shaping operations will support the achievement of an endstate that provides a global security environment favorable to U.S. interests. Figure IV-3 is an example of a phasing construct for Humanitarian Assistance. Note that this construct supports the phasing model; however, the phases have been modified to support an endstate in which the DOD is in support of OGOs (DOS/USAID).

(c) The Joint Force, as part of a larger multinational and interagency effort, conducts continuous, anticipatory shaping operations that build partnerships with governmental, non-governmental, regional and international organizations, and reduces the causes of conflict and instability in order to prevent or mitigate conflict or other crises and set the conditions for success in other operations - all aimed at a secure global environment favorable to U.S. interests.[10]

[10] *Military Support to Shaping Joint Operating Concept.*

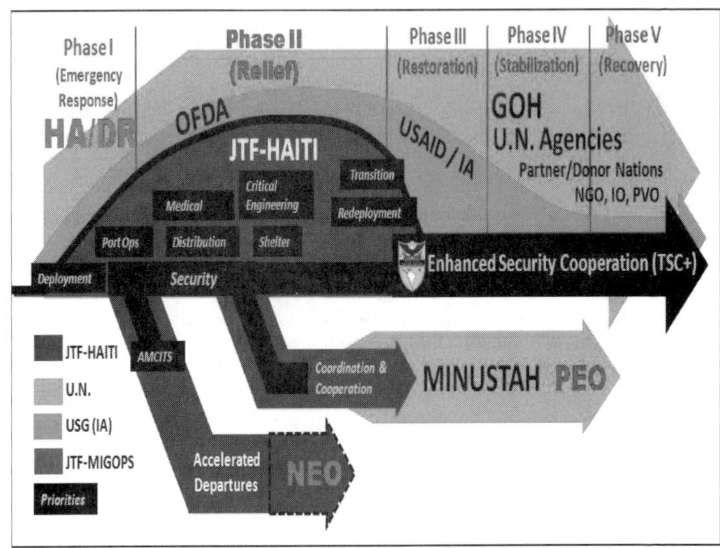

Figure IV-3. Example Humanitarian Phasing Construct

(d) Joint, interagency and multinational operations are executed continuously with the intent to enhance international legitimacy and gain multinational cooperation in support of defined national strategic and strategic military objectives. They are designed to assure success by shaping perceptions and influencing the behavior of both adversaries and allies, developing allied and friendly military capabilities for self-defense and coalition operations, improving information exchange and intelligence sharing, and providing U.S. forces with peacetime and contingency access. Shape phase activities must adapt to a particular theater environment and may be executed in one theater in order to create effects and/or achieve objectives in another. Planning that supports most "shaping" requirements typically occurs in the context of day-to-day security cooperation, and CCMDs may incorporate shaping activities and tasks into the SCP/Theater Campaign Plan. Contingency and crises requirements also occur while global and theater shaping activities are ongoing, and these requirements are satisfied in accordance with the CJCSM 3122 series. Moreover, the JOPP steps described in Chapter IX, "The Joint Operation Planning Process," are useful in planning security cooperation activities as well as developing OPLANs and OPORDs.

(2) Deter.[11] The intent of this phase is to deter undesirable adversary action by demonstrating the capabilities and resolve of the joint force. It differs from deterrence that occurs in the shape phase in that it is largely characterized by preparatory actions that specifically support or facilitate the execution of subsequent phases of the operation/campaign. Deterrence supports shaping by helping to reassure states that cooperative partnership with the United States will not result in an unacceptable threat. Insofar as the influencing of potential adversaries is concerned, deterrence deals with coercive forms of influence.

(a) Deterrence operations are designed to convince adversaries not to take actions that threaten the vital interests of the United States by means of decisive influence over their decision-making. Decisive influence is achieved by credibly threatening to deny benefits and/or impose costs, while encouraging restraint by convincing the actor that restraint will result in an acceptable outcome. Because of the uncertain future security environment, specific vital interests may arise that are identified by senior national leadership. Deterrence strategy and planning must be sufficiently robust and flexible to accommodate these changes when they occur.

[11] *Deterrence Operations, Joint Operating Concept.*

> *"These phases of a plan do not comprise rigid instructions, they are merely guideposts. Rigidity inevitably defeats itself, and the analysts who point to a changed detail as evidence of a plan's weakness are completely unaware of the characteristics of the battlefield."*
>
> *General Dwight D. Eisenhower*

> ***Coercion**: Increases the cost and/or reduces the benefits of defiance (decreasing the overall utility of defying our demands)*
>
> ***Deterrence**: Demand that the adversary refrain from undertaking a particular action linked to a threat to use force if it does not comply*
>
> ***Compellence**: Demand that the adversary undertake a particular action linked to a threat to use force if it does not comply*

(b) An adversary's deterrence decision calculus focuses on their perception of three primary elements:

1 The benefits of a COA.

2 The costs of a COA.

3 The consequences of restraint (i.e., costs and benefits of not taking the COA we seek to deter).

(c) Joint military operations and activities contribute to the "end" of deterrence by affecting the adversary's decision calculus elements in three "ways":

1 Deny benefits.

2 Impose costs.

3 Encourage adversary restraint.

(d) The "ways" are a framework for implementing effective deterrence operations. These "ways" are closely linked in practice and often overlap in their application; however, it is useful to consider them conceptually separate for planning purposes. Military deterrence efforts must integrate all three ways across a variety of adversaries and deterrence objectives. Deterrence ways are not either/or propositions. Rather, when properly leveraged to convince an adversary his best option is not taking a COA aimed against the vital interests of the United States, they are complementary and synergistic. Because future threats will be increasingly transnational, these military deterrence efforts will likely involve synchronized actions by multiple CCDRs worldwide.

(e) The central idea is implemented at the operational level by:

1 Tailoring deterrence operations to specific adversaries and contexts.

2 Conducting dynamic deterrence assessment, planning, and operations.

3 Deterring multiple decision-makers at multiple levels.

4 Characterizing, reducing, and managing uncertainty.

(f) The specific military "means" required to credibly threaten benefit denial and cost imposition, or otherwise encourage adversary restraint, will vary significantly by adversary and situation. Military objectives and means cannot be considered in isolation; these objectives may change over time and must be synchronized with the application of the other instruments of national power. Some aspects of these military means may contribute more directly to warfighting (i.e., defeat) than deterrence. However, it is possible to identify key joint capabilities (and deterrence-related attributes of those capabilities) that must be planned for regardless of their warfighting utility.

(g) The military means of "Deterrence Operations" fall into two categories: (1) those that directly and decisively influence an adversary's decision calculus, and (2) those that enable such decisive influence.

1 Direct means include:

- Force projection.
- Active and passive defenses.
- Global strike (nuclear, conventional, and non-kinetic).
- Strategic communication.

2 Enabling means include:

- Global situational awareness (ISR).
- Command and Control (C2).
- Forward presence.
- Security Cooperation and Military Integration and Interoperability.
- Deterrence assessment, metrics, and experimentation.

(h) Once a crisis is defined, these actions may include mobilization, tailoring of forces and other pre-deployment activities; initial deployment into a theater; employment of ISR assets to provide real-time and near-real-time situational awareness; setting up of transfer operations at enroute locations to support aerial ports of debarkation in post-chemical, biological, radiological, nuclear, and high-yield explosives attack configurations; and development of mission-tailored C2, intelligence, force protection, transportation, and logistic requirements to support the CCDR's concepts of operations.

(i) CCDRs continue to engage multinational partners, thereby providing the basis for further crisis response. Liaison teams and coordination with other agencies assist in setting conditions for execution of subsequent phases of the campaign or operation. Many actions in the deter phase build on security cooperation activities from the previous phase and are conducted as part of security cooperation plans and activities. They can also be part of stand-alone operations.

(3) Seize the Initiative. CCDRs and their subordinate JFCs seek to seize the initiative in combat and noncombat situations through the application of appropriate joint force capabilities. In combat operations this involves executing offensive operations at the earliest possible time, forcing the adversary to offensive culmination and setting the conditions for decisive operations. Rapid application of joint combat power may be required to delay, impede, or halt the adversary's initial aggression and to deny the initial objectives. If an adversary has achieved its initial objectives, the early and rapid application of offensive combat power can dislodge adversary forces from their position, creating conditions for the exploitation, pursuit, and ultimate destruction of both those forces and their will to fight during the dominate phase. During this phase, operations to gain access to theater infrastructure and to expand friendly freedom of action continue while the CCDR seeks to degrade adversary capabilities with the intent of resolving the crisis at the earliest opportunity. In all operations, the CCDR establishes conditions for stability by providing immediate assistance to relieve conditions that precipitated the crisis.

(4) Dominate. The dominate phase focuses on breaking the enemy's will for organized resistance or, in noncombat situations, control of the operational environment. Success in this phase depends upon overmatching joint force capability at the critical time and place. This phase includes full employment of joint force capabilities and continues the appropriate sequencing of forces into the Operational Area (OA) as quickly as possible. When a campaign or operation is focused on conventional enemy forces, the dominate phase normally concludes with decisive operations that drive an adversary to culmination and achieve the CCDR's operational objectives. Against unconventional adversaries, decisive operations are characterized by dominating and controlling the operational environment through a combination of conventional, unconventional, information, and stability operations. Stability operations are conducted as needed to ensure a smooth transition

to the next phase and relieve suffering. In noncombat situations, the joint force's activities seek to control the situation or operational environment. Dominate phase activities may establish the conditions for an early favorable conclusion of operations or set the conditions for transition to the next phase.

(5) Stabilize the Environment. The stabilize phase is required when there is no fully functional, legitimate civil governing authority present. The joint force may be required to perform limited local governance, integrating the efforts of other supporting/contributing multinational, Inter-Governmental Organizations (IGOs), Non-Governmental Organizations (NGOs), or USG agency participants until legitimate local entities are functioning. This includes providing or assisting in the provision of basic services to the population. The stabilize phase is typically characterized by a change from sustained combat operations to stability operations. Stability operations are necessary to ensure that the threat (military and/ or political) is reduced to a manageable level that can be controlled by the potential civil authority or, in noncombat situations, to ensure that the situation leading to the original crisis does not reoccur and/or its effects are mitigated. Redeployment operations may begin during this phase and should be identified as early as possible. Throughout this segment, the CCDR continuously assesses the impact of current operations on the ability to transfer overall regional authority to a legitimate civil entity, which marks the end of the phase.

(6) Enable Civil Authority.[12] This phase is predominantly characterized by joint force support to legitimate civil governance in theater. Depending upon the level of indigenous state capacity, joint force activities during Phase VI may be at the behest of that authority or they may be under its direction. The goal is for the joint force to enable the viability of the civil authority and its provision of essential services to the largest number of people in the region. This includes coordination of joint force actions with supporting or supported multinational, agency, and other organization participants; establishment of MOEs; and influencing the attitude of the population favorably regarding the United States and local civil authority's objectives. DOD policy is to support indigenous persons or groups promoting freedom, rule of law, and an entrepreneurial economy and opposing extremism and the murder of civilians. The joint force will be in a supporting role to the legitimate civil authority in the region throughout the enable civil authority phase. Redeployment operations, particularly for combat units, will often begin during this phase and should be identified as early as possible. The military endstate is achieved during this phase, signaling the end of the campaign or operation. Operations are concluded when redeployment is complete. CCMD involvement with other nations and agencies, beyond the termination of the joint operation, may be required to achieve the national strategic endstate.

d. As a general rule, the phasing of the campaign or operation should be conceived in condition-driven rather than time-driven terms. However, resource availability depends in large part on time-constrained activities and factors—such as sustainment or deployment rates—rather than the events associated with the operation. The challenge for planners, then, is to reconcile the reality of time-oriented deployment of forces and sustainment with the event-driven phasing of operations.

e. Effective phasing must address how the joint force will avoid reaching a culminating point. If resources are insufficient to sustain the force until achieving the end state, planners should consider phasing the campaign or operation to account for necessary operational pauses between phases. Such phasing enables the reconstitution of the joint force during joint operations, but the commander must understand that this may provide the adversary an opportunity to reconstitute as well. In some cases, sustainment requirements, diplomatic factors, and political factors within the host nation may even dictate the purpose of certain phases as well as the sequence of those phases. For example, phases may shift the main effort among Service and functional components to maintain momentum while one component is being reconstituted.

[12] *JP 5-0, Joint Operations Planning, Chapter III contains greater detail on Interagency, NGO, IGO and Enabling Civil Authorities*

f. Transitions. Transitions between phases are designed to be distinct shifts in focus by the joint force, often accompanied by changes in command or support relationships. The activities that predominate during a given phase, however, rarely align with neatly definable breakpoints. The need to move into another phase normally is identified by assessing that a set of objectives are achieved or that the enemy has acted in a manner that requires a major change in focus for the joint force and is therefore usually event driven, not time driven. Changing the focus of the operation takes time and may require changing commander's objectives, desired effects, measures of effectiveness (MOEs), priorities, command relationships, force allocation, or even the design of the OA. An example is the shift of focus from sustained combat operations in the dominate phase to a preponderance of stability operations in the stabilize and enable civil authority phases. Hostilities gradually lessen as the joint force begins to reestablish order, commerce, and local government and deters adversaries from resuming hostile actions while the US and international community take steps to establish or restore the conditions necessary for long-term stability. This challenge demands an agile shift in joint force skill sets, actions, organizational behaviors, and mental outlooks, and inter-organizational coordination with a wider range of interagency and multinational partners and other participants to provide the capabilities necessary to address the mission-specific factors.[13]

7. Joint Operation Planning Organization & Responsibility

Joint operation planning is an inherent command responsibility established by law and directive. This fundamental responsibility extends from the President and SECDEF, with the advice of the CJCS, to the CCDRs and their subordinate JFCs. Joint force Service and functional components conduct component planning that could involve planning for the employment of other components' capabilities, such as when the Joint Force Air Component Commander (JFACC) plans for the employment of all air assets made available. The CJCS transmits the orders of the President and the SECDEF to the CCDRs and oversees the CCMDs' planning activities. The JCS function in the planning process as advisers to the President, NSC, and SECDEF.

a. The CJCS, CCDRs, and subordinate JFCs have primary responsibility for planning the employment of joint forces. Although not responsible for directing the CCMDs Service forces in joint operations, the Military Departments participate in joint operation planning through execution of their responsibilities to: organize, train, equip, and provide forces for assignment and allocation to the CCDRs; administer and support those forces; and prepare plans implementing joint strategic mobility, logistic, and mobilization plans.

b. Headquarters, commands, and agencies involved in joint operation planning or committed to conduct military operations are collectively termed the Joint Planning and Execution Community (JPEC).[14] Although not a standing or regularly meeting entity, the JPEC consists of the CJCS and other members of the JCS, the JS, the Services and their major commands, the CCMDs and their subordinate commands, and the CSAs (Figure IV-4).

8. In-Progress-Reviews

IPRs are a continuous dialog rather than specific breaks in the process for socialization. IPRs can be held at any time in the planning process to update the SECDEF on the status of the plan. IPRs stress the importance of strategic communication between the SECDEF/CJCS, the CCDR's and other appropriate senior leaders. This dialogue will shape the appropriate campaign or contingency plan as it matures and give the SECDEF/CJCS visibility on the planning while the plan is being developed or reviewed. Further, IPRs expedite planning by ensuring that the plan addresses the most current strategic assessments and needs. They generate valuable feedback for planning staffs and provide a forum for guid-

[13] *JP 5-0, Joint Operations Planning*

[14] *JP 5-0, Joint Operations Planning*

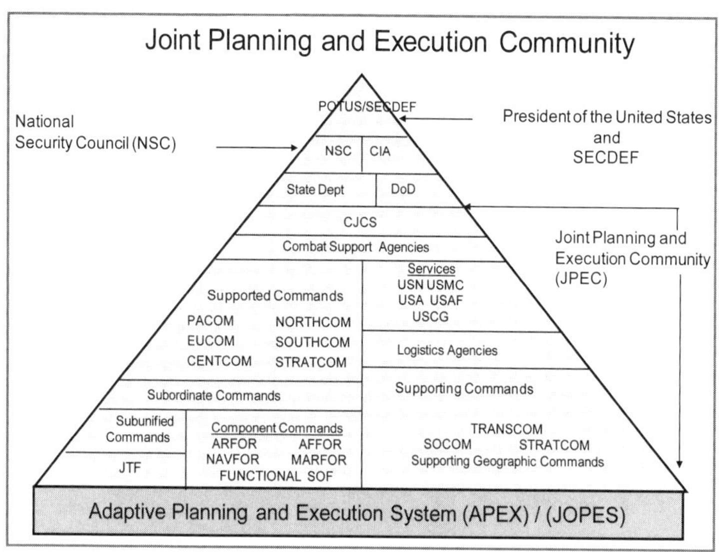

Figure IV-4. Joint Planning and Execution Community (JP 5-0)

ance on coordination with the interagency and multinational communities. IPRs provide the opportunity for discussion of key issues or concerns, identification and removal of planning obstacles, and resolution of planning conflicts. IPRs ensure that plans remain relevant to the situation and the SECDEF's intent throughout their development. Planning will include as many IPRs as necessary to complete the plan (Figure IV-5). For those plans not designated "top priority" IPR's will be conducted with the SECDEF's designated representative (see Chapter 8, IPRs).

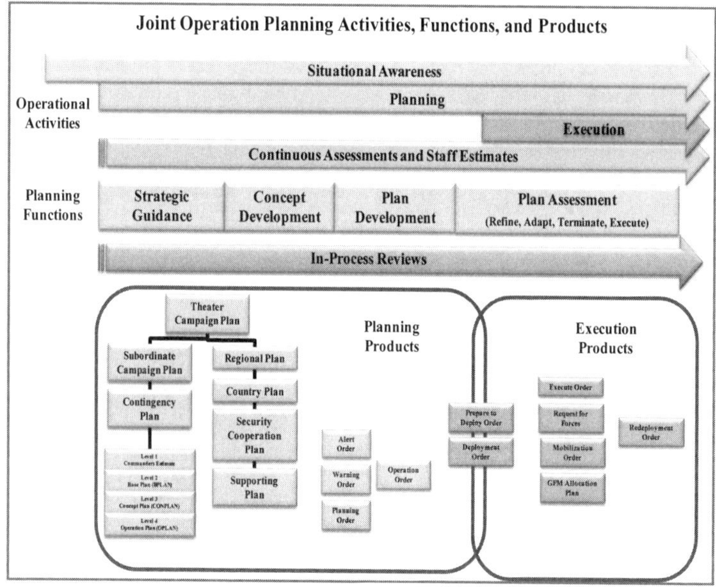

Figure IV-5. Joint Operation Planning Activities, Functions, and Products

9. Force Planning

- Force Planning at the National Strategic Level.
- Force Planning Construct.

a. National Strategic Level. Force planning[15] at the national strategic level is associated with creating and maintaining military capabilities. It is primarily the responsibility of the Military Departments, Services and USSOCOM and is conducted under the administrative control that runs from the SECDEF to the Secretaries of the Military Departments to the Service Chiefs. The Services recruit, organize, train, equip, and provide forces for assignment to CCMDs and they administer and support these forces. In areas peculiar to special operations, USSOCOM has similar responsibility for special operations forces (SOF), with the exception of organizing Service components.

b. Theater Strategic Level. At the theater strategic level, force planning encompasses all those activities performed by the supported CCDR, subordinate component CDRs, and support agencies to select, prepare, integrate, and deploy the forces and capabilities required to accomplish an assigned mission. Force planning also encompasses those activities performed by Force Providers to develop, source, and tailor those forces and capabilities with actual units.

c. Force Planning Construct. The force planning construct established by the 2006 QDR and affirmed in subsequent QDRs provides capstone guidance for DOD components to determine the overall size and composition of the joint force (capacity), the types of forces and systems (capabilities), and the levels of effort (steady-state or surge) needed to implement the NDS. DOD will maintain a robust force capable of protecting United States interests against a multiplicity of threats, including two capable nation-state aggressors, and must be capable of conducting a wide range of operations, from homeland defense and defense support to civil authorities, to deterrence and preparedness missions, to the conflicts we are in and the wars we may someday face. DOD components plan for force generation, sustainment and training activities as well as to size and shape their forces and capabilities to defend the United States Homeland, prevail in the War on Terror and Conduct Irregular Warfare, and conduct and win conventional campaigns.[16]

d. The planning conducted by the supported CCMD and its components, as well as by the Military Departments, Services, and Service component commands of the CCMDs, determine required force capabilities to accomplish an assigned mission, develop forces lists, order and tailor required force capabilities with actual units, identify and resolve shortfalls, and determine the routing and time-phasing of forces into the operational area.

[15] *Quadrennial Defense Review.*

[16] *JP 5-0, Joint Operations Planning.*

Planning Functions

I. Planning and Functions

Contingencies and crisis share the same planning activities and are interrelated. The joint operation planning process (JOPP) can be as detailed as time, resources, experience, and situation permit. The JOPP is detailed, deliberate, sequential, and time consuming. All steps and sub-steps are used when enough planning time and staff support are available. The JOPP is a planning model that establishes procedures for analyzing a mission, developing, analyzing, and comparing COAs against criteria of success and each other, selecting the optimum COA, and producing a plan or order. The JOPP applies across the spectrum of conflict and range of military operations (ROMO). The JOPP helps organize the thought process of commanders and staffs. It helps them apply thoroughness, clarity, sound judgment, logic, and professional knowledge to reach decisions.

a. Each JOPP step begins with inputs that build on previous steps. The outputs of each step drive subsequent steps. False assumptions and errors committed early affect later steps. While the formal process begins with the receipt of a mission and has as its goal the production of an order, planning continues throughout the operations process.

b. The four subordinate planning functions that embody joint operation planning are: (1) Strategic Guidance, (2) Concept Development, (3) Plan Development, and (4) Plan Assessment (Figure V-1).

c. Although the four planning functions are generally sequential, they often run simultaneously in the effort to accelerate the overall planning process. The Secretary of Defense or the CCDR may direct the planning staff to refine or adapt a plan by reentering the planning process at any of the earlier functions. The time spent accomplishing each activity and function depends on the nature of the crisis.

(1) In time-sensitive cases, activities and functions may be accomplished simultaneously and compressed so that all decisions are reached in open forum and orders are combined and initially may be issued orally.

(2) A crisis could be so time critical, or a single COA so obvious, that the first written directive might be a DEPORD or an EXORD.

(a) Commanders can alter the JOPP to fit time-constrained circumstances and produce a satisfactory plan. In time constrained conditions, commanders assess the situation; update their commander's visualization, and direct the staff to perform those JOPP activities needed to support the required decisions.

(b) Streamlined processes permit commanders and staffs to shorten the time needed to issue orders when the situation changes. In a time-constrained environment, many steps of the JOPP are conducted concurrently. To an outsider, it may appear that experienced commanders and staffs omit key steps. In reality, they use existing products or perform steps in their heads instead of on paper. They also use many shorthand procedures and implicit communication. Fragmentary orders (FRAGORDs) and warning orders (WARNORDs) are essential in this environment.

(3) Following each of the IPRs the Office of the Secretary of Defense (OSD) will publish a memorandum for record detailing SECDEF's guidance for continued planning.

Joint Operation Planning Activities, Functions, and Products

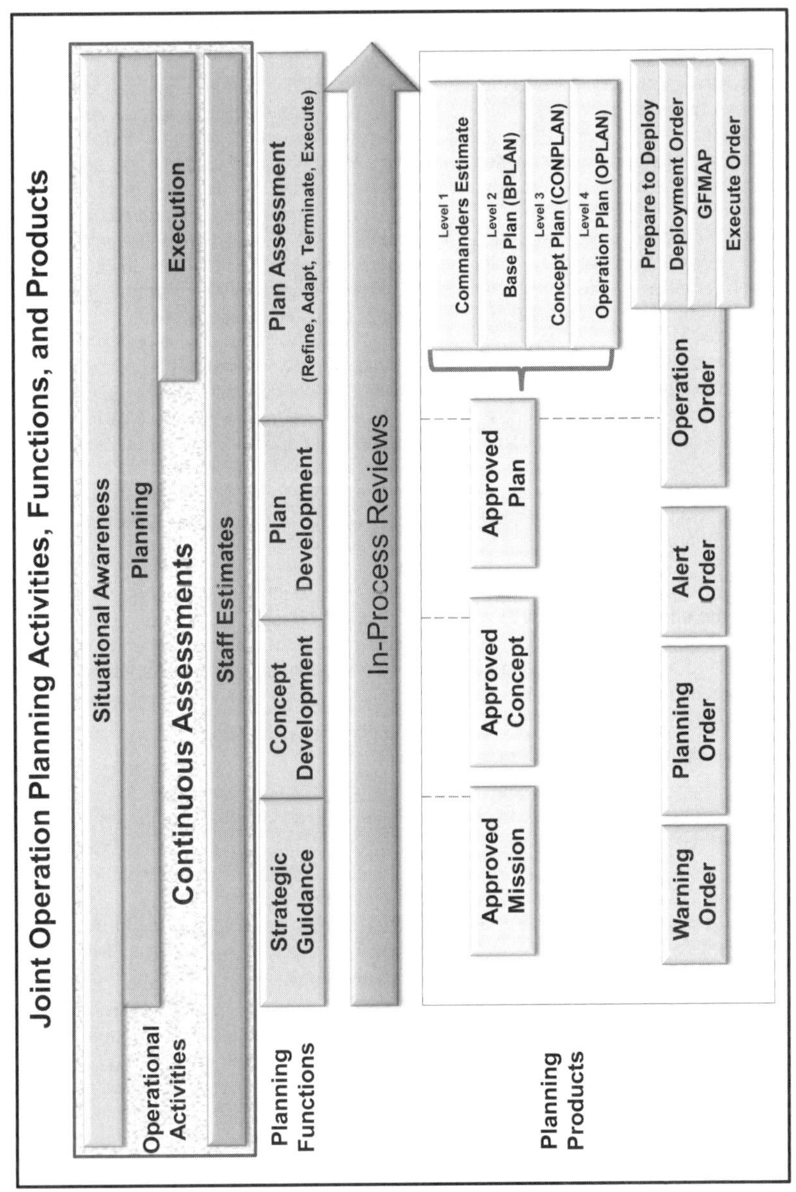

Figure V-1. Planning Functions (JP 3-35).

Planning Functions[1]

1. Strategic Guidance - Function I

The President, SECDEF, and the CJCS, with appropriate consultation, formulate suitable and feasible military objectives to counter threats. The CCDR may provide input through one or more CDR's Assessments. This function is used to develop planning guidance for preparation of COAs. This process begins with an analysis of existing strategic guidance (e.g., JSCP for contingency plans or a CJCS Warning Order, Planning Order or Alert Order for a crisis). The primary end product is a CDR's *Mission Statement* for deliberate planning and a *CDR's Assessment* (OPREP-3PCA) or CDRs *Estimate* in CAP.

2. Concept Development - Function II

During deliberate planning, the supported CCDR develops the CCDR's CONOPS for SECDEF approval, based on SECDEF, CJCS, and Service Chief planning guidance and resource apportionment provided in the JSCP and Service documents. In CAP, concept development is based on situational awareness guidance, resource allocations from approved contingency plans, and a CJCS Planning Order, or Alert Order. Using the CCDR's mission statement, CCMD planners develop preliminary COAs and staff estimates. COAs are then compared and the CCDR recommends a COA for SECDEF approval in a CDR's *Estimate*. The CCDR also requests SECDEF guidance on interagency coordination. The approved COA becomes the basis of the CONOPS containing conflict termination planning, supportability estimates, and, time permitting, an integrated time-phased database of force requirements, with estimated sustainment.

3. Plan Development - Function III

This function is used in developing an OPLAN, CONPLAN or an OPORD with applicable supporting annexes and in refining preliminary feasibility analysis. This function fully integrates mobilization, deployment, employment, conflict termination, sustainment, redeployment, and demobilization activities. Detailed planning begins with SECDEF approval for further planning in a non-crisis environment or a CJCS Warning Order, Alert Order or Planning Order in a crisis situation; it ends with a SECDEF-approved Plan or OPORD.

4. Plan Assessment – Function IV

During this function, the CCDR refines the complete plan while supporting and subordinate CCDRs, Services and supporting agencies complete their supporting plans for his/her review and approval. CCDRs continue to develop and analyze branches and sequels as required or directed. The CCDR and the JS continue to evaluate the situation for any changes that would trigger plan revision or refinement.

(a) The JS, Services, CCMDs, and Agencies monitor current readiness and availability status to assess sourcing impacts and refine sourcing COAs should the plan be considered for near-term execution.

(b) The CCDR may conduct as many IPR(s) as are required with the SECDEF during Plan Assessment. These IPR(s) could focus on branches/options and situational or assumption changes requiring major reassessment or significant plan modification/adaptation, but might also include a variety of other pertinent topics (e.g., information operations, special access programs, nuclear escalation mitigation).

[1] *CJCSM 3122.01*

II. Joint Operation Planning Process

Ref: JP 5-0, Joint Operation Planning (Aug '11), pp. IV-1 to IV-3 (fig. IV-2, p. IV-3).

Operational design and JOPP are complementary elements of the overall planning process. Operational design provides an iterative process that allows for the commander's vision and mastery of operational art to help planners answer ends–ways–means–risk questions and appropriately structure campaigns and operations. The commander, supported by the staff, gains an understanding of the operational environment, defines the problem, and develops an operational approach for the campaign or operation through the application of operational design during the initiation step of JOPP.

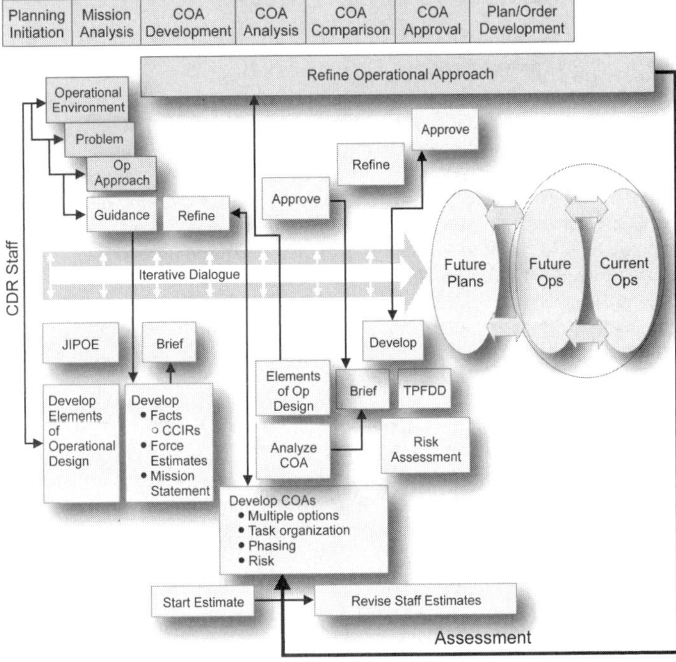

Commanders communicate their operational approach to their staff, subordinates, supporting commands, agencies, and multinational/nongovernmental entities as required in their initial planning guidance so that their approach can be translated into executable plans. As JOPP is executed, commanders learn more about the operational environment and the problem and refine their initial operational approach. Commanders provide their updated approach to the staff to guide detailed planning. This iterative process between the commander's maturing operational approach and the development of the mission and CONOPS through JOPP facilitates the continuing development of possible COAs and their refinement into eventual CONOPS and executable plans.

This relationship between the application of operational art, operational design, and JOPP continues throughout execution of the campaign or operation. By applying the operational design methodology in combination with the procedural rigor of JOPP, the command can help keep its aperture as wide as possible to always question the mission's continuing relevance and suitability while executing operations in accordance with the current approach and revising plans as needed. By combining the best aspects of both of these approaches, the friendly force can maintain the greatest possible flexibility and do so in a proactive vice reactive manner.

Operational Art & Design and the Joint Operations Cycle 6

> *"If I were given one hour to save the planet, I would spend fifty-nine minutes defining the problem and one minute resolving it."*
>
> *-Albert Einstein*

This Chapter gives a broad overview of Operational art and design. The primary sources for this chapter are JP 3-0, FM 3-0, FM 5-0 and JWFC Pamphlet 10 "Design in Military Operations." For a greater understanding of these critical concepts, review these documents in total.

I. Operational Art in the Strategic Context

1. Grand Strategy as the Basis for Operational Art

a. Military operations do not occur in a vacuum, and the preeminent aspect of any military campaign is the grand strategic context in which it occurs. Military effectiveness depends at least partly on the military's appreciation of and consistency with that context. Pursuing and maintaining the necessary levels of consistency is the realm of operational art.

b. Clausewitz's musing that "war is nothing but the continuation of politics by other means"[1] makes no inference that military campaigns are simple interruptions to grand strategy. To the contrary, war is a dynamic integral component of grand strategy for which standard civil planning systems are often insufficient. And so, it is through the application of operational art that military operations and engagements are objectively linked with the activities of those other elements of national power[2] in compliance with grand strategy. Figure VI-1 depicts this relationship.

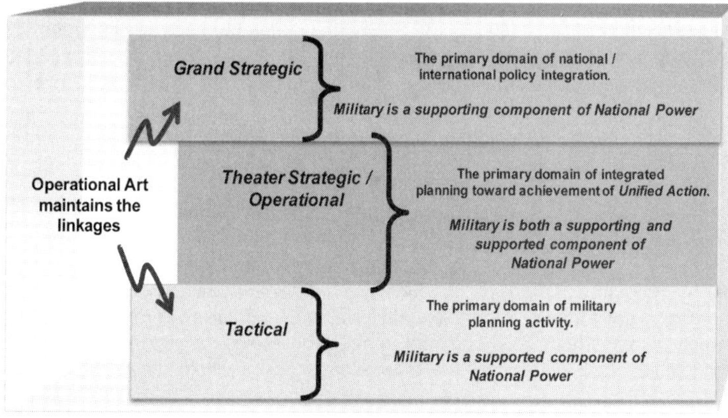

Figure VI-1. Operational Art and Linkages

[1] *Clausewitz, On War (Paret/Howard Translation), author's notes of 10 July, 1827.*

[2] *Or international - in the case of multi-national endeavors.*

c. How operational art is actually performed is a matter of important discussion in the joint forum. It is clear that it is not embodied in the processes of grand strategic policy development or integration. It is equally clear that it is not a component of classic military planning processes. Operational art occurs unevenly somewhere between the two.

d. When applied either in campaign or operational planning, operational art reveals itself through the elements of operational design, which demonstrate a cohesive central concept or 'big idea' behind the military action that is equal to the functional complexity, dynamicity and the temporal depth of the military challenges.

e. It is less necessary to understand which military activities do and do not constitute military campaigns, than it is to identify among them which may require the application of operational art.

II. Utilizing Operational Art and Design

1. Commanders and Operational Art

Operational art represents a creative approach to dealing with the direction of military forces. It expresses informed vision across the levels of war.

Operational art requires broad vision, the ability to anticipate the conditions necessary for success before, during and after the commitment of forces. It helps CDRs and their staffs order their thoughts and understand the conditions for victory before seeking battle, thus avoiding unnecessary battles.

a. *Operational art* is the application of creative imagination by CDRs and staffs—supported by their skill, knowledge, and experience—to design strategies, campaigns, and major operations and organize and employ military forces. Operational art integrates ends, ways, and means across the levels of war.[3] It reflects a holistic understanding of the operational environment and the problem. This understanding enables CDRs to develop end state conditions and an operational approach to guide the force in establishing those conditions for lasting success. The *operational approach* is a broad conceptualization of the general actions that will produce the conditions that define the desired end state.[4] Design assists CDRs in developing their operational approach. In visualizing an operation, CDRs determine which conditions satisfy policy, orders, guidance, and directives. Taken together, these conditions become the end state. CDRs devise and execute plans that complement the actions of the other instruments of national power in a focused, unified effort. To this end, CDRs draw on experience, knowledge, education, intellect, intuition, and creativity.

b. Applying Operational Art. CDRs use operational art to envision how to establish conditions that define the desired end state. Actions and interactions across the levels of war influence these conditions. These conditions are fundamentally dynamic and linked together by the human dimension, the most unpredictable and uncertain element of conflict. The operational environment is complex, adaptive, and interactive. Through operational art, CDRs apply a comprehensive understanding of it to determine the most effective and efficient methods to influence conditions in various locations across multiple echelons (see figure VI-2).

c. Operational art spans a continuum—from comprehensive strategic direction to concrete tactical actions. Bridging this continuum requires creative vision coupled with broad experience and knowledge. Operational art provides a means for CDRs to derive the

[3] *JP 3-0, Joint Operations.*

[4] *FM 5-0, The Operations Process*

Operational Art and Planning

Ref: ADRP 5-0, The Operations Process (Mar '12), p. 2-4 and ADRP 3-0, Unified Land Operations (Mar '12), pp. 4-2 to 4-9.

Conceptual planning is directly associated to operational art—the cognitive approach by commanders and staffs—supported by their skill, knowledge, experience, creativity, and judgment—to develop strategies, campaigns, and operations to organize and employ military forces by integrating ends, ways, and means (JP 3-0). Operational art is a thought process that guides conceptual and detailed planning to produce executable plans and orders.

In applying operational art, commanders and their staffs use a set of intellectual tools to help them communicate a common vision of the operational environment as well as visualizing and describing the operational approach. Collectively, this set of tools is known as the elements of operational art. These tools help commanders understand, visualize, and describe combinations of combat power and help them formulate their intent and guidance. Commanders selectively use these tools in any operation. However, their application is broadest in the context of long-term operations. These tools are useful when using Army design methodology and the military decision-making process and may frequently be used in both, simultaneously.

The elements of operational art support the commander in identifying objectives that link tactical missions to the desired end state. They help refine and focus the operational approach that forms the basis for developing a detailed plan or order. During execution, commanders and staffs consider the elements of operational art as they assess the situation. They adjust current and future operations and plans as the operation unfolds.

Elements of *Operational Design*

Within operational art, joint force commanders and staffs consider elements of operational design. Elements of operational design are individual tools that help the joint force commander and staff visualize and describe the broad operational approach.

- Termination
- Military end state
- Objective
- Effects
- Center of gravity
- Decisive point
- Lines of operations and lines of effort
- Direct and indirect approach
- Anticipation
- Operational reach
- Culmination
- Arranging operations
- Force and functions

Elements of *Operational Art*

As some elements of operational design apply only to joint force commanders, the Army modifies the elements of operational design into elements of operational art, adding Army specific elements. During the planning and execution of Army operations, Army commanders and staffs consider the elements of operational art as they assess the situation.

- End state and conditions
- Center of gravity
- Decisive points
- Lines of operations and lines of effort
- Operational reach
- Basing
- Tempo
- Phasing and transitions
- Culmination
- Risk

Refer to The Joint Forces Operations & Doctrine SMARTbook (Guide to Joint, Multinational & Interagency Operations) for complete discussion of operational art and design (from joint doctrine). Additional topics include joint operation planning, joint logistics, joint task forces, information operations, multinational operations, and IGO/NGO coordination.

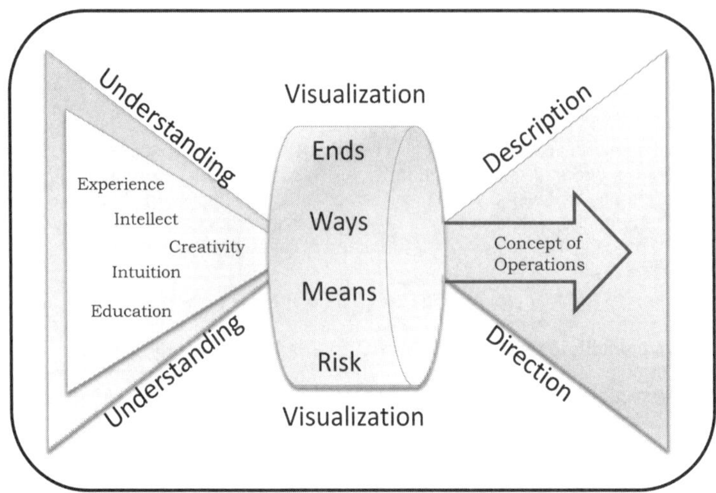

Figure VI-2. Operational Art (FM 5-0).

essence of an operation. Without it, tactical actions devolve into a series of disconnected engagements, with relative attrition the only measure of success. Through operational art, CDRs translate their concept of operations into an operational design and ultimately into tactical tasks. They do this by integrating ends, ways, and means and by envisioning dynamic combinations of the elements of full spectrum operations across the levels of war. They then apply operational art to array forces and maneuver them to achieve the desired end state.[5]

d. Every operation begins with a CDR's intent that guides its conduct. In almost all cases, a CDR's intent and concept of operations envision all the instruments of national power working toward a common end state. Using operational art, CDRs frame their concept by answering several fundamental questions:

- What is the force trying to accomplish (ends)?
- What conditions, when established, constitute the desired end state (ends)?
- How will the force achieve the end state (ways)?
- What sequence of actions is most likely to attain these conditions (ways)?
- What resources are required, and how can they be applied to accomplish that sequence of actions (means)?
- What risks are associated with that sequence of actions, and how can they be mitigated (risk)?

e. CDRs understand, visualize, describe, direct, lead, and assess all aspects of operations. Based on a comprehensive analysis of the operational environment, CDRs determine the centers of gravity (COG) around which to frame the plan.

f. When applying operational art, collaboration informs situational understanding. This collaboration involves an open, continuous dialog between CDRs that spans the levels of war and echelons of command. This dialog is vital in establishing a common perspective on the problem and a shared understanding of the operational environment's conditions. Effective collaboration enables assessment, fosters critical analysis, and anticipates opportunities and risk. Collaboration allows CDRs to recognize and react to changes in the situation.

[5] *FM 5-0, The Operations Process*

g. Practicing operational art requires a broad understanding of the operational environment at all levels. It also requires practical creativity and the ability to visualize changes in the operational environment. CDRs project their visualization beyond the realm of physical combat. They anticipate the operational environment's evolving military and nonmilitary conditions. Operational art encompasses visualizing the synchronized arrangement and employment of military forces and capabilities to achieve the desired end state. This creative process requires the ability to discern the conditions required for victory before committing forces to action.

h. Conflict is fundamentally a human endeavor characterized by violence, uncertainty, chance, and friction. Operational art helps CDRs integrate functions and capabilities. It also helps synchronize military actions with actions of other government and civilian organizations. Operational art supports design by providing a conceptual framework for ordering the thought process when planning operations. Operational art supports the CDRs' ability to seize, retain, and exploit initiative and achieve decisive results.

i. Operational art is a cognitive aspect of operations supported by design. While the character of conflict changes with time, the violent and chaotic nature of warfare does not. The essence of military art remains timeless. Operational art—the creative expression of informed vision to integrate ends, ways, and means across the levels of war—is fundamental to the Joint force's ability to seize, retain, and exploit the initiative.

2. Operational Design

a. Why Consider Design. Strategists, operational artists, and tacticians currently find it difficult to comprehensively understand and explain complex competitors situated in a dynamic, interactive environment. The complexity of many potential operational environments frustrates conventional thinking and makes it difficult to recognize patterns in an unconventional adversary's actions. This is compounded by the presence of multiple adversarial actors in the same setting, multiple partners with motives in tension with our own, and operations within a social mosaic that includes unique structures and norms, religions, classes, tribes, relationships, and other factors.

(1) Today's problems are not necessarily new. In today's environment, however, complexity is increased by the pace of events in the dynamic circumstances surrounding the problems. Modern technologies that allow nearly instantaneous sharing of information through communications networks and news services also allow closer coordination of criminal, insurgent, and trans-national actors. As a result, the new strategic context may require innovative approaches that effectively integrate all elements of U.S. and multinational power.[6]

b. Operational Design Defined. *Design* is a methodology for applying critical and creative thinking to understand, visualize, and describe complex, ill-structured problems and develop approaches to solve them. Critical thinking captures the reflective and continuous learning essential to design. Creative thinking involves thinking in new, innovative ways while capitalizing on imagination, insight, and novel ideas. Design is a way of organizing the activities of understanding, visualizing, and describing within an organization. Design occurs throughout the operations process before and during detailed planning, through preparation, and during execution and assessment.

c. Planning consists of two separate, but closely-related components: a conceptual component and a detailed component. The conceptual component is represented by the cognitive application of design. The detailed component translates broad concepts into a complete and practical plan. During planning, these components overlap with no clear

[6] *JWFC Pamphlet 10, Design in Military Operations.*

delineation between them. As CDRs conceptualize the operation, their vision guides the staff through design and into detailed planning. Design is continuous throughout planning and evolves with increased understanding throughout the operations process. Design underpins the role of the CDR in the operations process, guiding the iterative and often cyclic application of understanding, visualizing, and describing. As these iterations occur, the design concept—the tangible link to detailed planning—is forged.

d. Design enables CDRs to view a situation from multiple perspectives, draw on varied sources of situational knowledge, and leverage subject matter experts while formulating their own understanding. Design enables CDRs to develop a thorough understanding of the operational environment and formulate effective solutions to complex, ill-structured problems. The CDR's visualization and description of the actions required to achieve the desired conditions must flow logically from what CDRs understand and how they have framed the problem. Design provides an approach for how to generate change from an existing situation to a desired objective or condition.

e. Innovation, adaptation, and continuous learning are central tenets of design. Innovation involves taking a new approach to a familiar or known situation, whereas adaptation involves taking a known solution and modifying it to a particular situation or responding effectively to changes in the operational environment. Design helps the commander lead innovative, adaptive work and guides planning, preparing, executing, and assessing operations. Design requires agile, versatile leaders who foster continuous organizational learning while actively engaging in iterative collaboration and dialog to enhance decision-making across the echelons.

f. A continuous, iterative, and cognitive methodology, design is used to develop understanding of the operational environment, make sense of complex, ill-structured problems, and develop approaches to solving them. In contrast to detailed planning, design is not *process-oriented*. The practice of design challenges conventional wisdom and offers new insights for solving complex, ill-structured problems. While plans and orders flow down the echelons of command, new understanding may flow up from subordinate echelons where change often appears first. By enhancing and improving CDRs' understanding, design improves a higher authority's understanding of the operational environment and the problems CDRs are tasked to solve.

3. Design Goals

Successfully applying design seeks four concrete goals that, once achieved, provide the reasoning and logic that guide detailed planning processes. Each goal is an essential component to reshaping the conditions of the operational environment that constitute the desired end state. Collectively, they are fundamental to overcoming the complexities that characterize persistent conflict. The goals of design are—

- Understanding ill-structured problems.
- Anticipating change.
- Creating opportunities.
- Recognizing and managing transitions.

a. Understanding Ill-Structured Problems. Persistent conflict presents a broad array of complex, ill-structured problems best solved by applying design. Design offers a model for innovative and adaptive problem framing that provides leaders with the cognitive tools to understand a problem and appreciate its complexities before seeking to solve it. This understanding is fundamental to design. Without thoroughly understanding the nature of the problem, CDRs cannot establish the situation's context or devise approaches to effect change in the operational environment. Analyzing the situation and the operational variables provides the critical information necessary to understand and frame these problems

(FM 3-0, Operations, Change 1, discusses the operational variables. See FM 3-0, Chapter 2 for a discussion on the structure of problems).

(1) A CDR's experience, knowledge, judgment, and intuition assume a crucial role in understanding complex, ill-structured problems. Together, they enhance the cognitive components of design, enhancing CDRs' intuition while further enabling CDRs to identify threats or opportunities long before others might. This deepens and focuses CDRs' understanding. It allows them to anticipate change, identify information gaps, and recognize capability shortfalls. This understanding also forms the basis of the CDR's visualization. CDRs project their understanding beyond the realm of physical combat. They must anticipate the operational environment's evolving military and nonmilitary conditions. Therefore, design encompasses visualizing the synchronized arrangement and use of military and nonmilitary forces and capabilities to achieve the desired end state. This requires the ability to discern the conditions required for success before committing forces to action.

(2) Ultimately, understanding complex, ill-structured problems is essential to reducing the effects of complexity on full-spectrum operations. This understanding allows CDRs to better appreciate how numerous factors influence and interact with planned and ongoing operations. Assessing the complex interaction among these factors and their influences on operations is fundamental to understanding and effectively allows the CDR to make qualitatively better decisions under the most dynamic and stressful circumstances.

b. <u>Anticipating Change</u>. Applying design involves anticipating changes in the operational environment, projecting decision-making forward in time and space to influence events before they occur. Rather than responding to events as they unfold, design helps the CDR to anticipate these events and recognize and manage transitions. Through the iterative and continuous application of design, CDRs contemplate and evaluate potential decisions and actions in advance, visualizing consequences of possible operational approaches to determine whether they will contribute to achieving the desired end state. A thorough design effort reduces the effects of complexity during execution and is essential to anticipating the most likely reactions to friendly action. During detailed planning, these actions and sequences are often linked along lines of effort, which focus the outcomes toward objectives that help to shape conditions of the operational environment.

(1) Design alone does not guarantee success in anticipating change—it also does not ensure that friendly actions will quantifiably improve the situation. However, applied effectively and focused toward a common goal, design provides an invaluable cognitive tool to help CDRs anticipate change as well as innovate and adapt approaches appropriately. Performed haphazardly and without proper focus and effort, it may become time-consuming, ineffective, process-focused, and irrelevant. Iterative, collaborative, and focused design offers the means to anticipate change effectively in the current situation and operational environment as well as achieve lasting success and positive change.

c. <u>Creating Opportunities</u>. The ability to seize, retain, and exploit the initiative is rooted in effective design. Applying design helps CDRs anticipate events and set in motion the actions that allow forces to act purposefully and effectively. Exercising initiative in this manner shapes the situation as events unfold. Design is inherently proactive, intended to create opportunities for success while instilling the spirit of the offense in all elements of full-spectrum operations. Effective design facilitates command, ensuring that forces are postured to retain the initiative and, through detailed planning, consistently able to seek opportunities to exploit that initiative.

(1) The goals of design account for the interdependent relationships among initiative, opportunity, and risk. Effective design postures the CDR to combine the four goals to reduce or counter the effects of complexity using the initial CDR's intent to foster individual initiative and freedom of action. Design is essential to recognizing and managing the inherent delay between decision and action, especially between the levels of war and echelons. The iterative nature of design helps the CDR to overcome this effect, fostering

initiative within the initial CDR's intent to act appropriately and decisively when orders no longer sufficiently address the changing situation. This ensures CDRs act promptly as they encounter opportunities or accept prudent risk to create opportunities when they lack clear direction. In such situations, prompt action requires detailed foresight and preparation.

d. Recognizing and Managing Transitions. CDRs must possess the versatility to operate along the spectrum of conflict and the vision to anticipate and adapt to transitions that will occur over the course of a campaign. Design provides the cognitive tools to recognize and manage transitions by educating and training the CDR. Educated and trained CDRs can identify and employ adaptive, innovative solutions, create and exploit opportunities, and leverage risk to their advantage during these transitions, both friendly and adversary.

4. Fundamentals of Design

Today's operational environment presents situations so complex that understanding them—let alone attempting to change them—is beyond the ability of a single individual. Moreover, significant risk occurs when assuming that CDRs in the same campaign understand an implicit design concept or that their design concepts mutually support each other. The risks multiply, especially when a problem involves multiple units, Services, multinational forces, or other instruments of national power. CDRs mitigate these risks with collaboration and by applying the design fundamentals:

- Apply critical thinking.
- Understand the operational environment.
- Solve the right problem.
- Adapt to dynamic conditions.
- Achieve the designated goals.

a. Apply Critical Thinking. CDRs ensure that superiors and subordinates share a common understanding of the purpose behind intended actions. Initial guidance provided by a higher political or military authority may prove insufficient to create clearly stated, decisive, and attainable objectives in complex situations that involve political, social, economic, and other factors. After CDRs conduct a detailed study of the situation, they may conclude that some desired goals are unrealistic or not feasible within the limitations. These limitations stem from the inherent tension that often exists among different goals, historical tensions in the local population, interactions of different actors seeking to improve their own survivability and position, and limited resources and time available to achieve the mission. One can never fully understand the dynamics of a conflict in advance. Well-intentioned guidance without detailed study may lead to an untenable or counterproductive solution.

(1) Design helps mitigate the risk associated with guidance that does not fully account for the complexities of the operational environment by using a critical and creative approach for learning, innovation, and adaptation. Design helps to clarify objectives in the context of the operational environment and within the limits imposed by policy, strategy, orders, or directives. This does not imply that CDRs can arbitrarily disregard instructions. If, however, they receive unclear guidance or consider the desired conditions unachievable, CDRs engage in active dialog. Dialog clarifies guidance and enables CDRs to offer recommendations to achieve a mutual understanding of the current situation and the desired end state. Design can assist CDRs in leading the top-down/bottom-up approach at all echelons.

b. Understand the Operational Environment. Design challenges leaders to understand the impact of their decisions and actions on the operational environment. Gaining a deeper and more thorough understanding of the operational environment enables more effective decision-making and helps to integrate military operations with the other instruments of national power. In an environment characterized by the presence of joint, interagency, intergovernmental, and multinational partners, such understanding is essential to success. In this context, human variables, interactions, and relationships are frequently decisive.

Military force may be necessary to achieve national policy aims, but, by itself, force proves insufficient to achieve victory in these situations. More importantly, leaders must recognize the relationship between the character of conflict and the approach one takes to effect changes in the operational environment.

(1) Developing a thorough understanding of the operational environment is a continuous process. Even though this understanding will never be perfect, attempting to comprehend its complex nature helps identify unintended consequences that may undermine well-intentioned efforts. Deep understanding reveals the dynamic nature of the human interactions and the importance of identifying contributing factors . Leaders can gain this understanding by capitalizing on multiple perspectives and varied sources of knowledge. For example, intelligence knowledge generated as part of the intelligence process contributes to contextual understanding of the operational environment. Design encourages the CDR and staff to seek and address complexity before attempting to impose simplicity.

c. Solve the Right Problem. CDRs use design to ensure they are solving the right problem. When CDRs use design, they closely examine the symptoms, the underlying tensions, and the root causes of conflict in the operational environment. From this perspective, they can identify the fundamental problem with greater clarity and consider more accurately how to solve it. Design is essential to ensuring CDRs identify the right problem to solve. Effective application of design is the difference between solving a problem right and solving the right problem.

d. Adapt to Dynamic Conditions. Innovation and adaptation lead to capitalizing on opportunities by quickly recognizing and exploiting actions that work well while dismissing those that do not. Adaptation does not rely on being able to anticipate every challenge. Instead, it uses continuous assessment to determine what works and what does not. Adaptation occurs through the crucial process of assessment and subsequent changes in how one approaches problems. In the military domain, adaptation demands clearly articulated measures of effectiveness. These measures define success and failure along with after-action reviews that capture and implement lessons at all echelons.

(1) Effective use of design improves the ability to adapt. Adaptation in this sense involves reframing the situation to align with new information and experiences that challenge existing understanding. Through framing and reframing achieved through iterative collaboration and dialog, design provides a foundation for organizational learning and contributes to the necessary clarity of vision required by successful CDRs.

e. Achieve the Designated Goals. If the link between strategy and tactics is clear, the likelihood that tactical actions will translate into strategic success increases significantly. For complex, ill-structured problems, integrating and synchronizing operations to link sequences of tactical actions to achieve a strategic aim may prove elusive. Through design, CDRs employ operational art to cement the link between strategic objectives and tactical action ensuring that all tactical actions will produce conditions that ultimately define the desired end state. As understanding of the operational environment and problem improves, design adapts to strengthen the link between strategy and tactics, promoting operational coherence, unity of effort, and strategic success.

5. Leading Design

CDRs are the central figure in design. Generally, the more complex a situation is, the more important the CDR's role is in design. CDRs draw on design to overcome the challenges of complexity. They foster iterative collaboration and dialog while leveraging their collective knowledge, experience, judgment, and intuition to generate a clearer understanding of the conditions needed to achieve success. Design supports the CDR's ability to understand and visualize the operational environment.

a. In leading design, CDRs typically draw from a select group within the planning staff, red team members, and subject matter experts internal and external to the headquarters. The CDR selects these individuals based on their expertise relative to the problem. The CDR expects these individuals to gain insights and inputs from areas beyond their particular expertise—either in person or through reach back—to frame the problem more fully. Design serves to establish the context for guidance and orders. By using members of the planning staff to participate in the design effort, CDRs ensure continuity between design and detailed planning as well as throughout the operations process. These are purpose-built, problem-centric teams, and the CDR may choose to dissolve them once they complete the design effort.

b. CDRs leverage design to create and exploit opportunity, not just to ward off the risk of failure. Design provides the means to convert intellectual power into combat power. A creative design tailored to a unique operational environment promises—

- Economy of effort.
- Greater coherence across rotations among units and between successive operations.
- Better integration and coordination among the instruments of national power.
- Fewer unintended consequences.
- Effective adaptation once the situation changes.

c. Design requires the CDR to lead adaptive, innovative efforts to leverage collaboration and dialog to identify and solve complex, ill-structured problems. To that end, the CDR must lead organizational learning and develop methods to determine if reframing is necessary during the course of an operation. This requires continuous assessment, evaluation, and reflection that challenge understanding of the existing problem and the relevance of actions addressing that problem.

6. Design Methodology

Three distinct elements collectively produce a design concept as depicted in Figure VI-3. Together, they constitute an organizational learning methodology that corresponds to three basic questions that must be answered to produce an actionable design concept to guide detailed planning:

- Framing the operational environment—what is the context in which design will be applied?
- Framing the problem—what problem is the design intended to solve?
- Considering operational approaches—what broad, general approach will solve the problem?

a. During design, the CDR and staff consider the conditions, circumstances, and factors that affect the use of capabilities and resources as well as bear on decision-making. As an organizational learning methodology, design fosters collaboration and dialog as CDRs and staffs formulate conditions that define a desired end state and develop approaches that aim to achieve those conditions. When initial efforts do not achieve a thorough enough understanding of behaviors or events, CDRs reframe their understanding of the operational environment and problem. This cycle of inquiry, contextual understanding, and synthesis relies on continuous collaboration and dialog. Collaboration—especially with joint, interagency, intergovernmental, and multinational partners—is fundamental to success. Collaboration affords CDRs opportunities to revise their understanding or approaches so they can execute feasible, acceptable, and suitable approaches to achieve desired conditions or objectives.

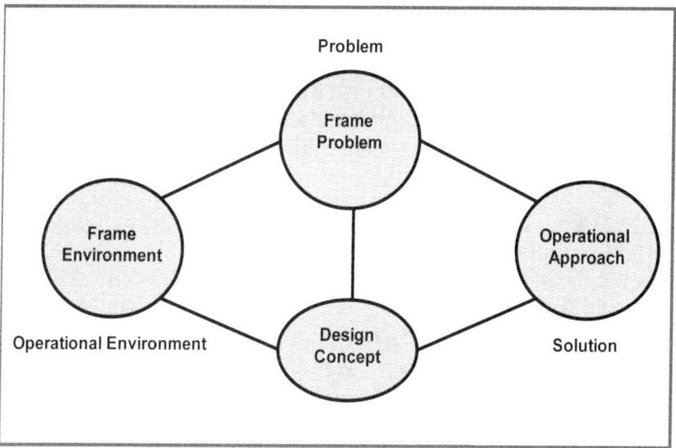

Figure VI-3. The Design Methodology.

b. Design is essentially nonlinear. It flows back and forth between environmental framing and problem framing while considering several operational approaches. No hard lines separate the efforts of each design element. When an idea or issue is raised, the CDR can address it in the appropriate element, even if the idea or issue is outside the current focus. The change in emphasis shifts from focusing on understanding the tendencies and potentials of actors in the operational environment, to understanding how they relate to and affect the problem, to understanding their likely contributions toward transforming existing conditions to a desired end state. As CDRs and staffs gain new knowledge or begin a new line of questioning, they often shift their focus among elements of design while building understanding and refining potential operational approaches to solve the problem.

c. Framing the Operational Environment. Framing involves selecting, organizing, interpreting, and making sense of a complex reality to provide guideposts for analyzing, understanding, and acting. Framing facilitates hypothesizing, or modeling, that scopes the part of the operational environment or problem under consideration. Framing provides a perspective from which CDRs can understand and act on a complex, ill-structured problem (see Chapter 13).

(1) Environmental Frame. The CDR and staff develop a contextual understanding of the situation by framing the operational environment. The environmental frame is a narrative and graphic description that captures the history, culture, current state, and future goals of relevant actors in the operational environment. The environmental frame describes the context of the operational environment—how the context developed (historical and cultural perspective), how the context currently exists (current conditions), and how the context could trend in the future (future conditions or desired end state). The environmental frame enables CDRs to forecast future events and the effects of potential actions in the operational environment. The environmental frame explains the actors and relationships within a system and surfaces assumptions to allow for more rapid adaptation. The environmental frame evolves through continuous learning, but scopes aspects of the operational environment relevant to higher guidance and situations.

(2) Within the environmental frame, CDRs review existing guidance, articulate existing conditions, determine the desired end state and supporting conditions, and identify relationships and interactions among relevant operational variables and actors. They analyze groupings of actors that exert significant influence in the operational environment knowing that individual actors rarely share common goals. By identifying and evaluating tendencies and potentials of relevant actor interactions and relationships, CDRs and their staffs formulate a desired end state that accounts for the context of the operational environment and higher directives.

d. Framing the Problem. Problem framing involves understanding and isolating the root causes of conflict—defining the essence of a complex, ill-structured problem. Problem framing begins with refining the evaluation of tendencies and potentials and identifying tensions among the existing conditions and the desired end state. It articulates how the operational variables can be expected to resist or facilitate transformation and how environmental inertia can be leveraged to ensure the desired conditions are achieved. The staff relies on text and graphics to articulate the problem frame.

(1) The Problem Frame. The problem frame is a refinement of the environmental frame that defines, in text and graphics, the areas for action that will transform existing conditions toward the desired end state. The problem frame extends beyond analyzing interactions and relationships in the operational environment. It identifies areas of tension and competition—as well as opportunities and challenges—that CDRs must address to transform current conditions to ultimately achieve the desired end state. Tension is the resistance or friction among and between actors. The CDR and staff identify the tension by analyzing the relevant actors' tendencies and potentials within the context of the operational environment.

(2) The CDR and staff challenge their hypotheses and models to identify motivations and agendas among the relevant actors. They identify factors that influence these motivations and agendas. The CDR and staff evaluate tendencies, potentials, trends, and tensions that influence the interactions among social, cultural, and ideological forces. These may include political, social, or cultural dispositions in one group that may hinder collaboration with another group.

(3) In the problem frame, analysis identifies the positive, neutral, and negative implications of tensions in the operational environment given the differences between existing and desired conditions. When the CDR and staff take action within the operational environment, they may exacerbate latent tensions. Tensions can be exploited to drive change, so they are vital to transforming existing conditions. If left unchecked, other tensions may undermine transformation and must be addressed appropriately. Because tensions arise from differences in perceptions, goals, and capabilities among relevant actors, they are inherently problematic and can both foster and impede transformation. By deciding how to address these tensions, the CDR identifies the problem that the design will ultimately solve.

(4) Identifying the Problem. A concise problem statement clearly defines the problem or problem set to solve. It considers how tension and competition affect the operational environment by identifying how to transform the current conditions to the desired end state—before adversaries begin to transform current conditions to their desired end state. The statement broadly describes the requirements for transformation, anticipating changes in the operational environment while identifying critical transitions. The problem statement accounts for the time and space relationships inherent in the problem frame.

7. Considering Operational Approaches

Considering operational approaches to the problem provides focus and sets boundaries for the selection of possible actions that together lead to achieving the desired end state.

a. The staff synthesizes and reduces much of the information and products created during the design to create the design concept and a shared understanding of a rationale behind it. The staff converges on the types and patterns of actions determining how they will achieve the desired conditions by creating a conceptual framework linking desired conditions to potential actions. The entire staff considers how to orchestrate actions to solve the problem in accordance with an operational approach.

b. The Operational Approach. The *operational approach* is a broad conceptualization of the general actions that will produce the conditions that define the desired end state. The operational approach enables CDRs to begin visualizing and describing possible combinations of actions to reach the desired end state given the tensions identified in the environmental and problem frames.

8. Reframing

Reframing is a shift in understanding that leads to a new perspective on the problems or their resolution. Reframing involves significantly refining or discarding the hypotheses or models that form the basis of the design concept. At any time during the operations process, the decision to reframe can stem from significant changes to understanding, the conditions of the operational environment, or the end state. Reframing allows the CDR and staff to make adjustments throughout the operations process, ensuring that tactical actions remain fundamentally linked to achieving the desired conditions.

a. Because the current operational environment is always changing and evolving, the problem frame must also evolve. Recognizing when an operation—or planning—is not progressing as envisioned or must be reconsidered provides the impetus for reframing in design. Reframing criteria should support the CDR's ability to understand, learn, and adapt—and cue CDRs to rethink their understanding of the operational environment, and hence rethink how to solve the problem. Generally, reframing is triggered in three ways: a major event causes a "catastrophic change" in the operational environment, a scheduled periodic review shows a problem, or an assessment and reflection challenges understanding of the existing problem and the relevance of the operational approach.

> *When existing understanding is no longer able to account for observed behavior, commanders must reframe their understanding and then assess their on-going actions for continued utility to achieving their goals.*

b. During operations, CDRs decide to reframe after realizing the desired conditions have changed, are not achievable, or cannot be attained through the current operational approach. Reframing provides the freedom to operate beyond the limits of any single perspective. Conditions will change during execution, and such change is expected because forces interact within the operational environment. Recognizing and anticipating these changes is fundamental to design and essential to an organization's ability to learn.

c. Reframing is equally important in the wake of success. By its very nature, success transforms the operational environment, creating unforeseen opportunities to exploit the initiative. Organizations are strongly motivated to reflect and reframe following failure, but they tend to neglect reflection and reframing following successful actions.

III. Joint Operations Cycle and Design (Nuts and Bolts)

a. The Joint Operations Cycle is both cognitive and physical and consists of several activities performed prior to, during and after operations to deliver on the benefits of operational art. The major formal activities are operational design, planning, preparation, execution and continuous assessment. These may or may not be further supported by formalized operational design activities which include diagnosis, dialog, design, learning, and re-design woven with continuous assessment into the joint operations cycle (Figure VI-4). These activities are overlapping and recurrent as circumstances demand.

b. The Joint Operations Cycle is utilized to assist CDRs in determining when and where to perform leadership actions such as making decisions, issuing guidance, providing command presence, and terminating the operations.

c. Following is a description for each of the processes and sub-processes inherent in the Joint Operations Cycle:

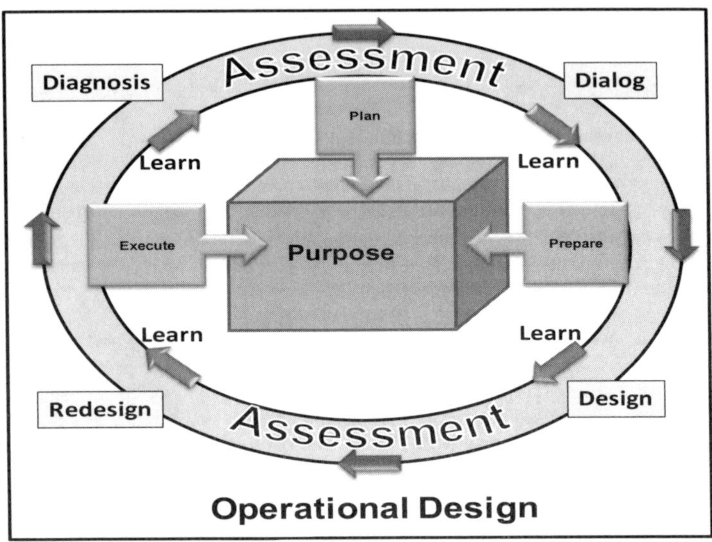

Figure VI-4. Joint Operations Cycle.

(1) Operational Design. Operational art is best applied through the conduct of operational design.

> *Design is not a function to be accomplished, but rather a living process.*

(a) As discussed, the purpose of operational design is to achieve a greater understanding, a proposed solution based on that understanding, and a means to learn and adapt. While operational art is the manifestation of informed vision and creativity, operational design is the practical extension of the CDR's creative processes whereby they synthesize their own intuition with the analytical and logical products of staff work to arrive at a single and comprehensive understanding of the Operational Environment (OE), and of what must be done militarily.

(b) Design does not stop once planning is concluded. Even during execution, CDRs and their staffs continue to consider design elements and through a learning and redesign process, adjust both current operations and future plans as the joint operation unfolds and central campaign concepts are refined.

(c) Operational design is particularly helpful during COA determination as the conceptual framework established through design provides an objective basis for evaluating and selecting COAs, and subsequently developing a detailed operational plan.

(d) Design and planning are qualitatively different yet interrelated activities essential for solving complex problems. While planning activities receive consistent emphasis in both doctrine and practice, design remains largely abstract and is rarely practiced through any distinguishable approach. Presented a problem, staffs may often rush directly into planning without clearly understanding the complex environment of the situation, purpose of military involvement, and approach required to address the core issues. Design informs and is informed by planning and operations. It establishes the intellectual foundation that aids in continuous assessment of operations of the OE. CDRs should personally lead relevant design processes and communicate the resulting framework to other CDRs for planning, preparation, and execution.

Campaign Design & Campaign Planning
Campaign Design
Campaign Planning
• Problem Setting (Framing)
• Conceptual
• Questions Assumptions & Methods
• Develops Understanding
• Paradigm-Setting
• Precedes Planning & Preparation
• Compliments Execution & Assessment
• Commander-Driven Dialog
• Supported by the Staff
• Supported by JOPP and JIPOE
• Problem-Solving
• Physical & Detailed
• Procedural (JOPP & JIPOE)
• Develops Products
• Paradigm-Accepting
• Patterns & Templates Activities
• Staff-Centered Process

Figure VI-5. Design and Planning Continuum (FM 5-0).

1 It is important to understand the distinction between design and planning (Figure VI-5, FM 3-24/JP5-0)). While both activities seek to formulate ways to bring about preferable futures, they are cognitively different.

2 Planning applies established procedures to solve a largely understood problem within an accepted framework. Design inquires into the nature of a problem to conceive a framework for solving that problem. In general, *planning is problem solving, while design is problem setting.* Where planning focuses on generating a plan—a series of executable actions—design focuses on learning about the nature of an unfamiliar problem.

3 When situations do not conform to established frames of reference—when the hardest part of the problem is figuring out what the problem is—planning alone is inadequate and design becomes essential. In these situations, absent a design process to engage the problem's essential nature, planners default to doctrinal norms - they develop plans based on the familiar rather than an understanding of the real situation. *Design provides a means to conceptualize and hypothesize about the underlying causes and dynamics that explain an unfamiliar problem.* Design provides a means to gain understanding of a complex problem and insights towards achieving a workable solution.

4 This insight is what Joint doctrine calls "CDR's visualization."[7] CDRs begin developing their design upon receipt of a mission. Design precedes and forms the foundation for staff planning. However, design is also continuous throughout the operation. As part of assessment, CDRs continuously test and refine their design to ensure the relevance of military action to the situation. In this sense, design guides and informs planning, preparation, execution, and assessment (Joint Operations Cycle). However, a plan is necessary to translate a design into execution.

5 Operational design may occur in two primary forms. The first and most basic is through the conduct of formalized military planning and merely reflects the intuition of the CDR communicated specifically to his staff in the form of inputs to planning. A second more intensive form is Campaign Design (CD), which while similar in its desired outcome, has a more limited applicability. This concept remains under development in Joint doctrine, but has occasionally been practiced as a distinct CDR-driven planning activity on key com-

[7] *JP 3-0, Joint Operations.*

plex operational problems. CD places much greater demands on the involvement of the CDR and other key participants above and beyond normal planning activities.

6 Given the difficult and multifaceted problems of operations today, dialog among the CDR, principal planners, members of the interagency team, and host-nation (HN) representatives helps develop a coherent design. This involvement of all participants is essential. The object of this dialog is to achieve a level of situational understanding at which the approach to the problem's solution becomes clear. The underlying premise is this: when participants achieve a level of understanding such that the situation no longer appears complex, they can exercise logic and intuition effectively. As a result, design focuses on framing the problem rather than developing courses of action.

(e) Operational Design Process. Design may occur by any means deemed appropriate by the CDR. The result of this process should be a framework that forms the basis for the coherent and complete joint campaign or operation plan and the conceptual linkage of ends, ways, and means. [8]

1 Conceptually, the design process will follow a logical progression through five sub-processes of Diagnosis, Dialog, Design, Learning, and Redesign. Figure VI-6, taken from FM 3-24, MCWP 3-33.5 - *Counterinsurgency*, depicts this progression for a high-level counterinsurgency challenge. The original diagnosis of objective OE factors on the left leads through CDR-driven dialogue toward the recognition of a central campaign purpose around which are designed conceptual approaches for interagency action. Learning occurs through execution leading to redesign where the CD cycle repeats itself. This establishes the design process that must be applied iteratively throughout the Joint Operations Cycle as assessments and/or changes in the OE might suggest.

2 As previously indicated, design normally occurs through the natural involvement of CDR's in their own planning processes. When it occurs through Campaign Design, it may occur either in advance, in-stride or following the conduct of formal operational planning.

> ***Designing*** *is creative and is best accomplished through discourse*
>
> ***Discourse*** *is the candid exchange of ideas without fear of retribution that results in a synthesis and a shared visualization of the operational problem.*

(f) Elements of Operational Design. The elements described in Figure VI-7 are tools to help CDRs and their staffs visualize the campaign or operation and shape the CONOPS.

1 CDRs and their staffs use a number of operational design elements to help them visualize the arrangement of actions in time, space, and purpose to accomplish their mission. These elements can be used selectively in any joint operation; however, their application is broadest in the context of a joint campaign or major operation.

2 Some elements (e.g., objectives, COGs, LOOs) can be described tangibly in the text or graphics of an operation order or plan. Other elements (e.g., balance, synergy, leverage) typically cannot be described in this manner. These elements will vary between COAs according to how the CDR and staff develop and refine the other elements of design during the planning process. For example, in the CDR's judgment, one COA could result in better *balance* and *leverage*, but not provide the *tempo* of operations that result from another COA. In the end, the CDR must be able to visualize these intangible elements and draw on judgment, intuition, and experience to select the best COA. Their detailed application to joint operation planning is provided in JP 5-0.

3 When produced apart from the effective application of operational art, these elements may be linked less coherently, or applicable only to a portion of the overall military approach.

(2) Plan.

[8] *FM 3-24, MCWP 3-33.5, Counterinsurgency.*

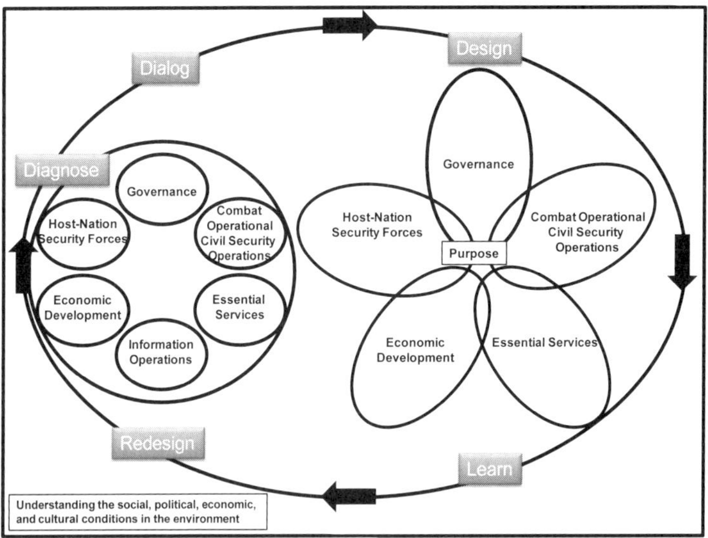

Figure VI-6. The Iterative Design Process.

(a) Planning focuses on the physical actions intended to directly affect the enemy or environment. Planners typically are assigned a mission and a set of resources; they devise a plan to use those resources to accomplish that mission. Planners start with a design (whether explicit or implicit) and focus on generating a plan—a series of executable actions and control measures. Planning generally is analytic and reductionist. It breaks the design into manageable pieces assignable as tasks, which is essential to transforming the design into an executable plan. Planning implies a stepwise process in which each step produces an output that is the necessary input for the next step.[9,10,11]

> *Though design precedes planning, it continues throughout planning, preparation and execution.*

(b) Planning is the process by which CDRs and staff translate the CDR's visualization or design into a specific COA for preparation and execution, focusing on the expected results. Planning involves having a desired endstate and describing the conditions and most effective methods to achieve it. It includes formulating one or more COA for accomplishing the mission. CDRs and staffs consider the consequences and implications of each COA. Once the CDR selects a COA, planning continues until the plan or order is published. Planning also continues through execution of an operation. At minimum, staffs refine plans for branches and sequels throughout an operation.

(c) *Plans forecast, but do not predict.* A plan is a continuous, evolving framework of anticipated actions that guides subordinates through each phase of the joint operation. Any plan is a framework from which to adapt, not an exact blueprint. The measure of a good plan is not whether execution transpires as planned, but whether the plan facilitates effective action in the face of unforeseen events. Good plans foster initiative, account for uncertainty and friction, and mitigate risk.

[9] *FM 3-24, MCWP 3-33.5, Counterinsurgency.*

[10] *FM 5-0, The Operations Process.*

[11] *MCDP 5, Marine Corps Doctrinal Publication (MCDP) 5, Planning.*

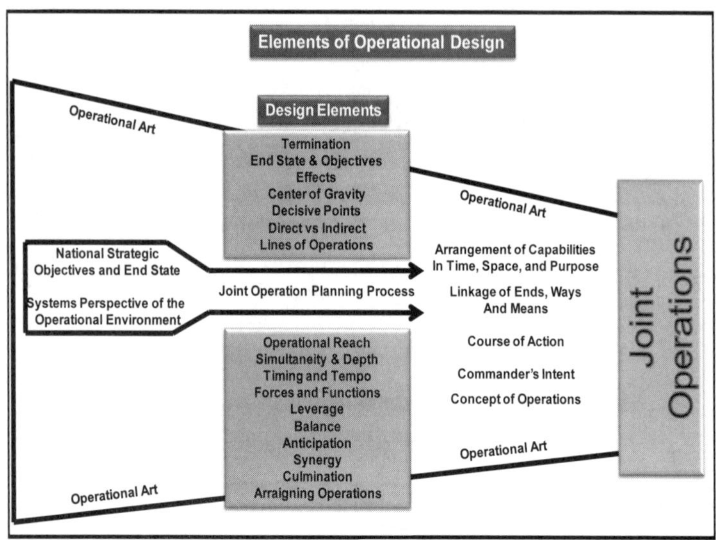

Figure VI-7. Operational Art and Design (JP 3-0).

(d) Scope, complexity, and length of planning horizons differ between strategic, operational and tactical planning. Campaign planning coordinates major actions across significant time periods. Planners integrate Service capabilities with those of joint, inter-agency, multinational and non-governmental organizations. Comprehensive, continuous, and adaptive planning characterizes successful Joint operations. Detailed discussion of planning through the JOPP are provided in Chapters 10 thru 22 of this document.

(3) Prepare.

(a) Preparation consists of activities by the joint community before execution to improve its ability to conduct the operation including, but not limited to, the following: subordinate/supporting plan refinement, possible rehearsals, ISR, coordination, and movement. It creates conditions that improve friendly forces' chances for success. It facilitates and sustains transitions, including those to branches and sequels.

(b) Preparation requires staff, supporting and subordinate coordination. Mission success depends as much on preparation as planning. Rehearsal of concept (ROC) drills help staffs, supporting commands and subordinates to better understand their specific role in upcoming joint operations.

(c) Several preparation activities begin during planning and continue throughout a joint operation. Many preparation activities continue during execution. Uncommitted forces prepare for identified contingencies and look to the joint operation's next phase or branch. Committed CDR's revert to preparation when they reach their objectives, conduct a branch plan or reach their termination criteria.

(4) Execute.

(a) Execution is putting a plan into action by applying all elements of national power to accomplish the mission and using situational understanding to assess progress and make execution and adjustment decisions. Whether the plan is a TCP or a subordinate CAP in the form of an Operations Plan, execution begins by focusing on a concerted action to shape, deter, seize, retain, and/or exploit the initiative. This represents Phase 1 of the current joint phasing model. In Phase II, joint forces seize the initiative as soon as possible and dictate the terms of action throughout a joint operation. Retaining the initiative requires constant effort. It enables CDRs to compel the adversary to accept action on terms established by friendly forces while maintaining our own freedom of action.

(b) Operationally, seizing the initiative requires leaders to anticipate events so their forces can see and exploit opportunities before the adversary does. This is when assessment, the J2, and IO planners will be critical. Once the initiative is seized, Joint forces exploit the opportunities it creates. Initiative requires constant effort to force adversaries to conform to friendly purposes and tempo while retaining friendly freedom of action. CDRs place a premium on audacity and making reasoned decisions under uncertain conditions. The CDR's intent and aggressiveness of subordinates create conditions for exercising disciplined initiative.[12,13,14]

(5) Assess.

(a) Assessment is the continuous monitoring and evaluation of the current situation and progress of a joint operation. It involves deliberately comparing forecasted outcomes to actual events to determine the overall effectiveness of force employment. Assessment measures the effectiveness of operations. More specifically, assessment helps CDRs determine progress toward accomplishing tasks and achieving objectives and the endstate. It helps identify opportunities, counter threats, and any needs for course correction. It results in modifications to plans and orders. This process of continuous assessment occurs throughout the joint planning process.

(b) Assessment and learning enable incremental improvements to the design. The aim is to rationalize the problem— to construct a logical explanation of observed events and subsequently construct the guiding logic that unravels the problem. The essence of this is the mechanism necessary to achieve success. This mechanism may not be a military activity—or it may involve military actions in support of nonmilitary activities. Once CDRs understand the problem and what needs to be accomplished to succeed, they identify the means to assess effectiveness and the related information requirements that support assessment. This feedback becomes the basis for learning, adaptation, and subsequent design adjustment. [15]

(c) Not all joint operations proceed smoothly toward the desired endstate. CDRs examine instances of unexpected success or failure, unanticipated adversary actions, or operations that simply do not progress as planned. They assess the causes of success, friction, and failure, and their overall impact on the force and the operation. CDRs and staffs continuously assess an operation's progress to determine if the current order is still valid or if there are better ways to achieve the endstate. Assessments by staff sections form the foundation of running estimates. Assessments by CDRs allow them to maintain accurate situational understanding and revise their visualization or operational design appropriately.

(d) Assessment precedes and guides every activity within the joint operation process and concludes each operation or phase of an operation. Assessment entails two distinct tasks: continuously monitoring the situation and the progress of the operations, and evaluating the operation against measures of effectiveness and measures of performance. Effective assessment requires criteria for evaluating the degree of success in accomplishing the mission. Criteria can be expressed as measures of effectiveness and a measure of performance (see Chapters 14, 21 and 22 for detailed information).

(e) Many aspects of operations are quantifiable. Examples include movement rates, fuel consumption and weapons effects. While not easy, assessing physical aspects of joint operations can be straightforward. However, the dynamic interaction among friendly forces, adaptable adversaries, and populations make assessing many aspects of operations difficult. For example, assessing the results of planned actions to change a group of people to support their central government is very challenging. In instances involving assessing

[12] *FM 3-24, MCWP 3-33.5, Counterinsurgency.*

[13] *FM 5-0, The Operations Process.*

[14] *MCDP 5, Marine Corps Doctrinal Publication (MCDP) 5, Planning.*

[15] *FM 3-24, MCWP 3-33.5, Counterinsurgency.*

change in human behavior, assessment relies on understanding trends and indicators over time to make judgments concerning the success of given actions.

(f) Just as CDRs devote time and staff resources to planning, they must also provide guidance on *what to assess and to what level of detail.* Depending on the situation and the echelon of command, assessment may be a detailed process (formal assessment plan with dedicated assessment cell or element). Alternatively, it may be an informal process that relies more on the intuition of the CDR, subordinate CDRs, and staffs.

(g) When assessing operations, CDRs and staffs should avoid excessive analysis. Excessive time and energy spent developing elaborate assessment tools and graphs squanders resources better devoted to other elements of the operations process. Effective CDRs avoid overburdening subordinates and staffs with assessment and collection tasks beyond their capabilities. As a general rule, the level at which a specific operation, task, or action occurs should be the level at which such activity is assessed. This focuses assessment at each level and enhances the efficiency of the overall assessment process.

d. Summary. While working through the JOPP within the next chapters, keep in mind that *operational design, planning, preparation, execution, and continuous assessment* make up the *Joint Operations* Cycle. Even though every operation is different, all operations follow the Joint Operations Cycle. These activities are cyclic, but not discreet. They overlap and recur as circumstances demand.

What are the ideas that underline the design methodology?

Design frames the problem to be solved in that it constructs a useful but provisional understanding of the relationships of the elements that comprise the situation so that an effective set of actions can be taken to transform the situation into a more acceptable state. Commanders should take the time to challenge existing mental constructs when faced with a new situation. If this is not done, commanders risk unknowingly anchoring their thinking based on preconceived notions that may not respond effectively to current reality.

The Joint Warfighting Center Joint Doctrine Series, Pamphlet 10, Design in Military Operations, A Primer for Joint Warfighters.

Global Force Management 7

> *"I believe that in 2020, we will still be the most powerful military in the world. More than 1 million men and women under arms — present in more than 130 countries and at sea — will still possess capabilities in every domain that overmatches potential adversaries. Enjoying alliances with a majority of the most powerful states, we will be the only nation able to globally project massive military power."* [1]
>
> *Quadrennial Defense Review*

The global security environment presents an increasingly complex set of challenges and opportunities to which all elements of U.S. national power must be applied. To protect U.S. national interests and achieve the objectives of the NSS in this environment, the Joint Force will need to continually recalibrate its capabilities and make selective additional investments to succeed in the following missions:[2]

> ***Counter*** *Terrorism and Irregular Warfare -* ***Deter*** *and Defeat Aggression -* ***Project*** *Power Despite Anti-Access/Area Denial Challenges -* ***Counter*** *Weapons of Mass Destruction -* ***Operate*** *Effectively in Cyberspace and Space -* ***Maintain*** *a Safe, Secure, and Effective Nuclear Deterrent -* ***Defend*** *the Homeland and Provide Support to Civil Authorities - Provide a Stabilizing Presence -* ***Conduct*** *Stability and Counterinsurgency Operations -* ***Conduct*** *Humanitarian, Disaster Relief, and Other Operations*

Likewise the Quadrennial Defense Review (QDR) affirms that our Armed Forces "be capable of conducting a broad range of several overlapping operations to prevent and deter conflict and, if necessary, to defend the United States, its allies and partners, selected critical infrastructure, and other national interests." The experiences with operations such as Operation Urgent Response in Haiti has demonstrated that even while sourcing major combat operations in one part of the world, we may be called upon to react to a crisis in disparate regions of the globe. An earthquake in one AOR can have a rippling impact on force sourcing for current operations and long-term security planning.

[1] Secretary Of Defense, Quadrennial Defense Review Report, (U.S. Department of Defense, 2014), 63.

[2] Department of Defense, Sustaining U.S. Global Leadership" Priorities for 21st Century Defense (January 2012)

The QDR also recognizes "simultaneously defending the homeland: conducting sustained, distributed counterterrorist operations; and in multiple regions, deterring aggression and assuring allies through forward presence and engagement. If deterrence fails at any given time, U.S. forces will be capable of defeating a regional adversary in a large-scale multi-phased campaign, and denying the objectives of, or imposing unacceptable costs on, a second aggressor in another region." [3] To remain dominant within this complex and uncertain security landscape the ability to dynamically align the force pool must improve and keep pace with the complexity of the operational environment. The *wicked problem* of balancing the force against global and institutional demand requires strict purposeful design to allow for timely and informed decisions and to ultimately satisfy the broadest array of objectives.

1. Purpose

This chapter provides an overview of the GFM process, which starts and ends with the SECDEF. In accordance with Title 10, United States Code (Title 10 USC), the SECDEF assigns forces/capabilities, allocates forces/capabilities, provides planning guidance to CCMDs, and provides overarching strategic guidance to the CJCS. The CJCS, in turn, develops strategic-level guidance including apportioned forces/capabilities to CCMDs for adaptive planning. CCMDs use apportioned forces as an assumption in developing plans and to coordinate execution force/capability requirements with the CJCS based on the SECDEF's guidance. The GFM process facilitates alignment of operational forces against known operational requirements, providing planning estimates in advance of planning, and deployment preparation timelines. The end result is a timely allocation of forces/capabilities necessary to execute CCMD missions (to include theater security cooperation tasks), timely alignment of forces against future requirements, and informed SECDEF decisions on the risk associated with allocation decisions.

2. Situation

The strategic environment will continue to be complex, dynamic and uncertain. The United States military will continue to be involved globally in named operations, Overseas Contingency Operations, Support to Campaign Plans, Theater Security Cooperation activities, exercises, and Security Force Assistance operations in support of National Security, National Defense and National Military Strategies.[4] Success in this environment requires a coherent use of the force pool (Figure VII-1) among the competing priorities in both planning and execution. This is achieved by the integrated use of assignment, allocation and apportionment. The goal of these processes are to provide CCDRs the forces to best support U.S. Military objectives outlined in the GEF using assigned and allocated forces to accomplish missions while mitigating military risk. To allow feasible plans to be developed, CCDRs are provided force planning assumptions based on analysis of the force pool. The number of forces that are reasonably expected to be available, should the plan be executed, are called apportioned forces.

a. CCDRs are directed by the Unified Command Plan (UCP), strategic guidance and various orders, to plan and execute operations and missions. CCDRs are assigned forces that are to be used to accomplish those operations and missions; however, in the dynamic world environment, competing missions may require adjusting the distribution of assigned forces among the CCDRs and Services through allocation. Each allocation decision involves tasking a CCDR, Secretary of a Military Department, or Director of DOD Agency to provide a force or individual to another CCDR. This involves risk to not only the providing

[3] *Secretary Of Defense, Quadrennial Defense Review Report, (U.S. Department of Defense, 2014), VI.*

[4] *SECDEF directed GFM procedures are contained in the Global Force Management Implementation Guidance (GFMIG) and the GEF.*

Service and/or CCDR, but also to contingency plans across the Joint Planning and Execution Community (JPEC).[5] The allocation process begins with the Supported CCDR identifying the forces or capabilities necessary to execute missions.

3. Scope

Title 10 USC, the UCP, GEF, GFMIG, JSCP, Forces For Unified Commands Memorandum ("Forces For"), and CJCSM 3130.06 *GFM Allocation Policy and Procedures*, are the baseline documents that establish the policy, guidance, doctrine and procedures in support of GFM. The SECDEF's GFMIG specifies guidance for the GFM assignment, apportionment, and allocation processes to manage the force pool from a global perspective. The GEF provides POTUS and SECDEF guidance for prioritizing planning, execution and GFM.

a. Title 10 USC. Title 10 USC Sections 161, 162, and 167 outline force assignment guidance and requirements. Section 162 gives the SECDEF the authority to allocate forces between CCDR's. Section 153 gives the CJCS authority to apportion forces to CCMDs basd on the SECDEF's GEF.

b. Unified Command Plan (UCP). Unified and Specified CCMDs were first described in the National Security Act (NSA) of 1947 and the statutory definition of the CCMDs has not changed since then. The NSA of 1947 and Title 10 USC provide the basis for the establishment of CCMDs. The UCP establishes the missions, responsibilities, and geographic areas of responsibility (AORs) for commanders of combatant commands (CCDRs). The UCP is approved by the President, published by the CJCS, and addressed to the CCDRs. The unified command structure generated by the UCP is flexible, and changes as required to accommodate evolving U.S. national security needs. Communications between the President or the SECDEF (or their duly deputized alternates or successors) and the CCDRs shall be transmitted through the CJCS unless otherwise directed. Title 10 USC, Section 161, tasks the CJCS to conduct a review of the UCP "not less often than every two years" and submit recommended changes to the President, through the SECDEF.

(1) Six CCDRs have geographic area responsibilities. These CCDRs are each assigned an AOR by the UCP and are responsible for all operations within their designated areas: U.S. Central Command, U.S. European Command, U.S. African Command, U.S. Pacific Command, U.S. Southern Command, and U.S. Northern Command. There are three CCDRs assigned worldwide functional responsibilities not bounded by geography: U.S. Special Operations Command, U.S. Strategic Command, and U.S. Transportation Command.

(2) Several key strategic documents provide direction for the execution of missions established in the UCP. Though not all-inclusive, they are: NSS, NDS, NMS, GEF, JSCP, and the GFMIG.

c. Guidance for Employment of the Force (GEF). The GEF transitions DOD planning from a contingency-centric approach to a strategy-centric approach. Rather than initiating planning from the context of particular contingencies, the strategy-centric approach requires commanders to begin planning from the perspective of achieving broad regional or functional objectives. Under this approach, planning starts with the NDS, from which the GEF derives theater or functional strategic endstates prioritized appropriately for each CCMD. The GEF, along with the JSCP, are the principal sources of guidance for CCMD steady state campaign, contingency and posture plans.

[5] *Joint Planning and Execution Community. Those headquarters, commands, and agencies involved in the training, preparation, mobilization, deployment, employment, support, sustainment, redeployment, and demobilization of military forces assigned or committed to a joint operation. See JP 1-02 and JP 5-0.*

d. Global Force Management Implementation Guidance (GFMIG). The purpose of the GFMIG is to integrate complementary force assignment, apportionment, and allocation information into a single GFM document allowing improvement in the DOD's ability to manage forces from a global perspective. The GFMIG compliments the GEF and Guidance for Development of the Force (GDF). The Forces For Unified Command Memorandum, commonly referred to as "Forces For," assignment tables are incorporated in Section II of the GFMIG. The apportionment tables are incorporated in Section IV and Annex C is the GFM Allocation Business Rules. In odd years, when the GFMIG is not updated, the "Forces For" and the two tables are published separately and posted on the JS J8 web site.

e. "Forces For." The "Forces For" Unified Commands Memorandum provides the SECDEF's direction to the Secretaries of the Military Departments for assigning forces to CCDRs. The assignment and apportionment tables are normally enclosures of the "Forces For" Memorandum. The assignment tables lists all the forces assigned to each CCDR. The apportionment tables list the forces apportioned for planning. In even years the "Forces For" assignment and apportionment tables are included in the GFMIG with the biennial update. In odd years, when the GFMIG is not updated, the "Forces For" and the two tables are published separately and posted on the JS J8 web site. Operational forces not assigned to a CCDR are retained under the Secretary of the Military Department and are referred to as "Service retained."

f. Joint Strategic Capabilities Plan (JSCP). The JSCP implements the strategic policy direction provided in the GEF and initiates the planning process for the development of campaign, campaign support, contingency, and posture plans. The JSCP contributes to the CJCS's statutory responsibility to assist the President and the SECDEF in providing for the strategic direction of the Armed Forces of the United States and directing contingency plans for the Nation. The goal is to provide CCMD and Service planners with meaningful, necessary guidance balanced between the details needed to conduct coordinated, sustainable steady-state activities and specific contingencies while still allowing commanders flexibility to respond to unanticipated events.

SECTION I - GFM GOALS AND PROCESSES

1. GFM Goals

The goals for GFM as defined by the GFMIG are:

a. Account for forces and capabilities committed to ongoing operations and constantly changing unit availability.

b. Identify the most appropriate and responsive force or capability that best meets the CCMD requirement.

c. Identify risk associated with sourcing recommendations.

d. Improve ability to win multiple overlapping conflicts.

e. Improve responsiveness to unforeseen contingencies.

f. Provide predictability for annual force requirements.

g. Identify forces and capabilities that are unsourced or hard-to-source (UHTS).

2. Processes

GFM is a compilation of three integrated processes; assignment, allocation and apportionment. These three processes are used to align U.S. forces in support of the NDS, joint force availability requirements, and joint force assessments. It provides comprehensive insights into the global availability of U.S. military forces/capabilities and provides senior decision makers a process to quickly and accurately assess the impact and risk of proposed changes in forces/capability assignment, apportionment, and allocation.

3. Force Pool

The three processes of assignment, allocation and apportionment are related to each other. Figure VII-1 shows the entire DOD force pool (every military unit, Soldier, Sailor, Airman and Marine) within the "Service Institutional" and "Operational Forces" box. This force pool is further divided by assigned forces to a CCDR, unassigned forces (Service Institutional), and Service retained forces. Most allocated forces come from operational forces assigned to a CCDR and Service retained, but in some instances, the Service may be directed to provide (allocate) forces from their Service Institutional forces (such as recruiters and schoolhouses).

__Service Retained Forces__ – AC and RC operational forces under the administrative control of respective Secretaries of the Military Departments, and not assigned to a CCDR. These forces remain under the administrative control of their respective Services and are commanded by a Service-designated Commander responsible to the Service unless allocated to a CCDR for the execution of operational missions. GFMIG

__Unassigned Forces__ – Forces not assigned to a CCDR IAW Title 10 USC, Section 162, and instead remain under Service control in order to carry out functions of the Secretary of a Military Department IAW Title 10 USC sections 3013(b), 5013(b), 8013(b). GFMIG

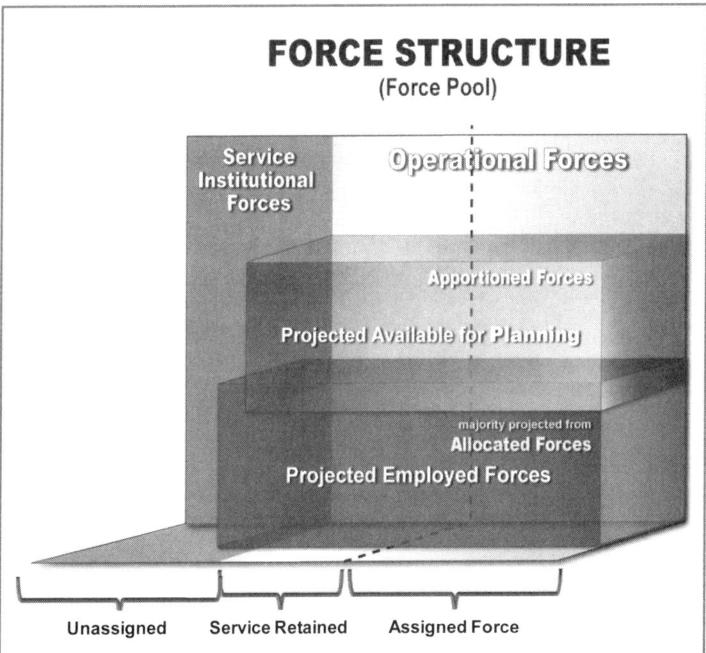

Figure VII-1. Force Pool (GFMIG).

a. Unassigned, Service retained and assigned forces may also be used by the Service to meet Service institutional requirements. This is the reason the Projected Employed Forces (shown in red) crosses the dashed line and into the Institutional Service Forces box. Apportioned forces are calculated by subtracting global demand from the assigned forces, and the fact that some assigned forces are employed performing Service institutional missions or are performing missions for their assigned CCDR, the employed and apportioned forces in Figure VII-1 overlap. CCDR force requests are constantly changing to respond to world events. To determine the Projected Employed Forces, analysis of the current CCDR force requests must be conducted in order to project the number of Employed Forces in order to calculate the number of forces left that can reasonably be expected to be available, or apportioned.

4. Global Demand

In order to distribute a limited number of forces among the competing CCDR demands, Joint Staff Joint Force Coordinator (JS JFC)[6], Joint Staff, Force Providers (FPs). In accordance with the GFMIG and the GFMIG Allocation Business Rules, FPs[7] include Secretaries of the Military Departments, CCDRs with assigned forces, the U.S. Coast Guard, DOD agencies, and OSD organizations that provide force-sourcing solutions to CCDR force requirements.

a. Operational Force Demand. Operational force demand encompasses all CCDR demand for forces in support of current operations (including shaping operations and OPLANS that require forces), activities in support of TCPs, theater security cooperation (TSC) events, activities in support of the TSC plan (including Joint Combined Exercise for Training (JCET) and Counter-Narco terrorism (CNT)) and requests for federal assistance (RFA). Operational force demand includes force requirements for conventional, SOF, mobility, DOD Agency and Intelligence, Surveillance and Reconnaissance (ISR) forces and capabilities in support of the above operations and activities. Excluded are exercise and JIA demand.

(1) Assigned Forces Demand. The demand on assigned forces is an integral part of operational force demand. CCDRs are assigned forces to perform operations and assigned missions. The CCDR to which forces are assigned often employs and deploys their assigned forces. Since the CCDR already has Combatant Command Authority (COCOM), and OPCON is inherent in COCOM, CCDRs can use their assigned forces for missions and operations they have authority to execute without additional SECDEF approval. Section II (*Assignment of Forces*) of the GFMIG provides details on authorities granted with assigned forces.

(a) CCDRs are directed to report the requirements for their assigned forces and assign each requirement a Force Tracking Number (FTN), just as they would when requesting a force to be allocated in either the annual or emergent process. Although the SECDEF does not allocate these forces back to the CCDR that has them assigned, this complete op-

[6] *CJCS, through the Director, J-3 (DJ-3), will serves as the Joint Force Coordinator (JFC) responsible for providing recommended sourcing solutions for all validated force and JIA requirements. In support of the DJ-3 the Joint Staff Deputy Director for Regional Operations and Force Management (J-35) assumes the responsibilities of the JFC. As such the JFC will coordinate with the Joint Staff J-3, Secretaries of the Military Departments, CCDRs, JFPs, JFM, and DoD Agencies. The Joint Force Coordinator (JFC) is referred to in current DoD GFM guidance and policy as the JFC. However, for clarity in this document, and distinction between the Joint Force Commander (JFC) and Joint Force Coordinator (JFC), the Joint Force Coordinator will be referred to as the Joint Staff Joint Force Coordinator (JS JFC).*

[7] *Force Providers (FPs). In accordance with the GFMIG and the GFMIG Allocation Business Rules, FPs include Secretaries of the Military Departments, CCDRs with assigned forces, the U.S. Coast Guard, DOD agencies, and OSD organizations that provide force-sourcing solutions to CCDR force requirements.*

erational demand provides the JS J3, JS JFC, JFPs, JFM, FPs, and OSD this vital visibility into global demand to help in determining risk when executing the allocation process.

b. Service Institutional Demand. Service responsibilities addressed in Title 10 USC include recruiting, training, organizing and equipping the force. To accomplish these Title 10 USC responsibilities, Services require resources and units organized towards these missions. Normally Service units assigned to a CCDR are organized for operational missions as are Service retained units. Unassigned forces are normally used by the Service to accomplish Service missions. These type units include recruiting commands, schoolhouses/ training commands, experimentation, system development and Service headquarters and are normally unassigned.

(1) Occasionally units assigned to a CCDR are tasked by a Service to support Service missions. An example of this may be to operationally test a weapon system.

(2) Services currently capture their institutional demand and the risks associated with meeting that demand. These risks to the Service are inserted into the allocation process as risk justifications that support the allocation decision process.

c. JIA Demand. Included in the GFM allocation process are JIA programs and procedures that support temporary Joint Headquarters requirements for personnel in support of Presidential or SECDEF-approved/directed operations. Joint and Combined Headquarters manning is documented on a Joint Manning Document (JMD). The individual billets for a JMD are each assigned a unique number in the electronic Joint Manpower and Personnel System (eJMAPS), the system of record to document JMDs. Each billet is classified as a coalition, contractor, unit fill, Other Government Agency (OGA), or JIA. The allocation process allocates U.S. military forces, so only the billets coded as JIA will be forwarded to the SECDEF for an allocation decision. Unit fills are force requirements for units that will be part of a joint headquarters. The billets designated as IAs are recorded as a unique demand signal on the forces. The individual billets for a joint headquarters are coded as JIAs.

(1) CCDRs will also enter the requirements for all billets JMDs, including those billets which are being filled with their assigned forces.

d. Exercise Demand. Exercise force demand encompasses all CCDR requirements for forces for CJCS directed and CCDR high priority exercises. CCDRs submit exercise force requests via the Joint Training Information Management System (JTIMS), a tool that automates and supports the global Joint Training System, directly to the JS JFC and supporting JFPs with JS J3 visibility. JS JFC/JFPs coordinate exercise requests with FPs, as able. Exercise sourcing is normally a lower priority than operational force demands. Once exercise forces are sourced in JTIMS, operational force demands may reallocate the forces previously sourced to an exercise to higher priority operational force requirement. Exercise forces deploy TACON for participation in (or in Direct Support of) a CCDRs Joint exercise.

e. Future Challenges. In addition to operational, institutional, JIA and exercise demand, which captures total global demand, the SECDEF also considers potential use of forces/ capabilities of future challenges. These future challenges are considered as potential usage of resources, but are not part of the global demand. Future challenges encompass all CCDR planning requirements for forces and capabilities required to support potential execution of a CJCS directed or SECDEF approved plan should execution be ordered. In accordance with the CJCSM 3122 series[8], CCDRs identify their force/capability requirements in Appendix 1 to Annex A of a plan. This list of force requirements is the Time Phase Force Deployment List (TPFDL), and should execution of a plan be directed, the forces required to support the plan would become an operational demand and be submitted into the GFM process for sourcing. Forces not assigned to the supported CCDR would have to be requested and allocated by the SECDEF.

[8] *CJCSM 3122.02 will be superseded with the publication of CJCSM 3130.04, Crisis Action TPFDD Development and Deployment Execution.*

5. Force Management Relationships

GFM aligns force assignment, allocation and apportionment in support of the NDS, joint force availability requirements, and joint force assessments. Authorities that govern the three processes are as follows:

a. Assignment. Title 10 USC Sections 161, 162, and 167 outline force assignment guidance and requirements. The President, through the UCP, instructs the SECDEF to document his direction for assigning forces in the Forces for Unified Commands Memorandum ("Forces For"). The Secretaries of the Military Departments shall assign forces under their jurisdiction to Unified and Specified CCMDs to perform missions assigned to those commands. Such assignment defines the COCOM and shall be made as directed by the SECDEF, including as to the command to which forces are to be assigned. Assignment is further explained in the biennial GFMIG within Section II.

(1) Unifying Concepts.

(a) *Force Assignment*. As mandated by Title 10 USC, forces are assigned to CCDRs to provide those CCDRs forces to meet their UCP-established missions and assigned responsibilities.

1 Forces permanently stationed overseas are generally assigned to the CCDR for the designated AOR, but CONUS based force's may be assigned to GCC's outside CONUS.

2 Special Operations Forces (SOF) are assigned to USSOCOM.

3 Conventional forces stationed within the continental U.S. (CONUS) are Service retained forces.

4 Historically, other CONUS forces have been assigned as follows:

a West Coast conventional combat forces: Army, Navy, and Marine Corps forces to USPACOM.

b Strategic nuclear and space forces to USSTRATCOM.

c Strategic defensive forces to U.S. Element NORAD (USELEMNORAD)

d Mobility forces to USTRANSCOM.

(b) *Service Components*. Service components provide:

1 Command relationships with Service components vary among CCDRs and are specified in assignment tables within the GFMIG. Service components can only be assigned COCOM to one CCDR. However, Service component commanders may support multiple CCDRs in a non-COCOM, supporting commander relationship.

2 The Services recommend the proper headquarters to provide support, and "Forces For" is the vehicle for the SECDEF to establish Service component support relationships.

(2) Scope. The "Forces For" is the SECDEF guidance to Service Secretaries to assign specific numbers of forces to CCDRs and CDRUSELEMNORAD.

(a) CCDRs exercise COCOM over assigned forces and are directly responsible to the President and SECDEF for performing assigned missions and preparing their commands to perform assigned missions. Although not a unified CCDR, USELEMNORAD exercises COCOM over assigned U.S. forces.

(b) The assignment of forces provided for within the GFMIG is separate from Title 32 USC provisions that deal with the National Guard and the parts of Title 10 USC that provide for ordering National Guard and reserve forces to active duty. CCDRs will exercise COCOM over assigned RC forces when mobilized or ordered to active duty (other than training). CCDRs may employ RC forces assigned to their commands in contingency operations only when the forces have been mobilized for specific periods in accordance with the law, or when ordered to active duty with the consent of the member and validated by their parent Service. During peacetime, CCDRs will normally coordinate with Service component commands on all matters concerning assigned or attached RC forces. CCDRs exercise Training and Readiness Oversight (TRO) for assigned RC forces when not on active duty or when on active duty for training. Service Headquarters exercise TRO for Service-retained RC forces.

(c) Force reductions or new assignments are required when force structure changes are programmed to occur. Reassignments, from one CCMD to another, may occur as directed by the SECDEF and are not necessarily tied to Presidential budget changes.

(3) Assignment and Transfer of Forces. A force is assigned in accordance with the guidance contained within the "Forces For" memorandum. Forces are allocated for execution through the GFM planning process specified in the CJCSM 3130.06 series[9], and Annex C of the GFMIG. Forces become attached when deployed via a SECDEF-approved deployment order.

(a) Forces, not command relationships, are transferred between commands. When forces are transferred, the command relationship that the gaining commander will exercise (and the losing commander will relinquish) will be specified by the SECDEF. The CCDR normally exercises OPCON over forces attached by the President or SECDEF. Forces are attached when transfer of forces will be temporary. Establishing authorities for subordinate unified commands and Joint Task Forces (JTFs) may direct the assignment or attachment of their forces to those subordinate commands and delegate the command relationship as appropriate.

(b) Transient forces do not come under the chain of command of the area commander solely by their movement across operational area boundaries, except when the CCDR is exercising TACON for the purpose of force protection.

(c) Unless otherwise specified by the SECDEF, and with the exception of the USNORTHCOM AOR, a CCDR has TACON for exercise purposes whenever forces not assigned to that CCDR undertake exercises in that CCDR's AOR. TACON begins when the forces enter the AOR and is terminated at the completion of the exercise, upon departure from the AOR. In this context, TACON is directive authority over exercising forces for purposes relating to that exercise only; it does not include authority for operational employment of those forces. In accordance with the UCP, this provision for TACON normally does not apply to U.S. Transportation Command (USTRANSCOM) or U.S. Strategic Command (USSTRATCOM) assets. When USTRANSCOM or USSTRATCOM forces are deployed in a geographic CCDR's AOR, they will remain assigned to and under control of their respective CCMD, unless otherwise directed.

(d) Subject to the two following exceptions, in the event of a major emergency in a geographic CCDR's (GCC) AOR requiring the use of all available forces, the GCC may temporarily assume OPCON of all forces in the assigned AOR, including those of another command. The CCDR determines when such an emergency exists and, upon assuming OPCON over forces of another command, immediately advises the CJCS, the appropriate operational commander, the Military Service Chief, and the Secretary of the Military Department of the forces concerned and the nature and estimated duration of employment of such forces. The CJCS shall notify the SECDEF, who will determine whether to permit the emergency exercise of OPCON to continue.

1 Exception 1: GCCs may not assume OPCON of all forces in the AOR for a major emergency if such use would interfere with those forces scheduled for or actually engaged in the execution of specific operational missions approved by the President or SECDEF.

2 Exception 2: CDRUSNORTHCOM's authority to assume OPCON during an emergency is limited to the portion of USNORTHCOM's AOR outside the United States. CDRUSNORTHCOM must obtain SECDEF approval before assuming forces not assigned to CDRUSNORTHCOM within the United States.

[9] *CJCSM 3130.06, Global Force management Allocation Policies and Procedures*

(4) Service Component Assignment. A CCMD's Service component consists of the Service component commander and the Service component command's forces (such as individuals, units, detachments, and organizations) that have been assigned to the CCMD.

(5) Supporting Service Component Commander. When a Service component is assigned to multiple CCMDs as a Supporting commander, the Service CDR and only that portion of the CDR's assets assigned to a particular CCDR are under command authority (Support) of that particular CCDR.

(a) Unless otherwise specified by the SECDEF, the CDR tasked as a supporting CDR to additional CCMDs maintains a general support relationship for planning and coordinating regarding the CCDR's assigned mission and forces.

b. Allocation. Pursuant to Title 10 USC, section 162, "[a] force assigned to a CCMD ... may be transferred from the command to which it is assigned only by authority of the SECDEF and under procedures prescribed by the Secretary and approved by the President." Under this authority, the SECDEF allocates forces between CCDRs. When transferring forces, the Secretary will specify the command relationship the gaining commander will exercise and the losing commander will relinquish. The allocation process provides forces to CCDRs. The Global Force Management Allocation Plan (GFMAP) will provide fully vetted and relevant sourcing solutions to CCDRs' force requirements to the SECDEF for approval in support of global and theater strategic objectives.

(1) Force allocation consists of three interrelated staffing cycles based upon urgency. The cycles are:

(a) Annual GFM force allocation cycle.

(b) Bi-weekly for emergent requests for forces.

(c) Special cycle for urgent (time-sensitive) emergent requests that cannot wait until the next bi-weekly cycle.

(2) GFMAP. Once approved by the SEDEF, the GFMAP is published on the JS J3 Joint Operations Division GFM website. There is a separate GFMAP for each FY. The GFMAP and attached Annexes are the SECDEF DEPORD for all allocated forces. The GFMAP Base Order is published 10-12 months prior to the applicable FY of execution and modified as necessary thereafter. The majority of rotational force and JIA requests are adjudicated and ordered in the GFMAP Base Order Annexes. Remaining requests from the Annual Submission and emergent requirements are added to the GFMAP Annexes in subsequent modifications. The GFMAP authorizes the transfer of forces from supporting CCDRs and Secretaries of the Military Departments and attachment to a supported CCDR.

(a) *GFMAP Annexes*. The SECDEF decisions to allocate forces are published within the four annexes of the CJCS GFMAP. A fifth annex, E, provides guidance specific to each CCDM. Annexes are published with the GFMAP and subsequent modification to the GFMAP. GFMAP Annexes A-D contains all the information inherent within a written order, and authorizes JFPs' to order forces in subsequent GFMAP Annex Schedules. Each Annex includes force schedules as follows:

- Annex A is Conventional Forces - JS J3 as the JFC.
- Annex B is Special Operations Forces (SOF) - USSOCOM as the JFP.
- Annex C is Mobility Forces - USTRANSCOM as the JFP.
- Annex D is Joint Individual Augmentees - JS J3 as the JFC.

(b) The GFMAP Annex serves as the deployment order (DEPORD) directing FPs to deploy forces at the specified dates. FPs implement the SECDEF orders in the GFMAP Annexes and JFP GFMAP Annex Schedules by issuing DEPORDs, through the chain of command, to the unit or individuals deploying. The JS JFC, JFPs, and JFM monitor the FP's progress in meeting the GFMAP orders. Since most operational forces are assigned, each allocation is a decision to take a force from one CCDR and deploy it to another. Each decision has risks that must be weighted to balance the operational necessity of deploying the force with the risk to the FP. These risks also include the financial cost, the stress on the Service and the stress on the Service men and women.

(3) Process. The JS J3 will seek SECDEF approval to allocate forces in support of CCDR, NORAD or USNMR requests via a bi-weekly cycle SECDEF Orders Book (SDOB), or if time-sensitive, via a special SDOB. Successful execution of the SDOB process requires supported and supporting commanders, JS JFC and JFPs, the JS and other agencies to closely adhere to the SDOB timeline.

(a) When a supported CCDR submits a force request, the JS J3 validates the force request and assigns the JS JFC to provide a recommended sourcing solution. Validation assesses CCMD requirements to determine viability (risk, priorities, and capabilities) for sourcing. The JS JFC and assigned JFP (working with their assigned Service components) develop recommended sourcing solutions. The JS JFC and JFP draft and staff the GFMAP Annex and forward the sourcing recommendations to the JS J3. The JS J3 staffs the draft GFMAP Annex with FPs, CCDRs, JS and OSD and the Chairman recommends the order to the SECDEF.

(b) When the SECDEF approves the GFMAP Annex, the CJCS publishes the modification to the GFMAP Annex to authorize allocation of forces from one command or Service to another.

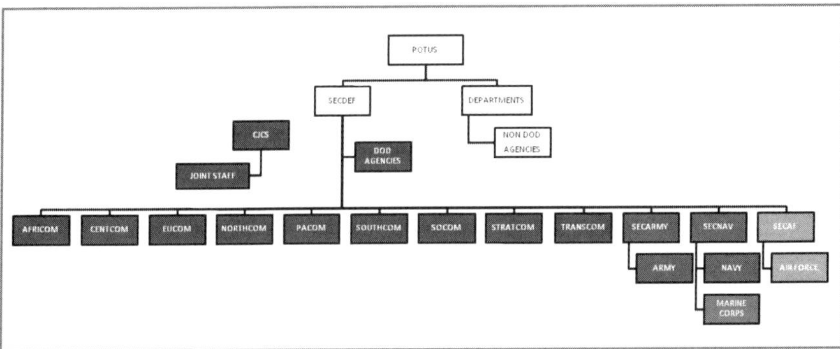

Figure VII-2. Chain of Command - Allocation Decision Process.

(4) Secretaries of Military Departments (for Service-retained forces), CCDRs and DOD agencies all work directly for the SECDEF. The allocation decision process is centralized with the SECDEF making the final decisions (Figure V11-2). When unassigned or Service-retained forces are allocated to a CCDR, they are normally attached. The providing CCDR is tasked to relinquish OPCON of that unit during its deployment and resume OPCON following the deployment. The providing CCDR normally retains COCOM of the unit. When assigned forces are allocated from one CCDR to another, the gaining CCDR is usually tasked to exercise OPCON, for the time the unit will be in the gaining CCDRs AOR. Specific command relationships are specified in the CJCS GFMAP and the GFMAP Annex. Objectives of the GFM allocation process are to:

- Globally prioritize CCDR operational tasks, including war plan response posture, SCP activities, or other missions as assigned by the CCDR -- taking into consideration ongoing operations -- and allocate annual forces to satisfy these tasks. Prioritization of CCDR operational tasks and the assessment of risks are essential as global demand for annual forces may exceed available supply and consequently affect the Military Department's ability to sustain rotation rates or available forces/capabilities.
- Optimize force management to reduce risk to achieve operational and strategic objectives, while balancing the Military Departments' responsibility to organize, train, and equip the force against CCDR operational tasks.
- Establish a mechanism, through the Global Force Management Board (GFMB) process, to provide joint solutions to CCDR's requirements, and enable -- as the technical capacity matures -- capability trade-offs among Military Departments.

- Provide predictability for Military Service Chiefs' annual deployment scheduling and CCDR's operational planning.
- Increase flexibility and options for senior leadership decision-makers (e.g., facilitate the Military Departments' ability to surge forces, as well as the ability, over the longer term as the capabilities to do so are fielded, to provide capability trade-off options among Military Departments, such as use of Navy Aegis Cruisers, Army Patriot assets, Navy or USAF fighters, or a mix of assets to provide defensive counter air capabilities).

(5) Allocation is divided into two force requirements, *annual and emergent:*

(a) Annual. This process lays out the roles, missions, and functions to support the sourcing of CCMD annual force requirements. The annual force allocation process provides guidance for the allocation of annual forces to support CCDR requirements. The Armed Forces of the United States provide overseas presence through a combination of annual forces and forward-based forces and the resources (infrastructure and pre-positioned equipment) necessary to sustain and maintain those forces. Forward-based forces are assigned to GCCs (e.g. USEUCOM, USPACOM, and USSOUTHCOM) in the assignment tables of the "Forces For" document. Annual forces are forces allocated to a CCDR in order to execute tasks in that CCMD's AOR, and are typically deployed for a specified period of time as outlined by Service deployment policies. Annual forces deploy as "units," sized from the Army Brigade Combat Team (BCT), Marine Corps Regimental Combat Team (RCT) or Marine Air Ground Task Force (MAGTF), Air Force Air and Space Expeditionary Task Force (AEF), or Navy Carrier Strike Group/Amphibious Ready Group (CSG/ARG) level or larger down to smaller sized capability packages (e.g., individual ships, squadrons, or mission unique teams), and deploy in support of CCDR missions. Annual forces are sourced globally -- all forces are The desired endstate for the annual force allocation process is a recommendation for filling validated CCDR capability requirements with appropriate joint forces to achieve an optimum level of risk in executing on-going combat operations and other missions and tasks called for in the National Defense and National Military Strategies.

1 Annual forces support the following activities:

a CCDR security cooperation activities undertaken in support of CCDR TCP.

b CCDR OPLAN response timelines in support of GEF and JSCP requirements.

c SECDEF-named operations with identified enduring requirements.

2 Annual allocation decisions are promulgated in the GFMAP:

a The SECDEF-approved GFMAP gives the designated JS JFC/JFP, CCMDs, and Military Departments authority for annual allocation of capabilities, forces, and units for the next fiscal year (FY+1). The annual force allocation plan will include, at a minimum:

- Type of force or capability required (i.e., CSG, Multi-role Fighter, etc.).
- Number of units or overall AOR presence.
- Operation or mission the annual force is tasked to support (e.g., TSCP or named exercise/operation).
- Start Date and End Date of required capability.
- FTN.
- Supported CCDR.
- Type of force or capability required.
- FP.
- Dates of deployment or PTDO status.

3 In the event of an emerging crisis, forces allocated to other operations may be considered by the SECDEF for allocation to the emerging crisis. Based on the risks, the SECDEF may allocate forces previously allocated elsewhere. However, the intent of the

GFM allocation process is to provide some measure of predictability for CCDRs, Military Departments, Defense Agencies, and individuals.

4 Annual Force Sourcing Timeline.

a The annual allocation process is an extended effort that typically spans a year in development with the goal of publishing a given FY's annual Deployment Order, known as the GFMAP Base Order, one year in advance of execution. The primary purpose of releasing the GFMAP one year in advance is to provide better predictability to Service members, families and units, and to better synchronize with the Planning, Programming, Budgeting and Execution (PPBE) process.

b The annual process is facilitated by periodic (Quarterly) GFMBs that have a specific purpose and associated product to guide the development of annual force requests and sourcing recommendations. The GFMIG details the Quarterly cycle (see paragraph 6 in this section).

5 Annual Force Planning. During execution, planning continues. The plan being executed is under constant review and the next step or phase of the operation is under review. During OIF and OEF, the SECDEF directed that force requirements be reviewed and re-validated annually. This re-validation became the basis for annual force planning. Today, all CCDRs review their ongoing operations and submit force requirements for the upcoming FY in their annual submission. The annual submission is, essentially, a consolidated RFF for the entire FY. CCMDs review every operation in progress and determine what forces are needed for each operation. The CCDR must also project the force requirements for engagement and shaping operations to the maximum extent possible. To determine the operational requirements from ongoing operations, a way to organize this task is to review the forces currently conducting the operation and validate the continuing need for each force in the coming FY for the phase of the operation that the plan will be in. The electronic force requirement for the current unit is refined, validated and submitted to the JS J3 to enter the annual allocation process. It is important to link current forces and their force requirements because the JS JFC and JFPs must schedule units into the GFMAP Annex in order to provide uninterrupted mission coverage. Specifying the evolving missions and tasks for specific units is imperative so the Services can train, organize and equip forces to be prepared to conduct those evolving missions. Once the annual force requirements have been submitted, newly identified, or refined, force requirements enter the emergent allocation process via a RFF.

Start Dates Defined:

Start Date *– The Start Date is the date the force, including personnel and equipment, arrive in the supported CCDRs AOR to begin JRSOI, normally at the port of debarkation (POD). Start Date is requested by the CCDR, nominated by the FP, recommended by the JFP, and ordered in the GFMAP Annex.*

Requested Start Date *– The date the CCDR requests the force, including personnel and equipment, to arrive in the supported CCDR's AOR to begin JRSOI normally at the POD. For most Naval forces this specifies the requested numbered fleet AOR IN-CHOP (AOR entry) date. For requests for Prepare to Deploy Order (PTDO) forces it is the date the CCDR requests the force be available to deploy within the designated PTDO response time.*

Ordered Start Date *– The date the force, including personnel and equipment, is ordered to arrive in the supported CCDR's AOR to begin JRSOI, normally the POD. For most Naval forces this specifies the requested numbered fleet AOR IN-CHOP (AOR entry) date. For forces on PTDO, it is the date the unit is to be available for deployment within the designated PTDO response time.*

(b) Emergent. The emergent process begins with the CCDR identifying a force or individual requirement that cannot be met using available assigned forces or forces already allocated. In order to optimize the timeliness of sourcing solutions and to ensure requested Start Dates are met the CCDRs must manage expectations and submit all emergent RFFs with as much lead time as possible. Emergent RFFs fall into three categories: routine, urgent, and immediate.

- *Routine RFFs.* Have requested Start Dates for forces that are 120 days (180 days if requesting non-standard forces which are forces not in the DOD inventory) or greater from the date, time, group (DTG) of the RFF message.
- *Urgent RFFs.* The CCDR places the greatest priority on these requests with the understanding that it will take precedence over all other sourcing requirements, except immediate RFFs, to include normal rotation of forces. Urgent RFFs may be sourced with units already assigned or allocated and require expedited handling.
- *Immediate RFFs.* Normally requested in response to a crisis situation and will be relayed in the most expeditious means possible.

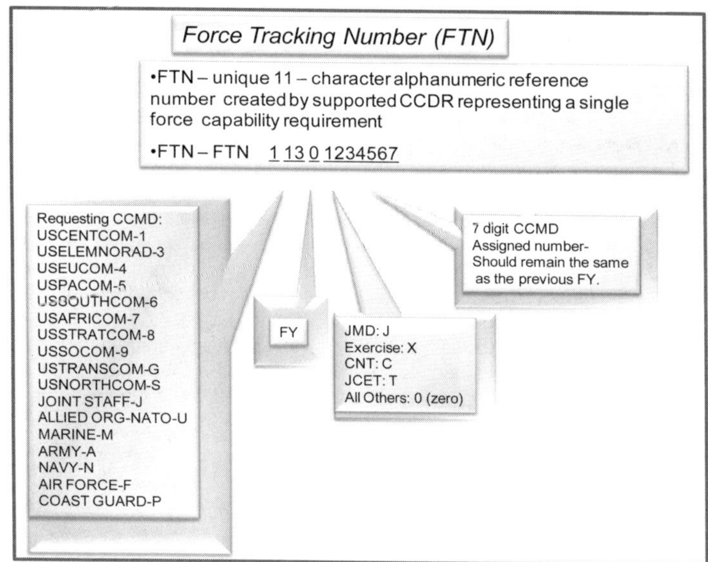

Figure VII-3. Force Tracking Numbers

1 *Force Tracking Numbers (FTNs).* The CCDR documents each force requirement, usually one unit per requirement. The force requirement contains information of what type of force is needed as well as the operational risk if the force is not provided. Each requirement is validated by the CCDR and assigned a FTN. The FTN is an identification number or the identification tracking number used by CCDRs and FPs to track force requirements from requesting document to unit deployment IAW the GFMIG (Figure VII-3).

a The FTN is a unique 11 character alpha-numeric reference number created by the supported CCDR and assigned to a single requested force capability requirement. Adherence to the assigned eleven character structure for each FTN position is mandatory in order to support data standardization between CCDRs and establish a joint process. A FTN cannot be reused or duplicated for additional requirements. Each force requirement has specified data fields attached that are used by the supporting commands in developing recommended sourcing solutions and informing the SECDEF of risks when making allocation decisions. Each force request is assigned a unique FTN and forwarded electronically to the JS J3 for validation.

b The force request is also transmitted to the JS J3 with a RFF message which requests sourcing of one or more force requirements. The JS J3 validates each force request, by FTN, provides priority based on the GEF and sourcing guidance, and assigns the JS JFC or one of the JFPs to provide a recommended sourcing solution. At the direction of the JS J3, the JS JFC and JFPs staff the force requirement with the FPs via their JFP assigned Service components to determine the recommended sourcing options and FP risk. The JS JFC and JFP generates a recommended sourcing solution, drafts a modification to their respective GFMAP Annex, staffs the draft Annex and forwards the recommendations, with the operational and FP risks to the JS J3. The JS J3 forwards the draft GFMAP Annex Mod to the CSA's (Services, CCMD's and Agencies) and OSD and briefs the solutions through the CJCS to endorse, and finally to the SECDEF for a decision. For some contentious issues, the GFMB may meet to review and endorse sourcing recommendations prior to the CJCS. Once the SECDEF decides to allocate, the JS J3 publishes the Modification to the GFMAP Annex.

2 *Voice Order of the Commanding Officer (VOCO)*. VOCO directs action without the requirements of a written order. SECDEF may approve CCDRs force requests via VOCO in order to expedite the sourcing and deployment process. During periods of rapidly-developing events such as natural disasters or other emergent crises, issuing requests and/or orders using the standard RFF/RFC process may not be practical. If the situation is sufficiently time sensitive, VOCO may be used to execute the steps contained in the emergent, crisis-based force allocation process. In all cases, information will be fully documented using the standard RFF process immediately following the VOCO. JS will record all VOCO approvals (date, time, and name of person relaying SECDEF's order), will track the progress of force sourcing and development, and will staff a special SDOB to obtain SECDEF signature for the record. Care must be taken to properly document all aspects of a VOCO authorization to ensure fiscal accountability is maintained, particularly in cases where DOD is not the Lead Federal Agency (LFA) managing a crisis or event. Additionally, the roles and responsibilities of all GFM principals, including command relationships, will remain unchanged. The VOCO process is not used merely as a convenience to supported or supporting commanders or as a method of bypassing the checks and balances inherent in the allocation decision process. It is available only in those circumstances where time constraints render the standard process impractical. As in the standard RFF/RFC process, all VOCO requests not otherwise addressed in a Standing EXORD are presented to the SECDEF via the CJCS. The SECDEF's initial VOCO approval of a request implies that, when time does not allow, VOCO may be used for each step of the allocation process, provided that the allocation decision and authorities granted are fully documented using the standard procedures as soon as practicable following the VOCO authorization. Requests received by entities other than SECDEF/CJCS should immediately be referred/forwarded to the JS J3 for consideration.

(6) Risk Assessment. This paragraph describes the two different assessments of allocation risk used during the annual and emergent force allocation processes.

(a) Overall risk assessment for the annual allocation process. Title 10 USC 153 requires the CJCS to assess the nature and magnitude of strategic and military risks in executing the missions called for in the NMS, and to provide a report of his assessment to Congress through the SECDEF. If the CJCS assesses the risk to be significant or higher, the SECDEF is required to submit a plan for mitigating that risk. Overall risk assessments performed in support of the annual force allocation process and emergent requirements inform the Chairman's Risk Assessment (CRA), and apply a common framework and assessment criteria. Services, CCMDs, and JS JFC/JFPs provide an overall assessment of the nature and magnitude of the risk associated with sourcing recommendations for a given fiscal year as it pertains to their ability to execute operations, activities, plans and functions, with amplifying information in each of the applicable risk categories.

1 The GFM process incorporates an assessment framework based on the NDS when discussing overall risk associated with annual allocation submissions and plans. Operational, future challenges, force management, and institutional risks are categories

used to express the overall risk associated with fiscal year requirements. When JS JFC/JFPs, FPs, and CCDRs present risk during the sourcing GFMB for fiscal year requirements, the overall risk categories of operational, future challenges, force management, and institutional are appropriate. CCDRs will present risks associated to operational and future challenges. In particular, supported CCDRs should focus their overall risk assessments to address the operational and future challenges risks associated with the four major areas of the FY sourcing effort (i.e., conventional forces, SOF, ISR and mobility forces). JS JFC/JFPs and FPs will present risks associated to future challenges, force management, and institutional risk categories. CDRUSSOCOM will present risks associated with all four categories. Overall allocation risk categories are summarized in the following sections.

Operational *-- Risk associated with the current force executing the GCCs strategy successfully within acceptable human, material, financial, and strategic costs. CCDRs risk of not getting forces requested.*

Future Challenges *-- Risk associated with the GCCs, JS JFC/JFPs or Services capacity to execute future missions successfully against an array of prospective future challengers.*

Force Management *-- JS JFC/JFP or Service risk associated with managing military forces and fulfilling the missions described in the NDS. The primary concern is recruiting, retaining, training, and equipping a ready force and sustaining that readiness.*

Institutional *-- JS JFC/JFP or Service risk associated with the capacity of new command, management, and business practices.*

2 <u>Risk Levels</u>. Provided is a separate set of categories and definitions for CCDR and FP use when presenting the impact of not sourcing or impact of sourcing risk. CCDRs and JS JFC/JFPs should use these risk categories throughout the GFM process when addressing the risk associated with sourcing a specific requirement. CCDRs and FPs should use the following definitions to identify their specific level of risk in each of the overall allocation risk categories, minus operational risk, when applicable.

Low Risk. *Success in achieving strategic objectives is assured, and can be achieved with planned resources and on planned timelines; unanticipated requirements can be easily managed with minimal impact on the force. FPs have the full capacity to source CCDR requirements.*

Moderate Risk. *Success in achieving strategic objectives is very likely, but timelines may be extended and execution may require additional resources from other plans or operations. Unanticipated requirements may necessitate adjustments to plans. FPs can source all requirements. Worldwide force allocation solutions may result in limited duration capability gaps. Intra-Service adjustments are required to source CCDR requirements.*

Significant Risk. *Success in achieving strategic objectives is likely, but timelines may require significant adjustments. Execution will require additional resources from other plans or operations, but some significant capability shortfalls still exist. Unanticipated requirements will necessitate adjustments to plans. FPs can source priority CCDR requirements. Worldwide force allocation solutions may result in extended duration capability gaps. Joint source solutions and force substitutions are required to source CCDR requirements.*

High Risk. *Success in achieving strategic objectives will require extraordinary measures. Timelines will require significant adjustments, but still may not achieve the commander's end-state. Execution will require significant resources from other plans or operations, and some resources are severely deficient or absent altogether. Unanticipated requirements cannot be managed. FPs require full mobilization to sustain sourcing solutions to achieve strategic objectives. CCDR requirements exceed joint force capacity to substitute capability.*

3 *Ad hoc and JIA* risk assessments will be completed by CCDR and FPs for un-sourced individual requirements as endorsed by the GFMB. The most current version as posted on the JS J3 web site will be used. These risk assessments will be compiled by job code, skill code and grade.

4 For additional Risk assessments see the GFMIG.

(7) Joint Individual Augmentee (JIA) Sourcing. The JIA execution sourcing allocation process mirrors the force sourcing process with the following exceptions; the format and information requirements of the JIA request is different, and different information technology (IT) tool is used, called *electronic* ***Joint Manpower and Personnel System*** (eJMAPS). The CCDR message requesting the JIA(s) is called an "Emergent JIA Request" rather than an RFF. JS J3 as the JFC for Annex D coordinates directly with the Services for identifying recommended sourcing solutions for JIA requirements. The Service may delegate the responsibility for generating JIA sourcing recommendations to specific commands within the Service.

(8) Allocation and the Inter-Agency Process. Although requests for interagency capabilities are not usually part of the allocation process, the interagency process is detailed in JP 3-08. There are cases in which a CCDR will request a capability via the allocation process and a non-DOD agency appears to offer the best solution. In these cases, the JS JFC coordinates between the agencies. The CJCS GFMAP Annex is used to relay the sourcing solution back to the JPEC; however, the SECDEF does not have authority to direct people and capabilities of other U.S. Government Agencies.

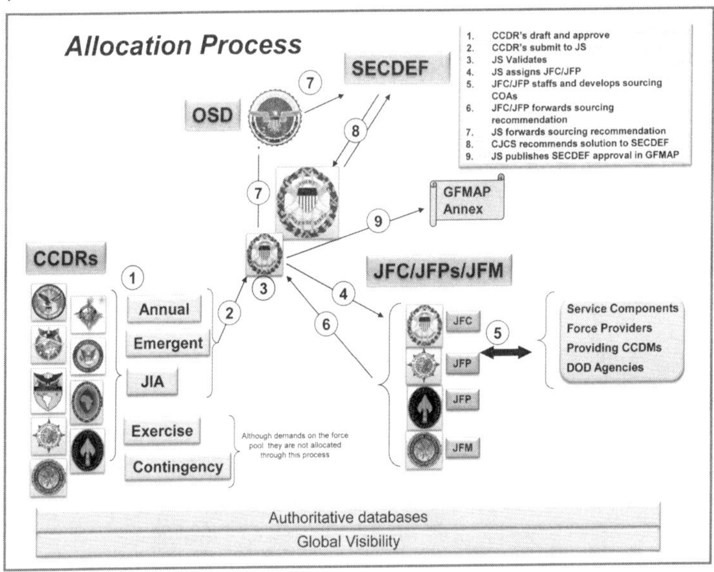

Figure VII-4. Allocation Process.

(9) Allocation Overview. As an allocation overview (Figure VI-4), the execution sourced forces are identified and recommended by JS JFC and JFPs, assisted by their Service components (the JS JFC will coordinate directly with the Service Headquarters. The Mobility and SOF JFPs continue to develop sourcing recommendations from the Service via their Service component). The recommended sourcing solution is reviewed through the GFM allocation process and becomes sourced when approved by the SECDEF for the execution of an approved operation of potential/imminent execution of an operation plan or exercise. The JS J3 provides specific guidance for the selection of forces in the execution sourcing message, including unit reporting requirements which will be done in accordance

with current GFM procedures per CJCSM 3130.06, GFM Allocation Policies and Procedures.

c. Apportionment. Apportionment is the distribution of forces and capabilities as a starting point for planning. Pursuant to Title 10 USC, Section 153, "The CJCS shall be responsible for preparing strategic plans, including plans which conform with resource levels projected by the SECDEF to be available for the period of time for which the plans are to be effective." Pursuant to the JSCP, "apportioned forces are types of combat and related support forces provided to CCDRs as a starting point for planning purposes only." Forces apportioned for planning purposes may not be those allocated for execution. The CJCS approves force apportionment based on the SECDEF guidance in the GEF. The apportionment tables are contained in Section IV of the GFMIG and is the process utilized to determine apportioned forces starting with assigned forces, and then subtracting the number of forces allocated to the high priority operations (global demand). The result is the number of apportioned forces which are published in the apportionment tables of the GFMIG. Apportioned forces are a quantity of a given type of force. During odd-numbered years, the apportionment tables are published separately and posted on the JS J8 web site. Throughout the planning process, planners consider the forces they have at their disposal to execute the plan as well as identifying additional augmentation forces that are not apportioned or provided as a planning assumption by the CJCS.

6. The Global Force Management Board (GFMB)

The GFMB is a flag officer-level body chaired by the DJS to provide senior DOD decision makers the means to assess operational effects of force management decisions and provide strategic planning guidance. The GFMB convenes periodically to address specific recurring tasks, and as required to address emergent issues.

a. Purpose. The purpose of the GFMB is to:

(1) Establish strategic guidance prior to developing force management options and recommendations.

(2) Serve as a strategic-level review panel to address issues that arise on recommended GFM actions prior to forwarding to the SECDEF for decision.

(3) Semi-annually identify forces/capabilities that are sporadically or persistently UHTS.

b. GFMB Membership. GFMB membership consists of flag/general officer (GO/FO) or equivalent representation from:

(1) The JS (including JS JFC).

(2) CCMDs (including JFPs).

(3) Services (FPs).

(4) OSD. Policy and Personnel & Readiness will attend at a minimum; others will attend as required.

(5) Defense agencies. Defense agency representatives may attend in a non-voting capacity as required.

SECTION II - FORCE SOURCING AND THE GFM PROCESS

> *Within the range of multiple sourcing[10] methodologies, execution and contingency are most prevalent because the force sourcing process generally results in an endstate in which JS JFC/JFPs identify units to satisfy a capability requirement for execution or planning.*
>
> *Global Force Management Implementation Guidance*

1. Force Planning, Assessment, and Execution and the GFM Process

a. Within the GFM process there are three methodologies utilized to provide CCDRs with requested capabilities for planning, assessment or execution. The intent is to provide the CCDR with the most capable forces based on stated capability requirements, balanced against risks (operational, future challenges, force management, institutional) and global priorities.

b. Force Providers (FPs). In accordance with the GFMIG FPs include Secretaries of the Military Departments, CCDRs with assigned forces, the United States Coast Guard, DOD agencies, and OSD organizations that provide force sourcing solutions to CCDR force requirements.

c. Force Providing. There are three types of forces provided within the GFM process depending upon the fidelity and endstate required; *preferred force identification* (planning only), *contingency sourcing* (plan assessment) and *execution sourcing* (allocation). Within the range of these methodologies, execution and contingency are most prevalent because the force sourcing process generally results in an endstate in which the JFPs identify units (UIC) to satisfy a capability requirement for either assessment or execution. The following paragraphs clarify and describe the meaning of these broad categories and related terms.

(1) Preferred Force Identification. Don't confuse preferred force identification with apportionment. The differences are discussed below and illustrated in Figure VII-5.

(a) Forces found in the Apportionment Tables are types of combat and related support forces *provided* to CCDRs as a starting point for planning purposes only. Annually the Services provide force updates to the JS who in turn staff the inputs and publish revised apportionment tables. The Services' submittals attempt to compensate for known deployments and average the availability of forces for that given year. The take away is that the apportionment tables are at best a good estimate. One could reasonably argue that the day the tables are published they are dated.

> *At various stages of the planning process, planners may require more detailed (unit level) information than is available in the Apportionment Tables.*

(b) While some directed plans are formally supported through the business process of contingency sourcing, the capacity of the Service and JFP planners limits the periodicity of these staffing efforts and cannot support all CCMDs plan assessment requirements. Therefore, CCMD planners, as directed by their command, work through Service, functional components, the JS JFC and the JFPs and their Service components to select preferred forces.

[10] *Only execution sourcing is actual force "sourcing" defined as, "identification of actual forces or capabilities that are made available to fulfill valid CCDR requirements." CJCSM 3130.06 GFM Allocation Policies and Procedures.*

(c) Most deliberate plans begin force planning utilizing the apportionment tables, and are later refined, as required, utilizing forces identified by the supported CCDR that place *specified forces* (unit level) into the plan, in lieu of apportioned forces (unit type). This allows the supported CCDR to continue employment, sustainment and transportation planning and assess risk and shortfalls with a greater fidelity. This is preferred force identification.

(d) CCDR planners, Service and functional components are encouraged to work with the JS JFC, the JFPs and their Service components to make the best possible assumptions with respect to generating preferred forces.

(e) The preferred forces identified for a plan by the CCDR should not be greater than the number of forces apportioned for planning unless the CJCS or his designee either grants permission or has provided amplifying planning guidance. To the degree the CCDR is able to make good planning assumptions when selecting preferred forces determines the feasibility of a plan, and may assist the JS JFC and/or the JFP in identifying forces should the plan be designated for contingency sourcing or transitions to execution.

> ***Preferred Forces*** - *Preferred forces are forces that are identified by the supported CCDR in order to continue employment, sustainment, and transportation planning and assess risk. These forces are planning assumptions only, are not considered "sourced" units, and do not indicate that these forces will be contingency or execution sourced.*
>
> - *To the degree the CCDR is able to make good assumptions with respect to preferred forces for planning, the JFPs will begin with a higher fidelity solution should the plan be designated for contingency or execution sourcing.*
> - *CCMD Service/functional components are encouraged to work with JFPs and their components to make the best possible assumptions with respect to preferred forces for planning.*
> - *The preferred forces identified for the plan by the CCDR should not be greater than the forces apportioned for planning unless granted permission to do so by the CJCS or designated representative.*

(f) Plan analysis achieves greater fidelity when utilizing preferred forces; however, preferred forces are still based on planning assumptions. Preferred forces *are* notional forces based on current information and *are not* contingency sourcing nor execution sourced. Preferred forces give the plan analysis a greater fidelity, however it is still based on planning assumptions; these units are still not considered "sourced."

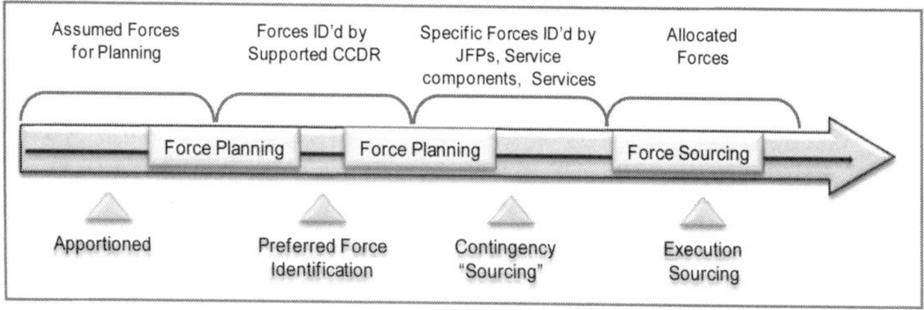

Figure VII-5. Force Sourcing.

(2) Contingency Sourcing. Contingency sourcing requests support the plan development and plan assessment phases of planning in three general categories: CCMD requests to support plan development/refinement; CJCS directed exercises with contingency souring requirements; and Joint Combat Capability Assessment Plan Assessment (JCCA-PA).

(a) The JS J5 Joint War Plans Division (JOWPD) is responsible for collecting all contingency sourcing requests; developing a proposed Contingency Sourcing Schedule Memorandum for staffing; and presenting the recommended memorandum to the JCCA Group (JCCAG) and Global Force Management Board (GFMB) for approval. Contingency sourcing is a manpower-intensive and time-consuming requirement that competes for resources with FP execution-level sourcing responsibilities. The GFMB is the final arbiter of all contingency sourcing requested events.

(b) Contingency sourced forces are not allocated forces. The term contingency "sourced" is misleading. Contingency sourced forces are specific forces which meet the planning guidance at a specified point in time identified by the JS JFC, ICW the FPs, and the JFP, ICW their assigned Service components. The JS J5 provides specific guidance through a list of sourcing assumptions and planning factors contained in a Contingency Sourcing Message. The JS JFC and JFPs have final approval of the sourcing solution and provide the approved solution to the supported CCDR in the CCDR. The JS JFC is responsible for consolidating all sourcing solutions and recommending global joint sourcing solutions.

(c) Contingency sourcing allows greater fidelity in force planning than preferred force identification discussed previously. However, because these forces are identified based on planning assumptions and planning guidance provided for the sourcing effort, there should be no expectation that forces sourced via contingency sourcing will be the actual forces sourced during execution sourcing. The frequency of contingency sourcing actions is, in part, dependent on the capacity of the JS JFC/JFPs and their FPs.

(d) Variables for contingency sourced units include but are not limited to: current disposition of forces, a specified as-of-date, categories of forces to be excluded from consideration, defined C-Day or C-Day window, substitution and mitigation options/factors, readiness reporting requirements, and training requirements.

(3) Execution Sourcing. The execution sourcing process is conducted by the JS J3. Execution sourced forces are considered allocated forces and are unavailable for use in other plans/operations unless reallocated by the SECDEF.

(a) Execution sourcing is the process of identifying forces identified and recommended by JS JFC and JFPs (via the JFPs Service components or Services for conventional forces) and allocated by the SECDEF to meet CCDR force requirements for the execution of an approved operation or potential/imminent execution of an operation plan or exercise. Execution sourcing of forces, including PTDO, are ordered in a Deployment Order (DEPORD). The GFMAP and its associated annexes are the global DEPORD for all allocated forces. For execution sourcing during allocation, the JS JFC/JFP is the supported commander for the force planning steps of identifying recommended sourcing solutions. All CCDRs and FPs are in support. The GFMAP Annex specifies the ordered FP. The GFMIG and GEF allows the FP and JS JFC/JFP some flexibility in identifying the appropriate unit for deployment.

(b) Units tasked must meet minimum readiness and availability criteria as directed by the tasking authority.

(c) Execution sourced forces are considered allocated forces and are unavailable for use in other plans/operations unless reallocated by the SECDEF.

(d) There are four execution force sourcing categories: standard, joint, in-lieu-of, and ad-hoc force sourcing solutions as listed in Figure VII-6.

1 *Standard Force Solution*. A mission ready, joint capable force with associated table of organization and equipment executing its core mission. This force will also have completed core competency training associated with the RFF's requested capability (s).

•Standard Force Solution. A mission-ready, joint capable force with associated table of organization and equipment executing its core mission
EX: Ship/Squadron/CSG/ESG
NON STANDARD
•Joint Force/Capability Solution. A Service providing a force/capability in place of another Service's core mission; the capability is performing its core mission.
EX: NMCB
•In-Lieu-of. A standard force, including associated table of organization and equipment, deployed/employed to execute missions and tasks outside its core.
EX: Air Ambulance (only one)
•Ad-Hoc. A consolidation of individuals and equipment from various commands/Services forming a deployable/employable entity, properly manned, trained and equipped to meet the supported CCDR's requirements.
EX: Provincial Reconstruction Team (PRT), Training/Transition Teams, other non-standard sourced with Navy individuals

Figure VII-6. Execution Sourcing Categories.

2 *Joint Force/Capability Solution*. Joint sourcing encompasses Services providing a force/capability in place of another Service's core mission. As in a standard force solution, the capability is performing its core mission. An example of this is sourcing an Army combat heavy engineer requirement as an enabler for an Army BCT with a Naval Mobile Construction Battalion (NMCB) unit. Navy is providing a like type capability, a capability that is performing its core competency mission, in the place of another Service's core mission. Joint sourcing may also encompass a force or capability composed of elements from multiple Services merged together to develop a single force/capability meeting the requested capability. Joint sourcing solutions will increase the time required to properly train, equip, and man the force/capability prior to deployment. Additional challenges include the fact that unlike a standard force, a joint sourcing solution may require movement of personnel and/or equipment from various locations to a single locality for consolidation and issuance of equipment. Second, once personnel and equipment are consolidated, familiarization, proper usage, and maintenance practices must also be incorporated into the training regimen to ensure that all members comprising the joint solution are well versed in required actions for sustaining operability.

3 *In-lieu-of (ILO)*. ILO sourcing is an overarching sourcing methodology that provides alternative force sourcing solutions when standard forces sourcing options are not available. An ILO/capability is a standard force, including associated table of organization and equipment that is deployed/employed to execute missions and tasks outside its core competencies. The force can be generated by normal FPs or be a result of a change of mission (s) for forward deployed forces. An example of this is taking an existing artillery battery, providing it a complete training and equipment package and then deploying it to fill a transportation company requirement. ILO/capability solutions will require retraining and in some instances will require re-equipping. ILO solutions will increase the time required to properly train, equip and man the force/capability prior to deployment. Additional challenges include the fact that unlike a standard force, an ILO sourcing solution may require movement of personnel and/or equipment from various locations to a single locality for consolidation and issuance of equipment. Second, once personnel and equipment are consolidated, familiarization, proper usage, and maintenance practices must also be incorporated into the training regimen to ensure that all members comprising the joint solution are well versed in required actions for sustaining operability.

4 *Ad-hoc*. An ad-hoc capability is the consolidation of individuals and equipment from various commands/Services and forming into a deployable/employable entity properly trained, manned and equipped to meet the supported CCDR's requirements. Ad-hoc solu-

tions will increase the time required to properly train, equip, and man the force/capability prior to deployment. Additional challenges include the fact that unlike a standard force, an ad-hoc sourcing solution will require movement of personnel and/or equipment from various locations to a single locality for consolidation and issuance of equipment. Second, once personnel and equipment are consolidated, familiarization, proper usage, and maintenance practices must also be incorporated into the training regime to ensure that all members comprising the ad-hoc solution are well versed in required actions for sustaining operability.

SECTION III - FORCE PLANNING AND THE GFM PROCESS

1. Force Planning and the GFM Process

a. GFM Process during Planning. The apportionment tables provide the number of forces reasonably expected to be available for planning. Theses tables should be used as a beginning assumption in planning. As the plan is refined, there may be forces identified that are required above and beyond those apportioned. Those forces should be requested, as required, to be augmented above the number apportioned for planning, or "augmentation forces." The CJCS may approve planning to continue with the revised assumption of using the identified augmentation forces. These augmentation forces are then allotted for planning. However, should the plan be executed, be prepared for the risk associated with the potential of those "augmentation forces" not being available.

(1) Throughout COA and plan development, planners assess the plan as it is refined. To enable the assessments, the planners must assume that units are allotted to the identified plan force requirements and to enable plan assessments, planners generate preferred forces. As the plan is refined, the level of analysis used to generate preferred forces usually increases. Since contingency plans rely on a foundation of assumptions, if an event occurs that necessitates execution of a contingency plan, the plans assumptions have to be re-validated. The plan will usually enter a phase of CAP to adapt it to the realities surrounding the event rather than transitioning directly to execution.

(2) As a contingency plan is either approved or nearing approval, the CJCS may direct the JS JFC/JFPs to contingency source a plan to support CJCS and/or SECDEF's strategic risk assessments. CCDRs may request contingency sourcing of specific plans. These requests are evaluated by the JS J5 and a contingency sourcing schedule is presented to the GFMB. The GFMB endorses the schedule and the CJCS orders the JS JFC/JFPs to contingency source specific plans per the schedule (see Contingency Sourcing).

b. GFM Process during CAP. The same planning steps that are used to develop contingency plans are used in CAP, but the time to conduct the planning is constrained to the time available. During CAP, preferred force identification is used the same as it was during deliberate planning. Contingency sourcing is rarely used during CAP due to the time constraints involved, but if time allows, the option exists for the CJCS to direct JS JFC/JFPs to contingency source a plan.

(1) For contingency plans and crsis, the difference in force planning is the level of detail done with the force requirements for the plan. With contingency plans the number of planning assumptions prevents generating the detailed force requirements needed by the JS JFC/JFPs to begin execution sourcing. During CAP, a known event has occurred and there are fewer assumptions. The focus of crisis planning is usually on transitioning to execution quickly. The detailed information requirements specified to support the execution sourcing process, either emergent or annual, preclude completion until most assumptions are validated.

(2) CCDRs usually have a good understanding of the availability of their assigned forces. Availability entails the readiness of the unit, as well as the unit's time in the deployment cycle and whether it meets SECDEF deployment-to-dwell (D2D) ratio requirements, and whether the unit is already allocated to another mission. The supported CCDR

generally reviews the force requirements for the deliberate plan and conducts a review of assigned and previously allocated forces to determine if the mission can be done without requesting additional forces. If forces are already assigned and/or allocated that can perform the mission, the CCDR may direct those forces to perform the mission, within the constraints of the allocation authorities in the GFMAP. If additional forces are required, the CCDR will forward a RFF with all the details necessary, both electronically and by message RFF. The emergent force allocation process (discussed above) is the process to identify force requirements in support of CAP.

(3) A CAP transitions to execution when the order is given to execute a mission or operation, although planning continues throughout execution. During CAP, the CCDR considers using assigned and already allocated forces to respond to the situation. If the CCDR identifies additional forces that are required, a force request is submitted. Once this request is approved by the CCDR, that force request is considered a CCDR requirement. The force request is sent from the CCDR to the SECDEF via the JS J3. The vehicle for the force request is a message called a RFF to the SECDEF and JS J3 info the JS JFC/JFPs, FPs, OSD, and all other CCDRs as specified in CJCSM 3130.06, *GFM Allocation Policies and Procedures.* Each individual force requested is serialized with a FTN. An RFF message may contain one or more FTNs. To request JIAs for a JTF Headquarters, the message is called a Emergent JIA Request. The initial force or JIA request to perform a mission is an emergent JIA request.

(4) For operations that are longer in duration, the SECDEF mandates that CCDR's validate their forces annually to determine the forces that require rotation. The process used to source annual forces is fundamentally the same as an emergent force request, but the annual process is necessarily modified to handle the large number of forces necessary to fulfill all CCDR requests for an entire FY. The annual submission is effectively the first RFF for a FY and should include all the forces for all the operations the CCDR anticipates executing.

2. GFM in Exercise Planning

Requests for forces to participate in exercises do not follow the same sourcing process as operational requests. Per reference CJCSI 3500.01, *Joint Training Policy and Guidance for the Armed Forces of the United States*, JS JFC/JFPs receive exercise force requests directly from the supported CCDRs. Supportability by JS JFC, the JFPs (and their Service retained conventional forces) is determined and the resulting sourcing solution is provided back directly to the supported CCDR. The SECDEF is not required to allocate forces for exercises, including exercises with other countries. Subsequent deployment of these exercise sourcing solutions is effected and tracked by the JS JFC/JFPs in concert with the supported CCDR. Under most circumstances, the GFMIG authorizes JS JFC/JFPs to transfer TACON of forces to support CCDR exercises and does not require a GFMAP mod to be approved by the SECDEF.

SECTION IV - DEPLOYMENT PLANNING

1. GFM and Four Levels of Planning

a. As discussed in Chapter III, deliberate planning, which occurs outside of crisis conditions, encompasses four levels of planning detail with an associated planning product at each level. It provides the JPEC the opportunity to develop and refine relevant plans in preparation of contingencies. The process is designed to uncover problems and issues. Contingency plans also provide a useful basis for crisis planning in response to a crisis when they occur. This process provides the planner with the time to accomplish detailed planning for operations in specific geographical areas, missions, and assumptions. As contingency plans are developed and refined over time, the detail and fidelity they contain as well as the value of the analytical process by which they are developed become extremely valuable tools in times of crisis. The four levels of planning detail include:

(1) Level 1- Commander's Estimate. The Commander's estimate provides the Secretary of Defense with military COAs to meet a potential contingency. Normally, these COAs will center on military capabilities in terms of forces available, response time, and significant logistic considerations. The estimate reflects the supported commander's analysis of the various COAs that may be used to accomplish the assigned mission and contains a recommended COA.

(2) Level 2 - Base Plan. This describes the CONOPS, major forces, concepts of support, and anticipated timelines for completing the mission. It normally does not include annexes or a TPFFD.

(3) Level 3 - Concept Plan (CONPLAN). A CONPLAN is an operation plan in an abbreviated format that may require considerable expansion or alteration to convert it into an OPLAN, Campaign Plan, or OPORD. It includes a base plan with selected annexes (usually A, B, C, D, J, K, S, V, and Z) and a CCDR's estimate of the plan's feasibility with respect to forces, logistics, sustainment, and transportation. It may also produce, if required, a time-phased force and deployment list (TPFDL) or a TPFDD.

(4) Level 4 - OPLAN. A OPLAN is a complete and detailed joint plan containing a full description of the CONOPS, all annexes applicable to the plan, and a TPFDD. It identifies the forces, functional support, and resources required to execute the plan and provide closure estimates for their flow into the theater. OPLANs can be quickly developed into an OPORD.

b. Key Service and Joint documents contribute to deliberate planning and help confirm availability of forces and resources for performing deployment operations. These documents along with the GEF and the JSCP combine to facilitate the joint planning process.

2. Deployment Planning

a. The primary objective of deployment planning is to provide ready, equipped, and integrated forces, equipment, and materiel when and where required by the supported CCDRs CONOPs and to retain visibility of those forces to support rapid adaption to changing operational conditions. Planning for joint force operations is guided by the procedures in the joint planning process. Both deliberate planning and CAP procedures require detailed analysis of the assigned mission to determine mission requirements for employment of the joint force. Employment planning considerations that directly impact deployment operations include: identification of force requirements, commander's intent for deployment, time-phasing of personnel, equipment, and materiel to support the mission, and closure of the forces required to execute decisive operations. These factors guide deployment planning and help determine mission requirements. Adaptive planning capabilities (such as collaboration and decision-support tools) will improve a transition from deliberate planning to CAP. The following paragraphs summarize the activities and interaction that occur during CAP. [11]

b. Military operations begin with an event that requires movement of forces somewhere in the world. This can be a planned or no-notice movement. Analyzing the mission leads to the development of COAs and selection of the desired COA, the development of orders, and their transmission, and continues through execution.

(1) Conduct Initial Mission Analysis and Develop COA's. Based upon information acquired, planners assess potential scenario developments, mission requirements, and develop COAs. Once a COA is selected by the commander, a thorough and continuous JIPOE to account for a changing operational environment is required.

(2) Commander's Intent for Deployment. The supported CCDR's intent for deployment may be detailed in the OPLAN, prepare to deploy order (PTDO), or OPORD. For unassigned forces, the CCDR may request forces to be placed on a PTDO by a RFF.

[11] Refer to the chapters on JOPP in this document, JP 5-0, and JOPES Volume I for detailed procedures.

If the SECDEF orders, the FP will be ordered to have a unit prepared to deploy in a DEPORD, usually that DEPORD is the GFMAP Annex. The PTDO may be later executed by requesting deployment from the SECDEF or the SECDEF may delegate authority to execute the PTDO to the supported CCDR. The supported CCDR's intent for deployment may direct the sequence for deployment of units, individuals, and materiel; identify immediate protection concerns; articulate specific force disposition requirements to support future operations; or identify general deployment timeline requirements needed for operational success. Within the breadth and depth of today's operational environment, effective decentralized control cannot occur without a shared vision. Without a commander's intent that expresses that common vision, unity of effort is difficult to achieve. In order to turn information into decisions, and decisions into actions that are "about right," commanders must understand the higher commander's intent. Successfully communicating the more enduring intent allows the force to continue the mission even though circumstances have changed and the previously developed plan/concept of operations is no longer valid. The supported commander's intent for deployment should, at a minimum, clearly articulate the CCDR's vision for how the deployment can best posture the joint force for decisive operations or operational success.

c. Time-Phasing of Forces. Once force requirements are identified, selected forces must be organized and time-phased to support the CONOPs.

(1) Time-phasing is the sequencing of the deployment and arrival and employment of forces based on the organization of forces to accomplish the mission, the commander's intent, the estimated time required to deploy forces from their point of origin to the operational area, and actual lift availability and port throughput for deployment. In addition to forces, support personnel, equipment, and materiel are time-phased to support the continuous operation of the joint force until the mission is accomplished.

(2) Finding the proper balance between projecting the force rapidly and projecting the right mix of combat power and materiel for the ultimate mission is critical. The commander must seek a balance that provides protection, efficient deployment, adequate support, and a range of response options to enemy activity. The availability of mobility assets is most often a constraining factor, so difficult trade-off decisions continuously challenge supported commanders.

(3) All movement priorities and phasing are based on the supported CCDR's required date for the deploying force capability. Movement data on the required delivery date, time-phasing of forces, and materiel is documented in the TPFDL or TPFDD. Ideally, forces and supporting materiel are time-phased in a manner that allows the CCDR to conduct decisive operations as quickly as possible. Closure of the minimum essential force required to accomplish the mission is a major factor in determining when decisive operations can be conducted.

(4) The JS JFC/JFP and FPs are responsible for providing forces as directed by the SECDEF. With the constraints imposed during OIF and OEF that maximum unit deployment length is specified, for rotational forces, the date the deploying unit must arrive is driven by the date the redeploying unit must be out of theater. The Latest Arrival Date is determined by the FPs and JS JFC/JFP based on the previous rotation End Date.

d. Force Tracking. Force tracking is the process of gathering and maintaining information on the location, status, and predicted movement of each element of a unit (including the unit's command element, personnel, and unit-related supplies and equipment) while in transit to the specified operational area. Force tracking is fundamental to effective force employment and C2. It provides information on transportation closure or physical moves of forces. CCDRs must be able to continuously monitor execution of the deployment operation and quickly respond to changing situations and unforeseen circumstances. Once basic mission requirements have been decided, joint force planners must review force tracking options to provide the supported CCDR with the requisite C2 means to monitor and control execution of the deployment.

e. Force Closure. Force closure is the point in time when a supported CCDR determines sufficient personnel and equipment resources are in the assigned operational area to carry out assigned tasks. Force closure is a function of several elements; most notably lift capacity, C2, POE and/or POD transshipment capability, port capacity, transit time for strategic lift, receipt of overflight and landing diplomatic clearances, and the effectiveness of deployment operations. As a planning factor, force closure is important because decisive operations cannot begin until sufficient forces and capabilities are available to execute the mission. Force closure includes force tracking, risk assessments, force readiness, and Mission, Enemy, Terrain and Weather, Troops and Support available-time available (METT-T). The supported commander's force closure decision must take into account the above factors and the commander's ability to integrate forces. Decisions affecting transportation mode or routing made prior to and during the early stages of a deployment may have a dramatic impact on joint force closure timelines. For this reason, planners must give careful consideration to the selection of POE and/or POD, scheduling of strategic lift modes, movement control, and JRSOI planning. Inherent in this complex process is the management of change and the ability to respond with accuracy and flexibility to dynamic conditions.

f. Identify DOD Agency, Host Nation, Contract, and Command Capabilities. Within each GCC's AOR, the United States organizations available to accomplish deployment operations vary significantly. Fundamental factors that cause this variance include geographical constraints such as the length of line of communications (LOCs), capability of HN infrastructure, acquisition and cross-servicing agreement, anticipated threat and mission. Each Service component possesses unique, specialized forces and capabilities to support various aspects of deployment operations. The supported CCDR must utilize this knowledge in assessing HN, contract, and command capabilities available to support key deployment functions. Depending upon the existing infrastructure, the HN, contract, and command capabilities that are available may greatly reduce the type and amount of support a CCDR must deploy from outside the theater. The inputs include requirements for:

- Transportation.
- Facilities.
- Security.
- Communications.
- Supplies.
- Services (life support).
- Labor service.
- POD support and other key functions.

g. There are many sources of support that may be applied to enhance deployment operations. These sources of support may combine in various ways under differing circumstances to make operations associated with deployment possible. How these combine will depend upon the condition of the HN infrastructure, what agreements exist (allied or otherwise), and how or if civil augmentation programs, or cross-Service logistics, are necessary.

(1) DOD Agency and Combat Support Agencies. CCDRs request capabilities from DOD Agencies (including CSAs) through the GFM allocation process per the GFMIG Section III and Annex C, *GFM Allocation Business Rules.* Once tasked, the JS JFC/JFP will staff these requests with the DOD agencies that have the capabilities. Once tasked, the DOD agency will provide FP nominations with their capacity to provide support for operating forces engaged in planning for, or conducting military operations, including support during conflict or in the conduct of other military activities related to countering threats to U.S. national security.

(2) Host-Nation Support. When available, HNS successfully assists in executing deployment operations. Maintaining current, comprehensive base support plans and conducting periodic site surveys are critical for validating HNS agreements required for

implementing specific OPLANs and CONPLANs. If HNS agreements do not exist, or have limited application, then the CCDR, in coordination with the DOS, should immediately start negotiation of HNS agreements and arrangements combined with an integrated contracting plan to obtain necessary support. It is recommended that counterintelligence teams be included for use in screening contractors.

(3) Multinational Support. Multinational support is another force multiplier. Many United States allies have capabilities or functional units similar to United States capabilities. The use of these units can enhance deployment operations, minimize U.S. support requirements, and ensure mission success. The joint planner should consider complementary multinational capabilities during COA development. However, during the planning phase, this capability should be balanced against the potential for competition for U.S. transportation assets to deliver those multinational units into the theater.

(4) Theater Contract Support. To optimize contractor support among Services, a central contracting authority (CCA) should be designated. The goal of the CCA is to achieve and maintain controls and optimize contracting resources. Contracting officers should make every effort to ensure that clauses excusing contractor performance in the event of hostilities or war are not included in contracts that augment U.S. forces during contingency or combat operations. *Surface Deployment and Distribution Command* (SDDC) and Military Sealift Command (MSC), for example, routinely use civilian contractors to perform or augment their operations. However, the joint planner should be aware that in some cases wartime exclusion clauses might prevent contractor personnel from delivering goods and services.

(5) Civil Augmentation Program.[12] Civil augmentation programs are separate Military Department contracting options most often used when HNS is insufficient or unavailable. They employ pre-existing contracts with United States and other vendors to provide support in many areas including facilities, supplies, services, maintenance, and transportation. Additionally, planners should consider initiating contracting services if status-of-forces agreements do not already contain those provisions. The goals of civil augmentation programs are to:

(a) Allow planning during peacetime for the effective use of contractor support in a contingency or crisis.

(b) Leverage global and regional corporate resources as facility and logistic force multipliers.

(c) Provide an alternative augmentation capability to meet facility and logistic services shortfalls.

(d) Provide a quick reaction to contingency or crisis requirements.

h. Prepare and send Deployment Directives. At this point in the planning process the supported CCDR has nearly completed the deployment planning process map's first functional area of mission analysis. Now the supported CCDR begins to give specific deployment guidance in the form of directives. These directives clarify the support that selected Services and nations should expect. Examples of the types of support directed by the supported CCDR may include information from the following agreements.

(1) Common-User Logistics. The term "common user logistics" defines materiel or service support shared with or provided by two or more Services, DOD agencies, or multinational partners to another Service, DOD agency, non-DOD agency, and/or multinational partner in an operation.

(2) Acquisition and Cross-Servicing Agreements. While HNS agreements provide United States pre-negotiated support for potential war scenarios, acquisition and cross-

[12] *Information concerning the Army logistics civilian augmentation program, Navy global contingency construction capabilities contract program and Air Force contract augmentation program may be found in the applicable Service publications, JP 4-0, Joint Logistic Support, JP 3-34, Joint Engineer Operations, and JP 4-10, Contracting and Contractor Management.*

servicing agreements (ACSAs) provide the legal authority for the United States military and other nation armed forces to exchange logistic goods and services during contingencies. Unlike HNS agreements, transactions under this program must be reimbursed, replaced in kind, or an exchange of equal value must take place.

3. Synchronizing and Balancing the Flow

Because JOPES is the system used to sequence and execute force movement, it is essential that movement data inputs are accurate. The TPFDD is a computer-supported database portion of an OPLAN or OPORD. It contains time-phased data for moving personnel, equipment, and materiel into a theater. The TPFDD reflects the requirements that strategic and intra-theater lifts are assigned against to ensure that the full scope of deployment requirements are identified and satisfied. Successful execution of the CCDR's plan depends on integrating deployment operations within JOPES[13] and the APEX enterprise. Planners are advised to consider sustainment when planning force projection and to include the sustainment requirement in the time-phased flow.

a. Confirm Deployment Data. Accuracy of data is essential prior to entry into the TPFDD.

b. TPFDD. As discussed earlier, the TPFDD establishes the flow of units into the theater. The supported CCDR must carefully balance the force mix and arrival sequence of combat forces and CSS units to ensure that deployment support and throughput requirements can be met. Service components responsible for deployment operations must validate and continuously review the TPFDD as changes may occur. As with any dynamic process, external changes in the environment, as well as those within the force, necessitate corresponding changes to the flow of forces (personnel, equipment, and materiel). The following recommended changes to the deployment force structure are requested by the Service component responsible to maintain balance and synchronization (IAW CCDR guidance) to accomplish the mission.

(1) Evaluate TPFDD Movement Requirements. The supported CCDR (or his components) continuously evaluates the movement progress of inbound units to ensure that mission requirements are met in a timely manner. Mission change requirements must be addressed in conjunction with the GFM process.

c. Analyze Capabilities. Dependable transportation feasibility analysis relies on accurate analysis of strategic lift capability and throughput capability. Port throughput data should consider not only port offload capability, but also the theater's ability to move and sustain forces away from the port. Matching the strategic TPFDD flow to the theater's reception, staging, and onward movement capability should prevent port saturation and backlogs that slow the build-up of mission capability. Intra-theater transportation feasibility may significantly impact upon port-to-port flow. It may show required changes to the type and sequence of strategic lift. It could also reveal whether the number, type, and sequence of units supporting deployment operations are adequate to deliver planned capabilities to the supported CCDR.

d. Review Historical Data. Theater infrastructure is studied during the concept development phase before the TPFDD is developed. Meanwhile, supported CCDRs can use historical data for a feasibility study of the flow of the TPFDD which can be based on the supporting Services components input. If historical data cannot be used, separate TPFDDs specifically for this type of analysis can be built without placing undue pressure on the sourcing units that are actually working on collecting real data.

[13] *CJCS Manual 3122.01, Series, Joint Operation Planning and Execution System (JOPES) Volume I: Planning Policies and Procedures.*

SECTION V - JOINT FORCE PROJECTION

> *Force projection, the ability to project the military instrument of national power systematically and rapidly deploys and integrates joint military forces and the associated sustainment material in response to requirements across the range of military operations.*
>
> *JP 1-02*

1. Joint Force Projection

Force projection, the ability to project the military instrument of national power, systematically and rapidly deploys and integrates joint military forces and the associated sustainment material in response to requirements across the range of military operations. Force projection allows a CCDR to strategically position and concentrate forces to set the conditions for mission success. Force projection, enabled by GFM, forward presence and agile force mobility, is critical to U.S. deterrence and warfighting capabilities. The President and/or SECDEF could direct CCDRs to resolve a crisis by employing immediately available forward-presence forces. However, when this response is not sufficient or possible, the rapid projection of forces from other locations may be necessary. Alternatively, responding to the full range of military operations may involve the deployment of forces and materiel within or outside the United States for humanitarian or disaster relief purposes. The requirement remains: to provide joint force capabilities that may include AC, RC, or civilian contract service in a timely and efficient and effective manner consistent with the CONOPS while retaining global force visibility.

a. Joint force projection, the use of the military instrument of national power is executed using APEX.

b. The CONOPS is the basis for force and deployment planning. The CONOPS details the phases of the operation, prioritizes the major missions within each phase, and identifies the forces required to provide those capabilities needed to meet mission requirements. The missions establish the force requirements and the prioritization determines the deployment flow of forces and support into the theater. The missions assigned to a force will determine the pre-deployment, support, and joint force integration requirements of that force. The CONOPS determines the size and scope of mobilization, deployment, JRSOI, sustainment, and redeployment activities.

c. Joint force projection encompasses a range of processes. The scope of these processes is dependent upon the joint operation. Planning for and execution of these processes normally occurs in a continuous, overlapping, and iterative sequence during each phase and for the duration of the operation. However, each joint operation or campaign usually differs in both sequence and scale. The following paragraphs on the facing page briefly describe each process. See Chapter 2 and JP 3-35 for greater details.

d. Joint Force Mobility. Mobility is the quality or capability of military forces, which permits them to move from place to place while retaining the ability to fulfill their primary mission. Mobility is a function of force, resource, operation, deployment, and sustainment planning. Mobility requires standard procedures; global force visibility; integrated employment and deployment planning; effective execution of pre-deployment actions; and movement execution supported by networked operation planning, deployment, and transportation information systems.

(1) The contingency plan and its TPFDL structure must be designed to ensure the Services acquire, train, and equip forces to meet the most probable operation requirements. The design must support rapid transition of the TPFDL to align forces and missions

Joint Force Projection

Joint force projection encompasses a range of processes.

(1) Mobilization. Mobilization is the process of assembling and organizing national resources to support national objectives in time of war or other emergencies by assembling and organizing personnel and materiel for active duty military forces, activating the RC including federalizing the National Guard, extending terms of Service, surging and mobilizing the industrial base and training bases, and bringing the Armed Forces of the United States to a state of readiness for war or other national emergency.

(2) Deployment. Deployment is the movement of forces within operational areas, the positioning of forces into a formation for battle, and/or the relocation of forces and materiel to desired operational areas. Deployment encompasses all activities from origin or home station through destination, specifically including intra-continental United States, inter-theater, and intra-theater movement legs, staging, and holding areas.

(3) Joint Reception, Staging, Onward Movement, and Integration. The last phase of deployment, JRSOI is the responsibility of the supported CCDR. This phase comprises the essential processes required to transition arriving personnel, equipment, and materiel into forces capable of meeting operational requirements. JRSOI is the critical link between deployment and employment of the joint forces. The time between the initial arrival of deploying forces and capabilities and operational employment is potentially the period of greatest vulnerability. During this transition period, deploying forces and capabilities may not fully sustain or defend themselves, or contribute to mission accomplishment because some elements may not have attained required mission capability. JRSOI planning is focused on the rapid integration of deploying forces and capabilities to quickly make them functioning and contributing elements of the joint force.

(4) Employment. Employment planning prescribes how to apply force and/or forces to attain specified national strategic objectives. The CONOPS establishes the phases, missions, and force requirements of a given operation. It is developed by the CCDR and the component commanders using the JOPP.

(5) Sustainment. Sustainment is the provision of personnel, logistic and other support required to maintain and prolong operations or combat until successful accomplishment or revision of the mission or of the national objective. Sustainment is ongoing throughout the operation. Sustainment operations must be closely linked to the phases and mission priorities of the CONOPS to ensure mission effectiveness without logistic shortages or excesses, which could reduce the efficiency of the force. Sustainment requirements are projected and planned based on the phases and missions of the operation. Consumption is monitored throughout the operation to support continuous projection of requirements. Force projection of sustainment operations may involve the establishment of support facilities in multiple sites outside the continental U.S. (OCONUS), including the crisis area. Logistics may be split-based between several theaters (ashore or afloat) and the U.S. The location and size of the base or bases supporting the operation is a key factor in operational reach.

(6) Rotation (Annual Force Requirements). As discussed in Section I of this chapter, CCDRs request their operational force requirements for all forces needed during an entire FY. The JS J3 works with OSD, JS JFC/JFPs, FPs, and CCMDs to forward for SECDEF approval FY allocation orders that best meets all CCDR force requests with available forces. Detailed sourcing guidance to CCDRs, JS JFC/JFPs, FPs and CSAs is given by the JS J3 in a published GFMAP PLANORD. The PLANORD includes strategic assumptions, timelines and specific directions for CCDRs to develop their force requirements submissions.

(7) Redeployment. Redeployment is the movement of forces out of an AO. This can be in support of another higher priority requirement in another AO, end of mission, or transfer of authority with an incoming unit. Redeployment is described in greater detail later within this chapter. The time for a unit to egress the AO and redeploy must be accounted for to schedule unit rotations.

for the phases of an actual crisis. An integration of sustainment forecasts with force projection plans is critical to success. Force alignment to missions within phases must be documented to enable rapid realignment with the actual mission requirements of a crisis. The OPLAN and the TPFDL structure for an OPORD must be designed to preserve unit integrity, agility, mobility, and security should events or conditions require execution of plan branches or sequels. The force structure must be understood by all supporting commands and enable global force visibility to ensure unity of effort and rapid response to actual operational events.

(2) Rapid force projection and force mobility are keystones of the NMS. Timely response to crisis situations is critical to deterrent and warfighting capabilities. The timeliness of response is a function of forward deployed forces and prepositioned assets, forces with organic movement capability, and adequate strategic and intra-theater mobility capability assets. Overseas presence, tailored to regional requirements, facilitates force projection by providing needed flexibility. The combination of organic force movement and rapid mobility capabilities, bolstered by pre-positioned assets, provides the supported CCDR with flexible mobility options that can be tailored to meet any crisis situation. Deployment operations normally involve a combination of organic and common-user lift supported movements using land (road and rail), sea, and air movement resources, as necessary. Successful movement depends on the availability of sufficient transportation capabilities to rapidly deploy combat forces, sustain them during an operation, and redeploy them to meet changing mission requirements or to return them to home and/or demobilization stations upon completion of their mission.

e. <u>The Strategic Mobility Triad</u>. Common-user airlift, sealift, and pre-positioned force, equipment, or supplies constitute the strategic mobility triad. Successful response across the range of military operations depends on sufficient port throughput capacity coupled with the availability of sufficient mobility assets to rapidly deploy combat forces, sustain them in an OA as long as necessary to meet military objectives, and reconstitute and redeploy them to meet changing mission requirements or to return to home and/or demobilization stations upon completion of their mission. To meet this challenge, USTRANSCOM's Transportation Component Commands (TCCs), Air Mobility Command (AMC), Military Sealift Command (MSC), and Military Surface Deployment and Distribution Command (SDDC), exercise OPCON of government-owned or chartered transportation assets for use by all DOD elements and, as authorized, other agencies of the USG or other approved users. Deployment operations normally involve a combination of land (road and rail), sea, and air movement augmented, as necessary, by pre-positioned assets.

(1) <u>Common-User Airlift</u>. The pool of common-user airlift consists of designated airlift assets from some or all of the following sources: AC and RC; the Civil Reserve Air Fleet (CRAF), when activated; contracted commercial assets; and foreign military or civil carriers, either donated or under contract.[14]

(2) <u>Common-User Sealift</u>. Sealift forces are those militarily useful merchant-type ships available to DOD to execute the sealift requirements of the Defense Transportation System (DTS) across the range of military operations. Called "common-user shipping," these ships will be engaged in the transportation of cargoes for one or more Services from one seaport to another or to a location at sea in the operational area pending a decision to move the cargo embarked ashore. The sealift force is composed of shipping from some or all of the following sources: active government-owned or controlled shipping; government-owned reserve or inactive shipping; United States privately owned and operated commercial shipping; United States privately owned, foreign flag commercial shipping; and foreign owned and operated commercial shipping.[15]

[14] *For additional information, see JP 3-17, Joint Doctrine and Joint Tactics, Techniques, and Procedures for Air Mobility Operations.*

[15] *For additional information, see JP 4-01.2, Sealift Support to Joint Operations.*

(3) DOD pre-positioned force, equipment, or supplies programs are both land and sea-based. They are critical programs for reducing closure times of combat and support forces needed in the early stages of a contingency. They also contribute significantly to reducing demands on the DTS.[16]

f. Operational and Tactical Deployment. There are numerous transportation resources available to a CCDR to support deployment operations. The type and number of sources vary by theater and by missions. Normally, operational and tactical deployment is executed through a combination of resources including: organic assets assigned to the CCDR for common transportation service; host-nation support (HNS) negotiated through bilateral or multilateral agreements; multinational civil transportation support organizations; or third-party logistics operations. When needed, theater lift resources and forces may be augmented by either assigning or attaching additional assets.

2. Integrated Planning and Execution Process

APEX, JOPES IT, GFM and the JOPP provide the processes, formats, and systems which link planning for joint force projection to the execution of joint operations. Standard processes that provide a common method for addressing force projection requirements enhance unity of effort. APEX and JOPP share the same basic approach and problem-solving elements, such as mission analysis and COA development. Planning for joint operations is continuous across the full range of military operations using two closely related, integrated, collaborative and adaptive processes – the APEX and the JOPP. The joint operation process supports the systematic, on-demand, creation and revision of executable plans, with up-to-date options, as circumstances require. A premium is placed on flexibility. The incorporation of collaboration capabilities, relational databases, and decision-support tools promotes planning with real-time access to relevant information and the ability to link planners and selected subject matter experts regardless of their location. The goal is shortened planning timelines and current, high-fidelity, up-to-date plans.

a. APEX is directed for all planning activities. JOPES IT use is directed for all joint force projections. APEX has three operational activities: situation awareness, planning, and execution. APEX provides the process, structure, reports, plans, and orders that orchestrate the JPEC's delivery of the military instrument of national power.

b. JOPP underpins planning at all levels and for missions across the full range of military operations. It applies to both supported and supporting CCDRs and to joint force component commands when the components participate in joint planning. This process is designed to facilitate interaction between the commander, staff, and subordinate headquarters throughout planning. The JOPP helps commanders and their staffs organize their planning activities, share a common understanding of the mission and commander's intent, and develop effective plans and orders. This planning process applies to developing contingency plans and crisis planning within the context of the responsibilities specified by APEX. Furthermore, JOPP supports planning throughout the course of an operation after the CJCS, at the direction of the President or SECDEF, issues the EXORD. In common application, JOPP proceeds according to planning milestones and other requirements established by the commanders at various levels.

3. Integrated Employment and Deployment. Deployment operations support employment of forces. Deployment positions mission-ready forces to execute the supported commander's OPLANs. Deployment plans align with the employment and the force requirements in the plans identify when and where specific forces are required. The GFM sourcing processes (preferred forces, contingency sourcing or execution sourcing) *link the requirements to the forces* that will ultimately be ordered to deploy via execution sourcing.

[16] *For additional information, see JP 4-01, Joint Doctrine for the Defense Transportation System.*

Employment is normally divided into a series of phases. Within those phases, missions are identified and prioritized. Forces are selected based on their ability to execute specific required missions. Any mission may require integrated or synchronized support from one or more Service forces. Deployment processes, systems, and plans must provide the supported commander with a joint force and the flexibility to redeploy forces as required to achieve the transition criteria for each phase of the operation and ultimately, mission success.

a. Operation Phase Model. The operation phase model is an element of operational design and is used to arrange operations.[17] It assists the CCDR to visualize and think through the entire operation to define requirements in terms of forces, resources, time, space, and purpose. Phasing enables commanders to systematically achieve objectives that cannot be attained all at once. It can be used to gain progressive advantages and assist in achieving objectives as quickly and effectively as possible. Phasing also provides a framework for assessing risk to portions of an operation or campaign and planning to mitigate that risk. Phasing the operation assists in framing commander's intent and assigning tasks to subordinate commanders. By arranging operations and activities into phases, the CCDR can better integrate and synchronize subordinate operations in time, space, and purpose; however, phasing is only part of a model and at any given time, the joint force will be operating across several phases simultaneously. As such, a phase represents a definitive stage during which a large portion of the forces and joint/multinational capabilities are involved in similar or mutually supporting activities. During planning, the CCDR establishes conditions, objectives, or events for transitioning from one phase to another and plans sequels and branches for potential contingencies. The CCDR adjusts the phases to exploit opportunities presented by the adversary or operational situation or to react to unforeseen conditions. The transitions between phases are designed to be distinct shifts in focus by the joint force, often accompanied by changes in command or support relationships. Changing the focus of the operation takes time and may require changing priorities, command relationships, force allocation, or even the design of the OA and should be thoroughly planned to avoid loss of momentum. This challenge demands an agile shift in joint force skill sets, actions, organizational behaviors, and mental outlooks. The operation phase model provides a flexible model to arrange combat and noncombat stability operations. This model can be applied throughout and across the range of military operations. Once the crisis is defined, deter phase actions may include mobilization, tailoring of forces and other pre-deployment activities; initial deployment into a theater; employment of intelligence, surveillance, and reconnaissance assets to provide real-time and near-real-time awareness of the operational environment; and development of mission-tailored command and control (C2), intelligence, force protection, and logistic requirements to support the CCDR's CONOPS.

b. Force Planning. The purpose of force planning is to identify all forces needed to accomplish the supported commander's CONOPS and phase the forces into the theater. It consists of determining the force requirements by operation phase, mission, mission priority, mission sequence, and OA. It includes major force phasing; integration planning; force list structure development; followed by force list development. Force planning is the responsibility of the CCDR supported by the Service component commanders, the Services, JS J3 and JS JFC/JFP. The primary objectives of force planning are to apply the right force to the mission while ensuring force visibility, force mobility, and adaptability. Force planning begins during CONOPS development. The supported commander determines force requirements, develops a TPFDD and LOI, and designs force modules to align and time-phase the forces in accordance with the CONOPS. Major combat forces are selected from those apportioned for planning and included in the supported commander's CONOPS by operation phase, mission, and mission priority. The supported CCMD Service components then collaboratively make tentative assessments of the combat support (CS) and combat service support (CSS) required IAW the CONOPS. During the planning stage, these forces

[17] *See JP 5-0 and Chapter 2 of this document for a description of Operational Phasing.*

are planning assumptions called preferred forces. The commander identifies and resolves or reports shortfalls. For some high-priority plans, the JS JFC/JFP may be tasked to contingency source the plan. Contingency sourcing is based on planning assumptions and contingency sourcing guidance. To the degree that the contingency sourcing planning assumptions and sourcing guidance mirrors the conditions at execution, contingency sourcing may validate the supported CCMD preferred forces. If the plan is directed to be executed, the force requirements are forwarded to the JS J3 for validation and tasked to the JS JFC/JFPs to provide sourcing recommendations to the SECDEF. When the SECDEF accepts the sourcing recommendation the FP is ordered to deploy the force to meet the requested force requirement. After the forces are sourced, the CCDR refines the force plan to ensure it supports the CONOPS, provides force visibility, and enables flexibility.

4. Deployment Process and Phases

a. Deployment Process. The joint deployment process is a dynamic process, beginning when force projection planning is initiated by a strategic or operational directive and ends when a force arrives at the prescribed destination, integrates into the joint force, and is declared ready to conduct operations by the CCDR. Deployment planning occurs during both deliberate planning and CAP and continues during execution. It is conducted at all command levels and by both the supported and supporting commanders. Pre-deployment activities are all actions taken by the JPEC, before actual movement, to prepare to execute a deployment operation. This includes training, organizing and equipping the force to be able to perform the mission specified in the force requirement. In the simplest form, the movement encompasses two primary nodes, Point of Origin and Destination. This simplest form occurs in the CONUS, operational inter-theater and intra-theater deployments, and during the movement of self-deploying forces. The more complex form encompasses four primary nodes, Point of Origin, Port of Embarkation (POE), Port of Debarkation (POD), and Destination. In the more complex form, there are three major movement "legs": Point of Origin to POE, POE to POD, and POD to Destination. The more complex form occurs when forces without organic lift capability deploy from one theater to another on strategic lift capability. The final phase, JRSOI, integrates the deploying forces into the joint operation. Joint force deployment has numerous process stakeholders and process seams, resulting from the multitude of organizations and functional processes involved in deployment planning and execution. Deployment stakeholders include the supported commander, operational commanders of forces, and supporting commanders of forces, Services, JS JFC/JFP, FPs, supporting and supported Service components and organizations with a deployment mission. Ideally, all process stakeholders will endeavor to ensure operational effectiveness (defined by successful mission accomplishment consistent with the supported commander's CONOPS and deployment efficiency (optimally phased force deployment and sustainment delivery with economical and effective use of available resources). Process seams may occur at functional or organizational interfaces when physical resources or information is transferred. Friction between operational and supporting stakeholders or process seams reduces the operational effectiveness and efficiency of the deployment process. The supported commanders reduce deployment execution friction by:

(1) Linking deployment to employment mission and force requirements.

(2) Minimizing CONOPS changes, but providing rapid dissemination of operation changes when they occur.

(3) Effective coordination and collaboration with other process stakeholders.

(4) Requiring accurate and early reporting of movement support requirements.

(5) Following APEX deployment procedures.

b. Deployment Phases. The joint deployment process is divided into four phases: planning, predeployment activities, movement, and JRSOI. They are iterative and often occur simultaneously throughout an operation. See facing page.

5. Sustainment Delivery

Sustainment delivery is the process of providing and maintaining levels of personnel and materiel required to sustain combat and mission activity at the level of intensity dictated by the CONOPS. Sustainment is ongoing throughout the entire operation and like deployment and redeployment, should be aligned with the mission and mission priorities of each phase. Sustainment delivery must frequently be balanced against force deployment or redeployment requirements because these operations share the same deployment and distribution infrastructure and other resources. However, deployment and force integration can be adversely affected by excess or insufficient sustainment support. Hence, operation planning must integrate deployment and sustainment operations.

6. Redeployment

Redeployment is the process of a unit egressing from an AO. In the deployment process a unit deploys, and at the end of the mission, redeploys. If the mission is enduring, the incoming unit arrives and conducts transfer of authority, and then, the outgoing unit redeploys. Redeployment operations may be conducted to re-posture forces and materiel in the same theater, to transfer forces and materiel to support another CCDR's CONOPS for employment, or to return personnel and materiel to home station and/or demobilization stations upon completion of their mission. Similar to deployment operations, redeployment planning decisions to re-posture, transfer, or return forces and materiel are based on the operational environment in the AOR or JOA at the time of redeployment.

a. In the GFM allocation process, CCDRs manage force requirements. No CCDR redeployment order is required to conduct a standard rotation or for a unit to redeploy at the end of a rotation ordered in the GFMAP Annex. However, when the unit and future rotations for that force requirement are no longer required due to end of mission, the CCDR must order the unit to redeploy and notify the force providing community that the unit is being redeployed and the requirement is no longer needed.

b. CCDRs manage forces deployed to their AOR as well as managing the force requirements for future forces that deploy. If the requirement for a deployed unit is no longer needed, the CCDR must order the unit to redeploy and notify the force-providing community that the unit is being redeployed and the requirement is cancelled. Likewise, if a CCDR endorsed force requirement has been submitted to the JS J3, the CCDR must alert the force-providing community that the requirement is no longer needed. This notification is called a redeployment order (REDEPORD) and is transmitted to the SECDEF, CJCS, JS J3, JS JFC/JFPs, and any ordered FPs via message traffic. The Supported CCDR initiates the redeployment process by releasing message traffic identifying units for curtailment (early redeployment) or requirements for off-ramping (identifying a FTN before the deployed unit initiates the deployment process). If a current unit is curtailed, future requirements will be off-ramped simultaneously. The supported CCDR REDEPORD is staffed by the JS JFC/JFPs and the resultant change in deployment schedules is transmitted as a change to the GFMAP Annex modifying orders to the FPs.

Deployment Phases

(1) Deployment planning occurs during both deliberate planning, CAP and continues during execution. It is conducted at all command levels and by both the supported and supporting commanders. Deployment planning activities include all action required to plan for a deployment and employment of forces. Deployment planning activities must be coordinated among the supported CCMD responsible for accomplishment of the assigned mission, the Services, JS JFC/JFPs, JS, OSD, and the supporting CCMDs providing forces for the joint force mission. Normally, supported CCDRs, their subordinate commanders, and their Service components are responsible for providing mission statements, theater support parameters, inter-theater lift requirements, applicable HN environmental standards, and pre-positioned equipment planning guidance during predeployment activities. The Services are responsible to organize, train, and equip interoperable forces for assignment to CCMDs and to prepare plans for their mobilization when required (see DODD 5100.1, *Functions of the Department of Defense and its Major Components*). Supporting CCDRs are responsible for providing trained and mission-ready forces to the supported CCMD. The JS JFC/JFPs are responsible to monitor the SECDEF ordered deployment timelines are met and to provide mitigation options if a force provided is unable to meet the timelines with trained and ready forces. Service and FP deployment planning activities should be coordinated with supporting CCMD planning activities to ensure that standards specified by the supported CCDR are achieved, supporting personnel and forces arrive in the supported theater fully prepared to perform their mission, and deployment delays caused by duplication of predeployment efforts are eliminated.

(2) Predeployment activities are all actions taken by the JPEC, before actual movement, to prepare to execute a deployment operation. It includes continued refinement of OPLANs, from the strategic to the tactical level at the supported and supporting commands. It includes completion of operation specific training, and mission rehearsals. As early as possible in pre-deployment, movement requirements are identified and lift support requirements are reported and scheduled.

(3) Movement includes the movement of self-deploying units and those that require lift support. It includes movements within CONUS, deployments within an AOR, and end-to-end origin to destination strategic moves. Though all forces are moved via the TPFDD, they may not always deploy in the order initially planned. During the initial and subsequent deployments, DOD leadership will use RFFs and DEPORDs as means to effect creation or sourcing of additional force requirements in the TPFDD. The supported commander issues the RFF message when he determines that additional forces, over and above those that have been previously approved by the SECDEF, are required to support military operations (note-SECDEF does not approve a TPFDD, he approves force requests). Deployments under RFFs or DEPORDs can significantly impact the subsequent flow of forces. Commanders and their staffs must understand the associated risks and impact of additional force flow that has not been previously planned. The TPFDD validation process ensures satisfaction of the deployment standards and controls deployments into and within the operation. Successful deployment movements are characterized by careful planning and flexible execution. Careful and detailed planning ensures that only required personnel, equipment, and materiel deploy, unit movement changes are minimized, and the flow of personnel, equipment, and materiel into theater supports the CONOPS. Deployment movement occurs in both inter-theater and intra-theater.

(4) The final phase, JRSOI, is the critical link between deployment and employment of the joint forces in the OA. It integrates the deploying forces into the joint operation and is the responsibility of the supported CCDR. Deployment is not complete until the deploying unit is a functioning part of the joint force. The time between the initial arrival of the deploying unit and its operational employment is potentially the period of its greatest vulnerability. During this transition period, the deploying unit may not fully sustain itself, defend itself, or contribute to mission accomplishment because some of its elements have not attained required mission capability. JRSOI planning is focused on the rapid integration of deploying forces and capabilities to quickly make them functioning and contributing elements of the joint force.

SECTION VI - RESPONSIBILITIES OF SUPPORTED AND SUPPORTING CCDRS

1. Responsibilities of Supported and Supporting CCDRS[18]

a. Geographic Combatant Commands (GCCs).

(1) GCCs are responsible for coordinating with USTRANSCOM and supporting CCDRs to provide an integrated transportation system from origin to destination during deployment operations.

(2) Responsibilities of Supported CCDRs. Supported CCDRs are responsible for deployment operations planned and executed during joint force missions in their AORs. This responsibility includes identification of the movement, timing, and sequence of deploying forces in the TPFDD, reception and integration of supporting units and materiel arriving in theater to support the operation, and assisting these units as required through to redeployment. Working through the U.S. Department of State (DOS), supported CCDRs negotiate HN diplomatic clearances and reception POD access when required, for deploying forces. For air movements, supported CCDRs must ensure that over flight and landing clearances are secured prior to the departure of forces from POEs. Supported CCDRs have four major responsibilities relative to deployment operations: build and validate movement requirements based on the CONOPS; determine predeployment standards; balance and regulate the transportation flow; and manage effectively.

(a) Determine Force Requirements and Build TPFDD Movement Requirements. Based on an approved CONOPS, the supported CCDR determines the forces or capabilities required for the mission and builds TPFDD movement requirements. Both activities encompass the supported CCDR providing detailed phase, mission, mission priority, and force tracking information. This effort permits force visibility and supports the CONOPS. Supported CCDRs designate latest arrival and required delivery dates. The supported CCDR specifies key employment information regarding when, where, and how forces will be employed by phase. The supported CCDR provides this information with the force requirements via the GFM process for allocation. Within the GFM process, JS JFC/JFPs, FPs, the supporting commands and DOD agencies assist in mission, force, training, sustainment and transportation planning. Service component commanders play a key role in advising the supported CCDR regarding the appropriate types and missions for the forces required. Employment information provided by the supported CCDR helps to ensure JPEC actions remain consistent with the CONOPS. To facilitate this process, the supported CCDR publishes a TPFDD LOI which provides specific guidance for supporting CCDRs, Services and agencies. The supported CCDR validates TPFDD movement requirements for all forces and agencies deploying or redeploying in support of an operation. Expeditious and thorough validation of movement requirements requires comprehensive, collaborative, and coordinated procedures.[19]

(b) Determine Pre-deployment Standards. Supported CCDRs establish predeployment standards for forces, capabilities, or personnel supporting their operations. Predeployment standards outline the basic command policies, training, and equipment requirements necessary to prepare supporting personnel and forces for the tactical, environmental, and/or medical conditions in theater. The Services' and CDRUSSOCOM's role is to ensure that designated forces for CCMDs are trained, maintained, and ready IAW these predeployment standards. Predeployment standards help ensure that all supporting personnel and forces arrive in theater fully prepared to perform their mission. See facing page for futher discussion.

[18] CCDRs operation planning responsibilities are described in the UCP, JP 1, Doctrine for the Armed Forces of the United States, and JP 5-0, Joint Operation Planning.

[19] Important note here: Remember that a planning TPFDD is completely under the GCC's control; an execution TPFDD implements the deployments ordered in the GFMAP.

Determining Pre-deployment Standards

Supported CCDRs establish predeployment standards for forces, capabilities, or personnel supporting their operations. Predeployment standards outline the basic command policies, training, and equipment requirements necessary to prepare supporting personnel and forces for the tactical, environmental, and/or medical conditions in theater. The Services' and CDRUSSOCOM's role is to ensure that designated forces for CCMDs are trained, maintained, and ready IAW these predeployment standards. Predeployment standards help ensure that all supporting personnel and forces arrive in theater fully prepared to perform their mission.

1. Preparation for Deployment

Preparation for deployment is primarily a Service responsibility. Specific responsibility for preparation for overseas deployment rests with the deploying unit or Service in the case of individual augmentees or replacements. For ad-hoc, ILO and joint sourcing solutions, the resourcing of preparation and training will be IAW Service to Service memorandums of agreement. Usually, the deploying unit commander must acknowledge completion of specified preparation for deployment requirements to the supported command during predeployment activities. In some instances, legal constraints may be in place that affects personnel readiness if specific training requirements have not been completed.

2. Pre-deployment Training

In addition to preparation for overseas movement, supported CCDRs may identify mission-specific training requirements that supporting individuals or units must complete before operational employment. These requirements are included in the force requests submitted in either the annual submission or in an emergent RFF. This training may be conducted at home station prior to deployment or en route to the OA. The supported CCDR should, at a minimum, specify theater-unique clothing and equipment requirements and mission-essential tasks to be trained prior to arrival in theater.

3. Balance and Regulate the Force Flow

The requested Start Dates on the request for each individual force or capability must support the overall CONOP. The SECDEF approved ordered Start Dates also support the CONOP while balancing operational with force providing risk. To the maximum extent possible, the supported CCDRs should balance and regulate the flow of forces with the flow of sustainment. Operational environment, available infrastructure, protection risk assessment, and the CONOPS are major factors in determining how to balance the flow of forces and capabilities. The supported CCDR manages the flow of forces via the TPFDD supported by strategic, operational, and tactical movement control organizations, to ensure effective interfaces between strategic and intra-theater movements.

4. Manage Effectively

Supported CCDRs balance the transportation flow of the joint force through effective employment planning. Balance is primarily a function of effective mission prioritization, alignment of forces, force composition, and force flow, but must also consider planned theater distribution (TD) and JRSOI requirements and capabilities. Force composition and transportation flow must accommodate mission requirements providing the supported CCDR with the operational flexibility and freedom of action required for successful mission execution.

c. Responsibilities of Supporting Combatant Commanders. Supporting missions during deployment of the joint force could include: the deployment or redeployment of forces from or to a supporting CCDR; sponsorship of en route basing or in-transit staging areas; or providing sustainment from theater stocks. Regardless of the supporting mission, the primary task for supporting CCDRs is to ensure that the supported CCDR receives the timely and complete support needed to accomplish the mission. Supporting CCDRs have five major deployment responsibilities: source, prepare, and verify forces; ensure units retain their visibility and mobility; ensure units report movement requirements rapidly and accurately; regulate the flow; and coordinate effectively.

(1) Source, Prepare, and Verify Forces. The GFMAP is a SECDEF-approved consolidated allocation order, by FY, that authorizes the transfer and attachment of forces to/from specified commanders and Services to supported CCDRs. Annexes within the GFMAP contain all the information inherent within a written order. The following Annexes and assigned JS JFC/JFP's are published in the GFMAP:

Annex A: Conventional Forces - (JS JFC).

Annex B: Special Operations Forces (SOF) (JFP) – (USSOCOM).

Annex C: Mobility Forces (JFP) - (USTRANSCOM).

Annex D: JTF-JMD Individual Augmentee - (JS JFC).

Annex E: CCDR Specific Coordination Instructions (Supported CCDR).

(a) The actual units sourced by the FP are not included in the GFMAP Annexes. Unit sourcing is available in the APEX and JOPES family of systems. Each unit line number (ULN) should reference the force tracking number (FTN) in order to trace back the force requirement to the originator as well as the SECDEF authority to allocate back through the GFMAP Annex.

(b) All CCDRs with assigned forces, Secretaries of the Military Departments, DOD Agencies, OSD organizations and U.S. Coast Guard support JS JFC/JFP planning and deploy forces specified in the JS JFC/JFP GFMAP Annexes.

(c) Any additional FY annual forces as well as emergent requirements are allocated in subsequent Annex modifications to the base GFMAP Annex upon SECDEF approval.

(d) JS JFC/JFPs publish modifications as required to their respective GFMAP Annexes in their entirety and supersede previous versions.

(3) Supporting and Supported CCDRs will publish DEPORDs implementing the orders in the JS JFC/JFP GFMAP Annex. All orders, down to the individual level, that implement the GFMAP Annexes will reference the FTN and FTN line number. Individual Marines, Soldiers, Sailors, and Airmen deploying should be able to trace their deployment orders back to the CCDR requirements via the FTN.

(a) Supporting and Supported CCDRs, Services and DOD agencies utilize APEX and JOPES as the systems of record for execution, to implement the deployment guidance in the GFMAP Annex.

(4) Command Relationships. Unless otherwise specified in the GFMAP or in a separate SECDEF approved order, supported commanders accept deploying forces and exercise OPCON upon arrival in their AOR, and relinquish OPCON of redeploying forces upon departure from their AOR. Supporting commanders that are providing forces transfer forces and relinquish OPCON to the supported commander upon arrival in the supported commander's AOR. CCDRs that provided forces exercise OPCON of their redeploying forces upon departure from the supported commander's AOR.

2. Planning

Deployment planning is based primarily on mission requirements and the time available to accomplish the mission. During deployment operations (mobilization, deployment, JRSOI, and redeployment), supported CCDRs are responsible for building and validating requirements, determining predeployment standards, and balancing, regulating, and effectively managing the transportation flow. Supporting CCMDs and DOD agencies source requirements not available to the supported CCDR as directed by the SECDEF and are responsible for: verifying supporting unit movement data; regulating the support deployment flow; and coordinating effectively during deployment operations. Based upon the supported CCDR's guidance, planners must assess the AOR's environment and determine deployment requirements for supporting the CCDR's CONOPS. Transportation feasibility must be included in the COA development.

SECTION VII - MUTUALLY SUPPORTING, INTERRELATED DOD PROCESSES

1. The GFM Process

The GFM process aligns force assignment, apportionment, and allocation methodologies in support of the NDS and joint force availability requirements. It provides DOD senior leadership with comprehensive insight into the global availability of forces and risk and impact of proposed force changes. The GFMB serves as a guiding body that provides complementary strategic focus and direction for the assignment, apportionment, and allocation process.

a. The mutually supporting, interrelated DOD processes are viewed in Figure VII-7 through a GFM lens. The stakeholders depicted include OSD, JS J3, Services (including theater Service Components), JFPs and the JFM (including their assigned Service Components and subordinate commands), CCDRs (including their assigned Service Components, JTFs, and other subordinate commands).

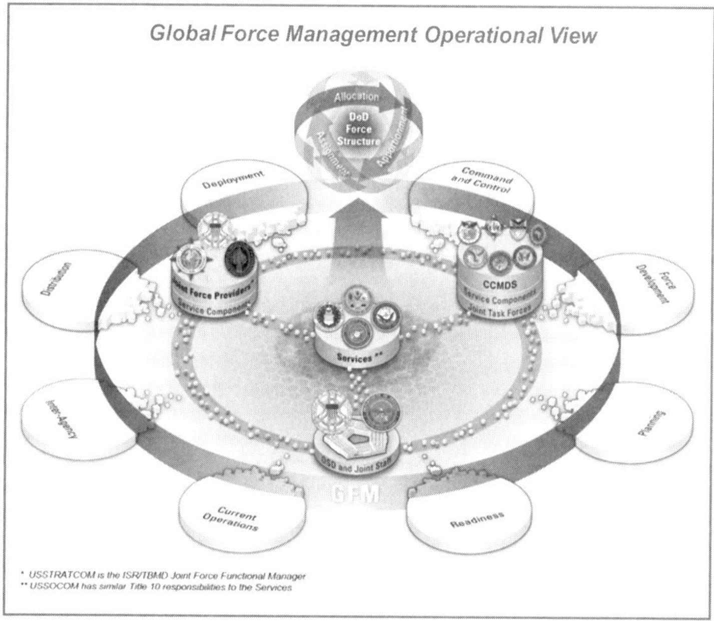

Figure VII-7. GFM Operational View (OV-1).

2. GFM Operational View

a. The GFM alignment (assignment, apportionment, and allocation) processes, tools, and data maintain synchronization across stakeholders and integration with related processes. This enhances the ability to efficiently and effectively align the force structure to respond to the complex, dynamic global environment.

b. Each stakeholder shares data and information to collaboratively determine the best use of the force structure to meet a situation. Impacts and risks of re-aligning the force structure are visible and all stakeholders collaboratively develop mitigation strategies. Planners obtain common force structure data directly from the entities responsible for building and maintaining the data.

c. Formally and rigorously specified force structure data contains unambiguously defined semantics implemented so GFM-related computer programs can readily exploit the data. Stakeholders share a common understanding of the meaning of GFM data. Changes in any of the processes depicted as overlapping and interacting with GFM influence not only GFM directly, but often influence changes in other processes. Seamless iterative interaction and integration between these related processes are necessary for the success of each of these processes as well as success of the missions these processes exist to support.

d. See facing page for discussion of GFM process interations.

SECTION VIII, GFM SUMMARY

a. The DOD continues to implement the most profound reordering of the U.S. military presence overseas since the start of the Cold War. The future operating environment has the potential to produce more challenges than the United States and its military forces can respond to effectively.[20] Therefore, resource management must enable the SECDEF and other senior civilian and military leadership to balance resources, to include forces, among the strategic priorities of on-going operations, shaping activities, and deliberate plans. Foremost in the resource management effort is the GFM enterprise which aligns forces using the assignment, allocation, and apportionment processes in support of strategic guidance and a common vision and agreed-upon end result. As a cornerstone to the APEX initiative, GFM will provide leaders at all levels with a clear picture of global force readiness, availability, and the risks associated with changing the alignment of the forces via the assignment, allocation and apportionment processes. The GFM enterprise provides the CJCS and SECDEF the decision support structure to make informed decisions on the use of forces. The metrics, information, and data requirements for these decisions can be situation dependent, making the GFM processes highly dynamic.

b. In the past, forces were apportioned for deliberate planning and presumed ready and available. Today's dynamic and demanding world requires us to challenge that assumption. As Operation Unified Response in Haiti was unfolding, planners recognized, yet again, that we must still be prepared to practice force management when a large percentage of forces may be committed to on-going operations and are unavailable. We must have the mechanisms available to rapidly mitigate the challenges and the risks this creates.

c. Doctrine and policy will continue to be challenged keeping current with the highly dynamic GFM processes. The many "independent" variables involved in GFM make data transmission and information gathering inconsistent and, at times, it is perceived as cumbersome. Compounding these issues is the fact that the number of individuals who fully

[20] *United States Joint Forces Command, The Joint Operating Environment (USJFCOM, 25 November 2008).*

GFM Process Interactions

The GFM process aligns force assignment, apportionment, and allocation methodologies in support of the NDS and joint force availability requirements. As displayed in Figure VII-5 the process interactions are:

Force Development

The Force Development Process takes input from many sources. Global demand for forces is one of those inputs. Likewise, GFM projects alignment of both current and projected force structure, so GFM is influenced by the changes taking place to the force structure through the Force Development Process.

Planning Process

The Planning Process is both influenced by and influences GFM. As the fidelity of a given plan is refined, the alignment of the force pool necessarily influences the plan. The forces assumed in planning must be aligned to the supported CCDR if/when a plan transitions to execution. This re-alignment, in turn, affects other planning efforts.

Command and Control

Command and Control by the Joint Force Commander includes establishing clear lines of command. Orders and direct planning assumptions dealing with the force structure (apportionment) communicate the alignment of forces. The orders and assumptions also direct near steady state Combatant Command authorities of forces (assignment) and re-alignment of forces and deployments to other CCMDs to respond to world events (allocation). GFM alignment processes both influence and are influenced by the Command and Control processes.

Deployment Process

The Deployment Process includes the tasks of preparing, deploying, sustaining and redeploying forces. Timelines for each of the tasks performed in the Deployment Processes inform the GFM alignment processes when planning the schedules to implement allocation and apportionment. Likewise, the Deployment Process receives and executes the deployment schedules generated from the GFM processes.

Distribution Process

The Distribution Process includes the tasks of moving units and sustainment worldwide. The distribution and transportation capabilities influence the schedules developed in the GFM process. Likewise, the GFM process deployment orders influence both planning and execution within the Distribution Process.

Interagency Process

While GFM does not manage the entire collection capabilities in all branches of government, GFM interacts with the Interagency Process by providing a conduit to non-DOD agencies to meet CCDR capability requests, for both planned and executed operations. As other (non-DOD) instruments of national power are committed to support CCDR capability requests, the GFM orders provide a vehicle to inform the JPEC of the directed sourcing solution.

Current Operations

Current operations, while formally defined as a subset of CAP, require forces. As the goals and strategies for these operations change, these changes directly impact the alignment of the force structure. Likewise, the capabilities resident in the force structure influence the strategies and plans to prosecute the operations.

Readiness

Readiness is a key indicator of availability when planning force deployments and aligning the force. The alignment of the force likewise influences the flow of efforts and resources to prepare forces to deploy.

understand the details of sourcing and force-providing from multiple perspectives is limited. The GFM enterprise is evolving rapidly and our senior Service schools, war colleges and universities must maintain pace with this evolution to ensure our graduates understand the principles and have a working knowledge of GFM and how GFM relates to the other processes. They will have the sword passed to them, and are expected to guide GFM into the next decade and beyond.

d. The dynamic complexity of GFM requires constant attention to provide comprehensive insights into the global availability of U.S. military forces and capabilities to quickly and accurately assess the impact and risk of force management decisions. With continued, focused attention, we will leverage the hard lessons of the past, capitalize on the savvy and knowledge of our people to operate today, and lead the evolution of GFM to be ready for tomorrow's challenges.

In-Progress Reviews (IPR) 8

If a strategy is the implementation of policy while balancing available ends, ways and means, a strategic planning dialog is the iterative conversation among civilian and uniformed senior leaders that considers potential contingencies and the application of military power to address them. A shared understanding of the problem, the goal, and the potential ways to apply military power, and the resources required is critical to presenting these leaders with credible choices. Military power may or may not be the last resort in influencing the problem or trying to bring it to resolution; its allocation is dependent upon senior decision makers having a clear understanding of what can be accomplished, the inherent risks, and how military power complements other elements of a USG response. This is the ultimate purpose of In-Progress Reviews (IPRs).

1. IPR Venue

IPRs provide a venue for senior military and civilian leaders to develop a shared understanding of the emerging problem (situation), how military power might address that problem to reach USG preferred resolution (strategic options), what resources can be applied across the government (USG unity of effort); what specific military actions might be taken (operational approach) – or COA); and what decisions, resources, and abilities the military needs in order to take specific actions.

2. IPR Objectives

The objective of a well-developed and maintained plan is twofold; first, to provide a mechanism for continuous strategic and operational discussion on using the military element of power to shape the current operational environment, including discussions of current U.S. policies and priorities and existing or emerging threats, and, secondly, when directed, transition to and from execution as smoothly as possible.

a. The APEX plan review process consists of four core planning phases. Strategic Guidance, Concept Development, Plan Development, and Plan Assessment.

b. During each phase CCDRs ensure planning is synchronized with the SECDEFs intent and consistent with current national objectives and assumptions. This is accomplished by maintaining an on-going dialog with the Joint staff and the OSD, as the guidance and assessments may change during the planning process and over the life of the plan.

c. The APEX system features early planning guidance and frequent iterative dialog in the form of reviews and updates between the OSD staff, the Joint Staff, Service staffs, and military commanders and planners. This dialog, known as socialization, facilitates an understanding of, and agreement on, the mission, planning and policy assumptions, strategic context, operational environment, threat courses of action, risks, and other key factors such as interagency (IA) and allied planning considerations between OSD leadership and the planners. In this sense, the plan is considered a “living” plan in terms of guidance, relevancy, and appropriateness because the CCMD, Joint Staff, and OSD frequently review and evaluate the plan throughout its development. For specified plans the socialization process culminates with an IPR to the SECDEF and CJCS, or Under secretary of defense for Policy (USD(P)) and the Vice Chairman of the Joint Chiefs of staff (VCJCS).

3. GEF and JSCP-tasked Plan Review Categories and Approval Authorities

a. GEF and JSCP-tasked plans are organized in four categories of review for purposes of IPR review and management. The four plan categories are:

(1) Plans briefed to SECDEF (identified in the GEF)

(2) Plans reviewed at other levels in OSD or CJCS/VCJCS level (to include review through the JPEC or Tank processes).

(3) Campaign plans, reviewed by the USD(P) and presented at a senior leader conference (SLC).

(4) Plans not requiring OSD/Joint Staff IPRs.

b. The SECDEF establishes the review requirements for all GEF- and JSCP-tasked plans. Guidance on which plans require reviews and the timing of those reviews are published by the SECDEF separately.

4. In-Progress Reviews with the SECDEF/Designated Representative[1]

a. DOD IPRs ensure that plans remain relevant and respond to the President and SECDEF direction throughout plan development. Each IPR is a key element of the APEX system to make plans more comprehensive and inclusive by integrating DOD senior leadership guidance via regular, deliberate participation, and interaction. IPRs ensure that plans remain relevant to the situation and the SECDEF's intent throughout their development.

b. The SECDEF establishes the review requirements for all GEF and JSCP tasked plans.

c. Bundled Plans. Separate plans, developed by a single or different CCMDs, are bundled to address a common GEF-identified problem set and are likely to be executed in concert with each other. CCMDs will usually breif bundled plans at the same time.

5. Plan Socialization and In-Progress Review Process

a. IPRs are an essential element of the APEX system, ensuring plans are more responsive to the President and SECDEF.

b. The IPR process should not be viewed as a set of briefings at discrete points in the planning process by the CCDR to DOD senior leadership but as a continuous and iterative dialog between the CCMD planners and OS, Joint Staff, and Service headquarters planning staffs to ensure the plan conforms to evolving national-level guidance, meets DOD and national-level leader needs, and reflects the changing operational environment.

c. IPRs are one tool DOD uses to ensure plans are developed consistent with the President's and SECDEF's strategic requirements. The IPR process:

(1) Enables the SECDEF and OSD staff to keep CCMD planners up-to-date with changes in strategic guidance, national priorities, and OSD's assessment of the operational and strategic environment;

(2) Provides the CCDRs the opportunity to inform DOD leadership on their plans, address planning and execution concerns, and get decisions on key planning issues;

[1] *CJCSI 3141.01 series, Management and Review of Campaign and Contingency Plans.*

(3) Ensures DOD senior leadership and the CCDR have a shared understanding of the strategic and operational guidance and parameters and have a common vision of mission execution and success;

(4) Generates feedback for planning staffs and provide guidance on coordination with IA and multinational communities;

(5) Inform key leadership and staffs of progress and effectiveness in achieving a plan's goals; identify changes in the strategic or operational environments, assumptions, or objectives; identify alternatives; and identify risk associated with plans that are being planned or executed.

d. For many plans, the President and SECDEF may require a range of military options that can be integrated with the other elements of national power (diplomatic, information, military, and economic or DIME), rather than a single military solution. The IPR process allows the CCMDs to have this conversation with the policy makers in OSD to identify a range of approaches ahead of a crisis. For these plans, the primary purpose of the IPRs is to create a shared understanding to: (1) inform DOD leadership of military COAs, limitations, costs, and risk; and (2) inform CCMD planners of policies and politics involved in decision making and execution of the plan.

e. The IPR process is conducted at two levels.

(1) Plan socialization is the ongoing, continuous dialog between CCMD planners and members of the Joint and OSD staffs to ensure a common understanding of the plan, its status, and issues. Most issues can be resolved at this level.

(2) Formal IPRs are scheduled discussions between the CCDR and the SECDEF or USD(P) to ensure the CCDR and SECDEF agree on the plan's objectives and operational approach and to address any unresolved issues. IPRs can be held at any time in the planning process and may be requested by either the SECDEF or the CCDR.

(3) Plans requiring greater specificity (i.e. CONPLANs with TPFDD (level 3T) and OPLNs (level 4) will likely be scheduled for more in person reviews than level 1-3 plans due to the greater amount of detail required.

f. Existing Plans. A number of the GEF – and JSCP-tasked planning requirements have existing plans that address the planning task or a previous version of the task. Often these plans require updates due to changes in capabilities, capacity (U.S., threat, allies), operational environment, and national level guidance, assumptions, and objectives. IPRs for existing plans provide the CCDR the opportunity to refresh DOD leadership's knowledge of the plan, its objectives, and assumptions. It allows both the CCDR and DOD leadership to discuss possible changes in plans.

(1) Campaign Plans. TCPs and FCPs are existing campaign plans that are in continuous execution. They describe the near- to mid-term objectives of a CCMD; the foundational activities the command undertakes to achieve those objectives; and the metrics by which the command defines and measures success. These campaigns require regular assessment and annual IPRs with OSD. TCPs and FCPs will usually be reviewed by the USD(P).

(2) Contingency Plans. IPRs for existing and revised contingency plans are conducted on an as-directed or as-needed basis. The SECDEF will provide guidance under separate cover on which plans will be reviewed in the upcoming year. At a minimum, all plans should be submitted for a paper review every two years. Should a GEF directed priority plan require a major update due to changes in friendly or enemy capabilities or in the operational environment, the CCDR will notify the Joint Staff that an IPR should be conducted.

g. New Plans. Newly tasked planning requirements necessitate more frequent interactions between the CCMD planners, OSD staff, the Joint Staff, and Service and force providing HQs. Where an existing plan has already been reviewed by DOD leadership resulting in a common starting point, a new task presents the possibility of divergence between DOD leadership's intent and how the task was perceived at the CCMD. To avoid this confusion, CCMD planners must ensure they are addressing the same issue and in the same frame as DOD leadership. Planners should ensure they understand:

- The problem the plan is intended to address.
- The strategic context and operational environment
- The nesting of DOD objectives with the President's national security objectives and the range of acceptable outcomes.
- How the military element of national power supports the other elements and what the other elements do to address the issue.
- The operational approach (are there any pre-established limitations or constraints).
- The time frame for possible execution and the decision timeline for the President to execute the plan.
- What other plans are related (should this plan be bundled with another CCMD plan, and should another CCMD's plan be bundled with this).

h. It is unlikely that not all can be answered at the start of planning, but by asking them, planners can reduce the likelihood of conducting a mission analysis for the wrong problem set. Much of this discussion can be done through action officer level coordination (email and phone calls). However, it is possible that during mission analysis, an issue could be discovered that requires the CCDR to personally discuss the problem set with OSD senior leadership.

6. Joint Planning and Execution Community (JPEC) Review

To improve the quality of planning with greater efficiency, the APEX process features early and detailed planning guidance and frequent dialog during the four planning functions in the form of IPRs between senior leaders and planners to promote an understanding of, and agreement on, the mission, planning assumptions, threat definitions, interagency, and allied planning cooperation, risks, courses of action, and other key factors.

a. JPEC Review. GEF- and JSCP-tasked plans should be continuously reviewed and refined by CCMDs, Services, NGB, and CSAs planning staffs. They will be more comprehensively reviewed at distinct times as noted below:

(1) Plans that have been directed by the SECDEF for IPRs shall be reviewed by the JPEC prior to the completed plan being presented to the SECDEF or USD(P). JPEC reviews are not required prior to IPRs during plan preparation, unless requested by the SECDEF, USD(P), CJCS, or CCDR.

(2) Plans directed in the JSCP for CJCS review will be reviewed by the JPEC.

(3) Plans that are not directed by the GEF or JSCP for an IPR or CJCS review, and campaign support plans, do not require JPEC reviews and can be reviewed at the CCMD level.

(4) Plans submitted to the JPEC for review should identify all related CCMD-level plans (functional plans, supporting/adjacent CCMD plans) that could reasonably be expected to be executed in conjunction with or following that plan under review.

b. JPEC Overview.

(1) In general, the JPEC will review plans using the five criteria in JP 5-0:

(a) *Adequacy*. The scope and concept of planned operations can accomplish the assigned mission and are within the planning guidance. Planning assumptions must be reasonable, valid, and comply with strategic guidance.

(b) *Feasibility*. The assigned mission can be accomplished using available resources within the time contemplated by the plan.

(c) *Acceptability*. This criterion is used in conjunction with feasibility to ensure the missions assigned can be accomplished with available resources and that the plan is proportional and worth the expected cost. It focuses on the level of risk to mission accomplishment, is consistent with domestic and international law, including the law of war, and is militarily supportable.

(d) *Completeness*. The plan incorporates all assigned tasks to be accomplished and to what degree they include forces required, deployment concept, employment concept, sustainment concept, time estimates for achieving objectives, description of the end state, mission success criteria, and mission termination criteria.

(e) *Joint Doctrine Compliance*. The plan complies with joint doctrine to the maximum extent possible.

(2) To complement the five review criteria and requisite subject matter expertise, principal respondents of the JS Directorates (JDIRs) will also review all plans in light of specific Focus Areas that have been identified as crosscutting issues for all GEF- and JSCP-tasked plans in addition to the functional areas identified. The Focus Areas, which are broken down by JS Directorate, can be found within each JDIRs section outlined in CJCSI 3141.01 series.

(3) For level 3T and level 4 planning, the logistics supportability analysis (LSA) will be presented to the Director for Logistics (DJ-4) as part of JPEC review.

c. JPEC Membership.

(1) JPEC membership consists of OSD, the CCMDs, the Services, the NGB, all JS directorates (including OCJCS Legal Counsel, Public Affairs, and National Guard and Reserve Matters), and the following DOD CSAs:

(a) Defense Intelligence Agency (DIA).

(b) Defense Information Systems Agency (DISA).

(c) Defense Logistics Agency (DLA).

(d) National Security Agency (NSA)/Central Security Service (CSS).

(e) National Geospatial-Intelligence Agency (NGA).

(f) Defense Threat Reduction Agency (DTRA).

(g) Defense Contract Management Agency (DCMA).

(2) Additionally, the following sub-unified commands are also included: U.S. Forces Korea, U.S. Forces Japan, and U.S. Cyber Command.

(3) DASD(P) serves as a single OSD point of contact (POC) to consolidate OSD comments during JPEC. OSD plan review may include the IA when appropriate and is reflected in Figure VIII-1.

d. JPEC Process.

(1) On receipt of a CCMD plan, J-5/JOWPD initiates a plan review directive to the JPEC. The plan review directive (JS Form 136) establishes review timelines, plan review level, comment formats, and other administrative directions for conducting the review.

(2) If feasible, J-5/JOWPD will coordinate an IPR style briefing at the AO level by the supported CCMD to the JPEC to familiarize the community with the plan early in the review process. If a briefing by CCMD planners is not possible, an IPR style brief may be included in the JPEC review documents.

(3) Planner-level or O-6 level plan review comments will be provided as execution-critical, substantive, or administrative as defined below. Execution-critical comments require a general officer/flag officer/Senior Executive Service endorsement.

(4) J-5/JOWPD consolidates review comments and provides a response to the CCMD planning staff with an information copy to OUSD(P) Plans. J-5/JOWPD will ensure all execution critical comments meet the definition of critical and oversee the adjudication of comments between the comment originator and the CCMD planning staff in coordination with the plan SME as necessary and appropriate.

(5) The CCMD planning staff reviews the consolidated JPEC and Focus Area review comments and then replies to J-5 /JOWPD regarding each execution-critical comment within 15 working days after receipt. J-5/JOWPD oversees adjudication of all critical comments with support as required from plan SMEs. Every effort must be made to adjudicate all critical comments, either through action officer, planner, or general/flag officer channels or, if required, through the JCS Tank process.

(6) The CCMD, will prepare a joint LSA for each fully developed OPLAN and CONPLAN with TPFDD by combining the Services, USTRANSCOM's, DLAs, and Defense Contract Management Agencies (DCMA's) assessments. The CCDR will make the TPFDD available to the Services, USTRANSCOM, and DLA prior to Force Flow, Transportation Feasibility, and Logistics Conferences, allowing time to use the TPFDD data to run analyses to produce their own assessments.

e. JPEC Completion.

(1) Once the supported command, Service, or CSA has completed their final adjudication of the plan, they will provide a planner level memo to J-5/JOWPD stating the JPEC review is complete and all execution-critical comments have been adjudicated.

(2) Functional CCMDs are responsible for the aligning planning and execution of other CCMDs, Services, Defense Agencies and activities, and, as directed, appropriate USG agencies within their specified global domain to facilitate coordinated and decentralized execution across geographic or other boundaries. Those plans that undergo a JPEC review will be reviewed by the respective functional command for that specific mission/area. Once complete, the functional CCMD endorses the plan via a letter to the SECDEF routed through JS J5/JOWPD, ensuring the alignment of specified planning and related activities.

7. IPR Attendance

CCDR participation in IPRs to the SECDEF is required. For IPRs with the USD(P) and VCJCS, the Deputy CCDR is appropriate.

a. Attendance of the SECDEF IPR is strictly limited by the SECDEF's executive staff and will be established prior to each IPR. The following will generally attend SECDEF IPRs (in addition to the SECDEF and the supported CCDR):

- USD(P) and/or Principal Deputy USD(P).
- Under SECDEF for Intelligence (USD(I)).
- DOD General Counsel or Principal Deputy General Counsel.
- DASD(P).

Completed Plan Review Process (Fig. VIII-1)

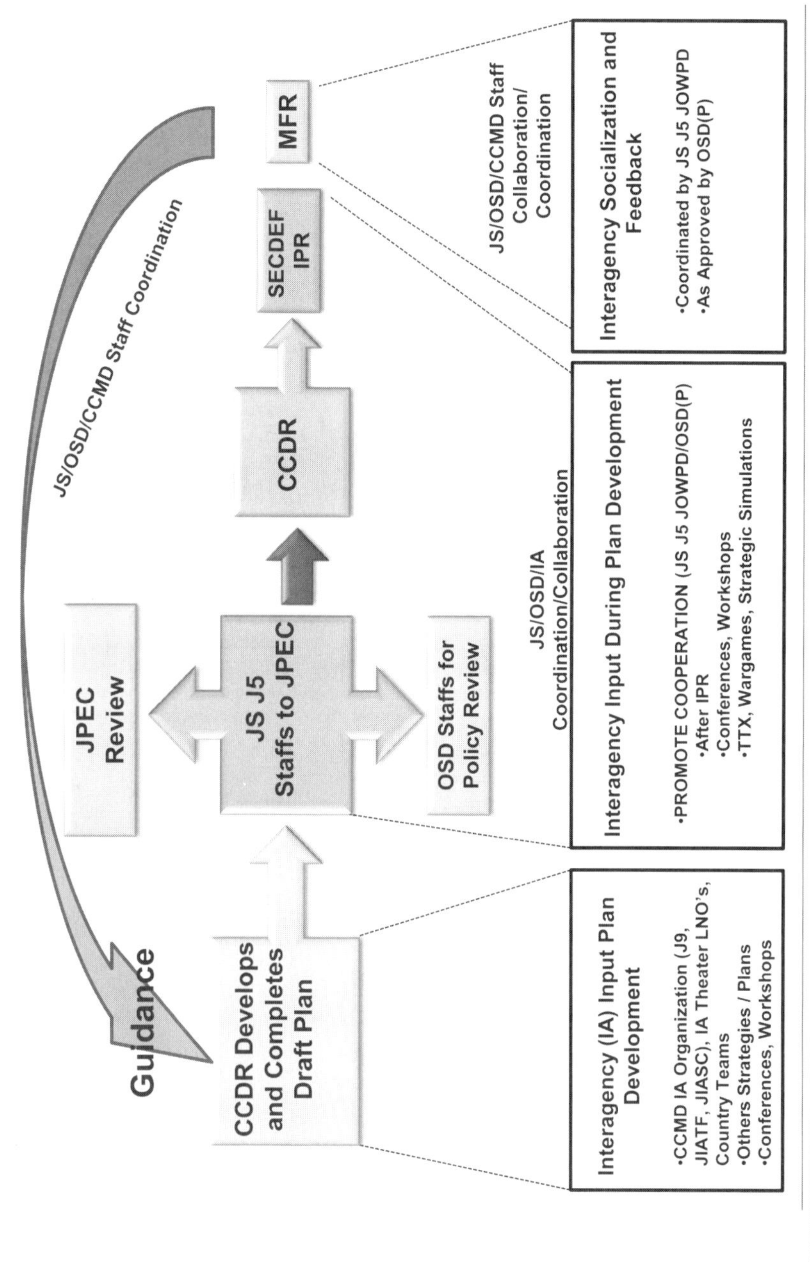

- Designated military assistants [as required].
- Applicable regional or functional ASD, DUSD, or DASD.
- Other key personnel as appropriate, approved, and or directed.

Adaptive Planning and Execution (APEX) and the JOPP

I. Adaptive Planning And Execution - Overview

1. Adaptive Planning and Execution (APEX)

At this writing all CJCSM 3122 series (JOPES) documents are being replaced by APEX. As the CJCSM 3122 series manuals are re-written they will be transformed into APEX manuals. APEX is defined as "the Joint capability to create and revise plans rapidly and systematically, as circumstances require." (CJCSG 3130, Adaptive Planning and Execution (APEX) Overview and Policy Framework). APEX is the Department level enterprise of joint policies, processes, procedures, and reporting structure supported by communications and information technology (IT). It incorporates a joint enterprise for the development, maintenance, assessment, and implementation of global campaign plans, theater campaign and related contingency plans and orders prepared in response to Presidential, SECDEF, or Chairman direction or requirements. It is utilized by the JPEC to monitor, plan, assess, and execute joint operations. APEX formally integrates the strategic-theater and operational planning activities of the JPEC and facilitates the seamless transition from planning to execution during times of crises. APEX activities span many organizational levels, including the interaction between the SECDEF, CCDRs, coalition, and interagency which ultimately assists the President and SECDEF to decide when, where, and how to commit U.S. military forces. APEX is applicable across the range and spectrum of military operations to achieve United States policy objectives.

a. APEX Enterprise. The APEX enterprise provides the processes, formats, and systems which link resource informed planning to the execution of joint operations.

(1) Well-developed plans have two objectives; first, provide a mechanism for continuous strategic and operational discussions on using the military element of power to shape the current operational environment, including discussions of current U.S. policies and priorities and existing or emerging threats, and, second, when necessary, execute those plans as effectively and efficiently as possible.

(2) APEX is a collaborative planning model (Figure IX-1) that facilitates the development and maintenance of plans and, when necessary, their transition to execution. APEX melds the best characteristics of joint operational planning process (JOPP) and departmental experience with planning and the execution process.

(3) The APEX enterprise is an iterative process. Each activity and function influences and is influenced by activities and functions which are performed and reviewed at multiple echelons of commands in overlapping timeframes. Facilitating communication and understanding of strategic guidance between these echelons of command takes place in several formats: formal strategy and policy documents; the plans review process; and via specific, individual communications with CCMDs. CCDR planning may also influence strategic direction and guidance, either during planning or execution.

b. APEX leverages CCDR design and the JOPP framework for planning functions that form the basis for planning. APEX also leverages continuous situational awareness, planning, assessments and staff estimates which form the basis for execution functions and enable the commander to direct, monitor, assess and adjust the plan or order as required throughout execution. Figure IX-1 further depicts this framework and some of the mutually supporting interrelated sub-processes that when properly integrated provide a dynamic, responsive APEX process. The activities and functions are continuous and iterative throughout, adapting to situational changes and governmental policy decision. The APEX process is abbreviated below, refer to CJCSM 3130 series documents for greater detail.

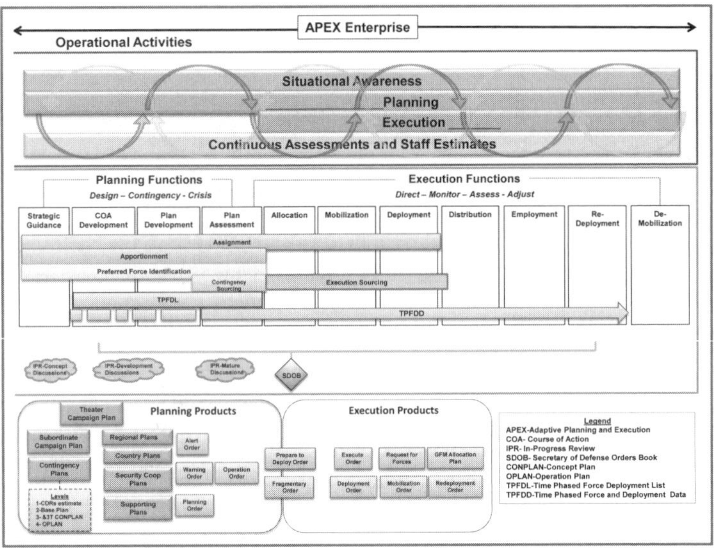

Figure IX-1. APEX Enterprise

(1) Operational Activities:

(a) Situational Awareness. Within APEX, situational awareness is the foundation supporting the continuous cycle of planning, execution, and assessment activities. CCDRs must know when strategic and operational facts and conditions change. Situational awareness is both internal (CCDR CCIRs/PIRs) and external (national interest, political environment). Situational awareness comprises the strategic environment including threats to national security while continuously monitoring the national and international political and military situations. Through situational awareness the determination of timely, relevant, and accurate information concerning the status of adversary, friendly forces, and resources is ascertained.

(b) Planning. APEX formally integrates military planning – campaign, contingency, and crisis – into one construct to better facilitate unity of effort and the transition from planning to execution. APEX establishes routine interactions between the SECDEF and CCDRs in the form of in-process reviews (IPRs). IPRs refine strategic direction and guidance, discuss military options, assist in understanding strategic risks and decision points, and address CCDR issues and concerns within potential contingency scenarios. Importantly, the Office of the Secretary of Defense (OSD) civilian leadership considers the review process as an iterative dialogue that matures over time (Chapter VIII – IPRs).

(c) Execution.

1. Theater Campaign Plan. The Theater Campaign Plan is always in execution. Consequently, theater campaigning is a continuous cycle of plan implementation – assessment – modification –and continuous implementation as theater operations, activities, events, and investments are applied to achieve regional IMOs.

2. Contingency Plans. For contingencies execution begins with the first activity in support of an approved operation order (OPORD) after the President or SECDEF authorize the Chairman to issue an execute order (EXORD) which directs the supported commander to initiate military operations, defines the time to initiate operations, and conveys guidance not provided earlier. Execution continues until the operation is terminated, forces are redeployed, or the mission is accomplished or revised. Detailed discussions on the execution process and the roles and responsibilities during execution of the joint community are contained in the CJCSM 3122 and CJCSM 3130 series of documents.

(d) Assessments. Assessment is the continuous monitoring and evaluation of the current situation and progress of a joint plan or operation toward mission accomplishment. It encompasses APEX at all levels and is iterative throughout and enables commanders to review and where necessary make the recommendations for additional resources or to recommend reallocating available resources to the highest priorities. Assessment and learning also enable incremental improvements to the commander's operational approach and the campaign or contingency plan. Once commanders understand the problem and what needs to be accomplished, they identify the means to assess effectiveness and the related information requirements that support assessment. This feedback becomes the basis for learning, adaptation, and subsequent adjustment. Assessment involves deliberately comparing forecasted outcomes to actual events to determine the overall effectiveness of force employment. In general assessments should answer two basic questions: Are we doing things right? Are we doing the right things? (Chapter XXII has greater details on assessment).

(e) Staff Estimates. Staff estimates are absolutely vital to establish and maintain an overall high degree of coordination and cooperation, both internally and with staffs of higher, lower, and adjacent units. The staff estimates continue throughout situational awareness, planning execution and assessment. Accurate and timely staff estimates directly affect the commanders ability to make well informed resource and risk based decisions (Chapter XVI has greater details on staff estimates).

(3) Planning Functions. The APEX process consists of core planning functions; Strategic Guidance, Concept Development, Plan Development, and Plan Assessment. During each function CCDRs ensure planning is synchronized with the SECDEF's intent and consistent with current national objectives and assumptions. This is accomplished by maintaining an on-going dialog with the Joint Staff and OSD, as the guidance and assessments may change during the planning process and over the life of the plan (Chapter IX has greater details on planning functions and the JOPP).

(4) Execution Functions. The APEX process consists of several execution functions. During each function, CCDRs continue to direct, monitor, assess and adjust as the plan moves forward within execution. CCDRs continue to review progress as needed during execution with the SECDEF, Chairman, and OSD. This on-going dialog is maintained to ensure guidance and intent is consistent with current national objectives and assumptions as the order is executed (Chapter XXIII has greater details on execution).

(5) In-Process Reviews. The APEX system features early planning guidance and frequent iterative dialog in the form of reviews and updates between the OSD staff, the Joint Staff, Service staffs, and military commanders and planners. This dialog, known as socialization, facilitates an understanding of, and agreement on, the mission, planning and policy assumptions, strategic context, operational environment, threat courses of action, risks, and other key factors such as interagency (IA) and allied planning considerations between OSD leadership and the planners. In this sense, the plan is considered a "living" plan in terms of guidance, relevancy, and appropriateness because the CCMD, Joint Staff, and OSD frequently review and evaluate the plan throughout its development. For specified plans the socialization process culminates with an IPR to the SECDEF and CJCS, or Under secretary of defense for Policy (USD(P)) and the Vice Chairman of the Joint Chiefs of staff (VCJCS) (Chapter VIII has greater details on IPRs).

(6) Secretary of Defense Orders Book (SDOB). The SDOB is a process used to route actions through Joint Staff Directors (JDIRs) and OSD to the SECDEF for approval for actions to include, but not limited to: EXORD, DEPORDs, Global Force Management Allocation Plan (GFMAP) and modifications, ALERTORD, NATO Force Preparation (FORCEPREP, Alert and mobilization orders for RC forces, and modification of CCDR authorities or previous SECDEF decisions.

c. APEX Evolution. APEX will continue to evolve to meet the challenges faced when applying the joint force to address global and regional challenges. The current approaches within APEX are intended to assist commanders in preparing and responding with improved understanding of contingency operations, apply military power when directed, and assess those military actions throughout planning and execution. Through the APEX enterprise CCDRs are able to coordinate, synchronize, and integrate TCPs and military operations with the activities of USG organizations and when directed, non-governmental organizations (NGOs), and coalition partners.

II. JOPP Overview

1. Joint Operation Planning Process (JOPP)

JOPP is an orderly, analytical process, which consists of a set of logical steps to examine a mission; develop, analyze, and compare alternative COAs; select the best COA; and produce a plan or order. Operational art and the application of operational design provide the conceptual basis for structuring campaigns and operations discussed in Chapter VI. JOPP provides a proven process to organize the work of the commander, staff, subordinate commanders, and other partners, to develop plans that will appropriately address the problem to be solved. It focuses on defining the military mission and development and synchronization of detailed plans to accomplish that mission. Commanders and staffs can apply the thinking methodology to discern the correct mission, develop creative and adaptive CONOPS to accomplish the mission, and synchronize those CONOPS so that they can be executed. Together with operational design, JOPP facilitates interaction between the commander, staff, and subordinate and supporting headquarters throughout planning.

a. JOPP is a four-function, seven-step process that culminates with a published Operations Order (OPORD) in CAP and results in an OPLAN, Concept Plan (CONPLAN), Base Plan or CDRs Estimate during contingency planning.

b. The four functions of the JOPP are: Strategic Guidance, Concept Development, Plan Development, and Plan Assessment. Each of these functions are further broken down into steps (Figure IX-2 and IX-3).

(1) Function I – Strategic Guidance consists of two steps: Planning Initiation and Mission Analysis.

(2) Function II – Concept Development consists of four steps: COA Development, COA Analysis and Wargaming, COA Comparison and COA Approval.

(3) Function III – Plan Development consists of Plan or Order Development.

(4) Function IV – Plan Assessment.

e. JOPP underpins planning at all levels and for missions across the full range of military operations. It applies to both supported and supporting CCDRs and to joint force component commands when the components participate in joint planning. This process is designed to facilitate interaction between the CCDR, staff, and subordinate headquarters throughout planning. JOPP helps CCDRs and their staffs organize their planning activities, share a common understanding of the mission and CCDR's intent, and develop effective plans and orders.

f. Operational design and JOPP are complementary elements of the overall planning process. As discussed in Chapter VI operational design provides an iterative process that allows for the commander's vision and mastery of operational art to help planners answer ends–ways–means–risk questions and appropriately structure campaigns and operations. The commander, supported by the staff, gains an understanding of the operational environment, defines the problem, and develops an operational approach for the campaign or operation through the application of operational design during the initiation step of JOPP. Commanders communicate their operational approach to their staff, subordinates, supporting commands, agencies, and multinational/nongovernmental entities as required

[1] *JP 5-0, Joint Operation Planning.*

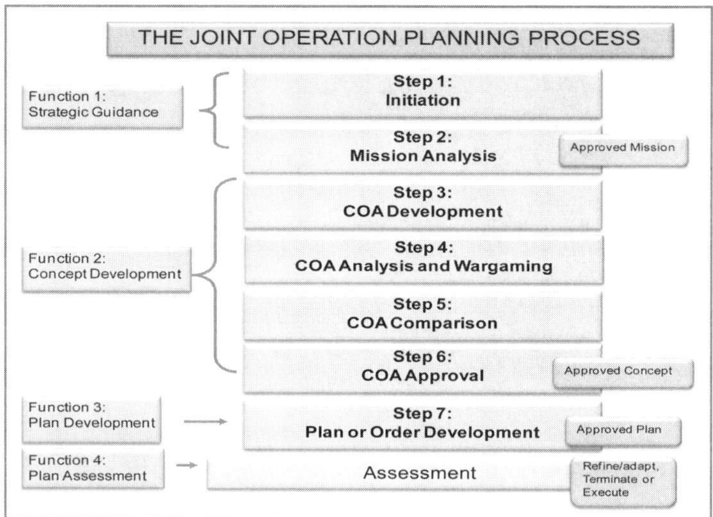

Figure IX-2. The Joint Operation Planning Process

***Example**: A CCMD might prepare a level 4 OPLAN per JOPES to meet a CPG requirement. The CCMDs Service components might provide input to the plan (TPFDD and other information), but will develop their component plans based on tasks given to them in the CCDR's plan. They will do their mission analysis and COA development, conduct their COA analysis, wargaming, and comparison, etc. They will consider various elements of operational design in conjunction with JOPP as they develop their plans. The JOPP is the fundamental process for all joint planning. Deliberate and CAP are structured using the JOPP.*

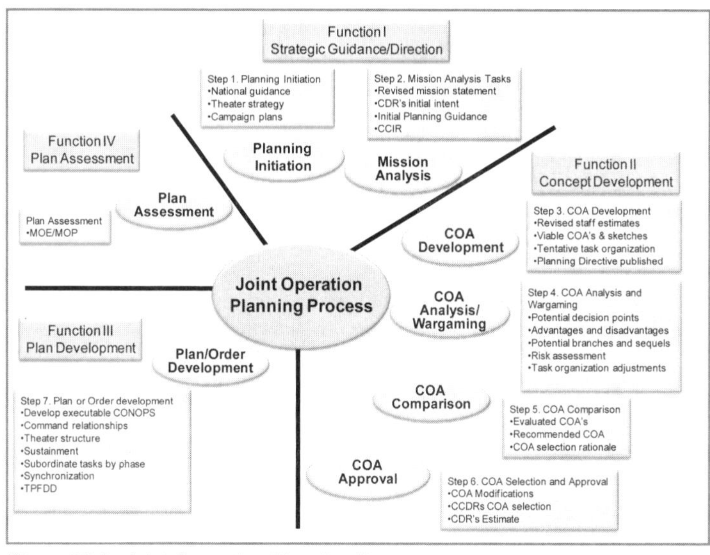

Figure IX-3. Joint Operation Planning Process

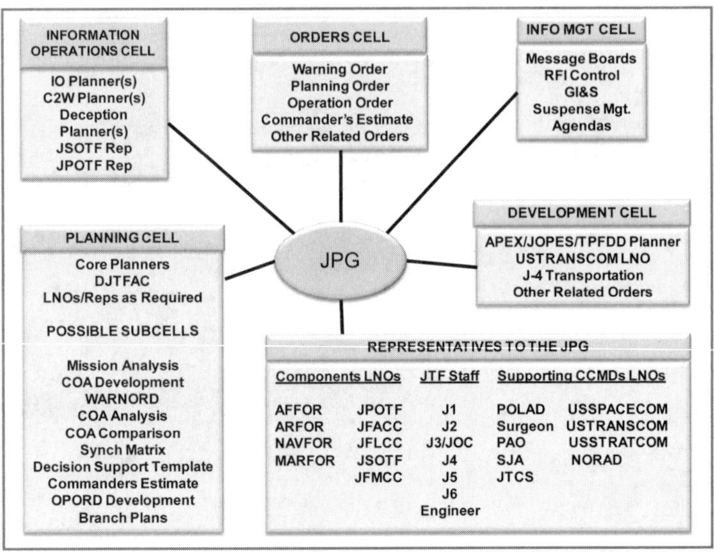

Figure IX-4. Example, Joint Planning Group

in their initial planning guidance so that their approach can be translated into executable plans. As JOPP is executed, commanders learn more about the operational environment and the problem and refine their initial operational approach. Commanders provide their updated approach to the staff to guide detailed planning. This iterative process between the commander's maturing operational approach and the development of the mission and CONOPS through JOPP facilitates the continuing development of possible COAs and their refinement into eventual CONOPS and executable plans.

2. Joint Planning Group

The Joint Planning Group (JPG) is typically organized within the J-5 Directorate. The JPG is responsible to the J-5 and CCDR for driving the command's planning effort. Effectiveness of the JPG will be measured, in part, by the support provided to it by the principal CCMD staff officers (J-1 through J-6). The composition of the JPG is a carefully balanced consideration between group management and appropriate representation from the JTF staff and components. JPG membership will vary based on the tasks to be accomplished, time available to accomplish the tasks, and the experience level of the JPG members. Representation to the JPG should be a long-term assignment to provide continuity of focus and consistency of procedure (Figure IX-4 is an example of a JPG).

3. Staff Estimates[2]

Staff Estimates are absolutely vital to establish and maintain an overall high degree of coordination and cooperation, both internally and with staffs of higher, lower, and adjacent units. The staff continues throughout situational awareness, planning and execution. The CCDR's staff must function as a single, cohesive unit—a professional team. Each staff member must know his own duties and responsibilities in detail and be familiar with the duties and responsibilities of other staff members.

[2]Staff estimate format. The staff estimate format contained in CJCSM 3122.01A, Appendix T, standardizes the way staff members construct estimates. The J2 (with input assistance from all staff members) will still conduct and disseminate the initial Joint Intelligence Preparation of the Operational Environment as a separate product. The Commander's Estimate format is located as enclosure J of the same document.

a. The staff's efforts must always focus on supporting the CCDR and on helping support the subordinate units. CDRs can minimize risks by increasing certainty. The staff supports the CCDR by providing better, more relevant, timely, and accurate information; making estimates and recommendations; preparing plans and orders; and monitoring execution.

b. The primary product the staff produces for the CCDR, and for subordinate CCDRs, is understanding, or situational awareness. True understanding should be the basis for information provided to CCDRs to make decisions. Formal staff processes provide two types of information associated with understanding and decision making. All other staff activities are secondary. The first is situational awareness information, which creates an understanding of the situation as the basis for making a decision. Simply, it is understanding oneself, the enemy, and the terrain or environment. The second type of information, execution information, communicates a clearly understood vision of the operation and desired outcome after a decision is made. Examples of execution information are conclusions, recommendations, guidance, intent, concept statements, and orders.

c. Mission analysis, facts and assumptions, and the situation analysis (of the area of operations, area of interest, adversary, friendly, and support requirements, etc.) furnish the structure for the staff estimate. The estimate consists of significant facts, events, and conclusions based on analyzed data. It recommends how to best use available resources. Adequate, rapid decision-making and planning hinge on good, timely command and staff estimates. They are the basis for forming viable courses of action. Failure to make estimates can lead to errors and omissions when developing, analyzing, and comparing COAs, developing or executing plans.

d. Essential Qualities of Estimates.

(1) CCDRs control tempo by making and executing decisions faster than the adversary. Therefore, CCDRs must always strive to optimize time available. They must not allow estimates to become overly time-consuming. However, they must be comprehensive and continuous and must visualize the future.

(2) Comprehensive estimates consider both the quantifiable and the intangible aspects of military operations. They translate friendly and adversary strengths, joint weapons systems, training, morale, and leadership into combat capabilities. The estimate process requires a clear understanding of the operational environment and the ability to visualize the operational or crisis situations requiring military forces or interagency support. Estimates must provide a timely, accurate evaluation of the operation at a given time.

(3) The demand on the C2 system is continuous as opposed to cyclical. Estimates must be as thorough as time and circumstances permit. The CCDR and staff must constantly collect, process, and evaluate information. They update their estimates:

- When the CCDR and staff recognize new facts.
- When they replace assumptions with facts or find their assumptions invalid.
- When they receive changes to the mission or when changes are indicated.

(4) Estimates for the current operation can often provide a basis for estimates for future missions as well as changes to current operations. Technological advances and near-real-time information estimates ensure that estimates can be continuously updated.

(5) Estimates must visualize the future and support the CCDR's operational visualization. They are the link between current operations and future plans. The CCDR's vision directs the endstate. Each subordinate unit CCDR must also possess the ability to envision the organization's endstate. Estimates contribute to this vision.

Types of Estimates

The CCDR and his staff make estimates that apply to any operational situation and all levels of command. They use estimates to look at possible solutions to specific operational missions and requirements. These estimates can form the cornerstone for staff annexes to orders and plans. The coordinating staff and each staff principal develop facts, assessments, and information that relate to their functional field or operating system. Types of estimates generally include, but are not limited to:

- CDR's estimate.
- Operations estimate.
- Personnel estimate.
- Intelligence estimate.
- Logistics estimate.
- Civil-military operations estimate.
- Signal estimate.
- Special staff estimates.

1. Commander's Estimate

The CDR's estimate, like the operations estimate, is an analysis of all the factors that could affect a mission. The CDR integrates his personal knowledge of the situation, his analysis of METT-T factors, the assessments of his subordinate CDRs, and any relevant details he gains from his staff. Once the CCDR has made a decision on a selected COA, provides guidance, and updates his intent, the staff completes the CDR's Estimate. The CDR's Estimate provides a *concise narrative statement* of how the CCDR intends to accomplish the mission, and provides the necessary focus for campaign planning and OPLAN/OPORD development. Further, it responds to the establishing authority's requirement to develop a plan for execution.

Estimate analysis includes risk assessment, force protection, and effective utilization of all resources. The estimate also includes visualizing all reasonable COAs and how each COA would affect friendly forces.

The CCDR's and operations estimates generally follow the same format. The CCDR uses his personal estimate as a cross-check of his staff's estimates.

2. Operations Estimate

The J3 prepares the operations estimate, which considers all elements that can influence the current operations and feasible future courses of action. It results in recommendations to the CCDR. To prepare this estimate, the J3 must *understand*:

- Strategic direction.
- CDR's intent (one and two echelons above).
- Risk assessment.
- Current task organization (two echelons below).
- Joint operational status of supporting commands/components, such as locations, capabilities (including level of training, effectiveness, degree of mobility, type of equipment, and limitations), and current or pending missions.
- Availability and capabilities of joint assets, such as air and space support, naval or amphibious assets.
- Other information, such as location, status, and mission of Transportation Command and other supporting commands.

Ref: FM 100-2-3 The Soviet Army: Troops, Organization and Equipment, p. 4-8.

3. Personnel Estimate

The J1 prepares the personnel estimate, which is an analysis of how all human resources and personnel factors impact effectiveness before, during, and after the operation. It includes a current overall personnel status of the joint organization, its subordinate commands, and any attached or supporting elements. Personnel status includes assessments of the following tangible and intangible factors:

- Medical evacuation and hospitalization.
- Command-strength maintenance.
- Replacements.
- Readiness.
- Organizational climate.
- Cohesion.
- Discipline, law and order.

The personnel estimate predicts losses (where and when losses could occur) and when, where, and if such losses cause the culmination of an operation. It contains the J1's conclusions and recommendations about the feasibility of supporting the operation.

4. Intelligence Estimate

Intelligence plays a critical role across the range of military operations. CCDRs use intelligence to anticipate the battle, visualize and understand the full spectrum of the battlespace, and influence the outcome of operations. Intelligence enables CDRs at all levels to focus their combat power and to provide full-dimensional force protection across the range of military operations.

Intelligence focuses on enemy military capabilities, centers of gravity (COGs), and potential COAs to provide operational and tactical CDRs the information they need to plan and conduct operations. It enables the CCDR to visualize, understand, and identify when and where to apply combat power to exploit enemy vulnerabilities and capitalize on opportunities with minimum risk. The J-2 must modify and tailor intelligence support to meet the unique challenges presented in each operation. In addition, the nature and intensity of a potential threat can change suddenly and dramatically. For example, a peacekeeping operation may abruptly transition to a combat peace enforcement operation should any of the belligerents fail to honor the terms of the truce. Therefore, intelligence resources at every echelon should be structured to provide support that is proactive, aggressive, predictive, and flexible.

Joint intelligence operations begin with the *identification of a need for intelligence regarding all relevant aspects of the battlespace, especially the adversary.* These intelligence needs are identified by the CCDR and all joint force staff elements, and are formalized by the J-2 as intelligence requirements early in the planning process. Those critical pieces of intelligence the CCDR must know by a particular time to plan and execute a successful mission are identified as the CDR's PIRs. PIRs are identified at every level and are based on guidance obtained from the mission statement, the CDR's intent, and the endstate objectives. *Intelligence requirements* provide the basis for current and future intelligence operations, and are prioritized based on consumer inputs during the planning and direction portion of the intelligence process. The J-2 provides the focus and direction for collection requirements to support the CCMD or subordinate joint force. The J2 prepares the intelligence estimate. Both the J2 and the J3 examine the area of interest to identify intelligence-collection needs. It's important to note that Joint Intelligence Preparation of the Operational Environment (JIPOE) is a staff process, not just a J2 process, and should be driven by the chief of staff.

Continued on next page.

Continued on next page

Continued on next page

Continued from previous page

Types of Estimates (cont.)

5. Logistics Estimate

The J4 prepares the logistics estimate, which provides an accurate and current assessment of logistical capabilities, environment, subordinate command capabilities, and any attached or supporting elements. The logistics estimate is an analysis of how logistical factors can affect mission accomplishment. It contains the J4's conclusions and recommendations about the feasibility of supporting the operation. This estimate includes how the functional areas of supply, transportation, services, maintenance, labor, facilities, and construction affect various COAs.

6. Civil-Military Operations (CMO) Estimate

The J5 prepares the Joint CMO estimate in relation to the operation and environment.

7. Signal Estimate

The J6 prepares the communications estimate in relation to the operational requirements and environment.

8. Special Staff Estimates

Each special staff officer creates their own staff estimate in relation to the situation and their functional responsibilities.

Staff estimate format. *The staff estimate format contained in CJCSM 3122.01A, Appendix T, standardizes the way staff members construct estimates. The J2 (with input assistance from all staff members) will still conduct and disseminate the initial Joint Intelligence Preparation of the Operational Environment as a separate product. The Commander's Estimate format is located as enclosure J of the same document.*

Continued from previous page

Refer to The Joint Forces Operations & Doctrine SMARTbook (Guide to Joint, Multinational & Interagency Operations) for further discussion of staff estimates. Additional topics include joint doctrine fundamentals, joint operations and unified action, the range of military operations joint operation planning, joint logistics, joint task forces, information operations, multinational operations, and IGO/NGO coordination.

Strategic Guidance & Strategic Direction 10

> "Strategic thought is inevitably highly pragmatic. It is dependent on the realities of geography, society, economics, and politics, as well as on other, often fleeting factors that give rise to the issues and conflicts war is meant to resolve."
>
> Peter Paret, Makers of Modern Strategy, 1986, Princeton University Press, p.3

1. Function I — Strategic Guidance/Strategic Direction

Strategic Guidance/Strategic Direction is the common thread that integrates and synchronizes the activities of the JS, CCMDs, Services, and CSAs with those of the other national agencies and departments. The NMS is informed by the NSS, QDR/NDS. While all true strategies have an ends-ways-means-risk approach, these national strategies (Figure X-1) tend to focus on one aspect of strategy more than others:

- The NSS is heavily weighted towards ENDS.
- The QDR/NDS focuses on MEANS.
- The NMS usually focuses on the WAYS and RISK.
- The commonality between them is national interests.

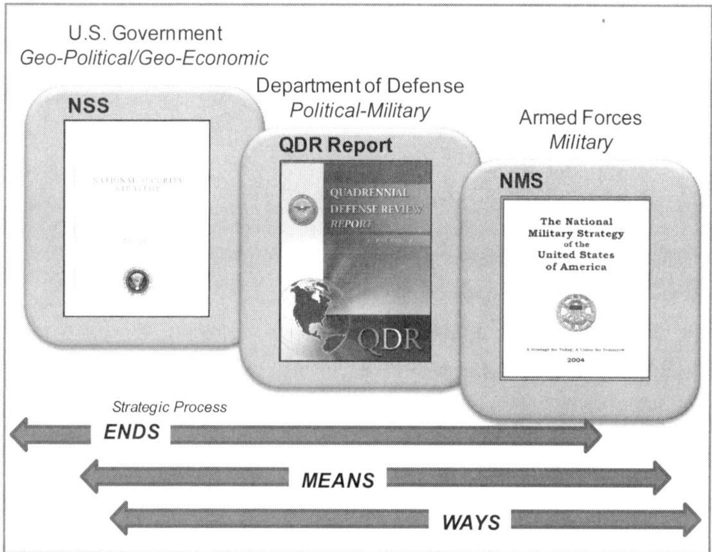

Figure X-1. Strategic Context and Linkages.

a. As an overarching term, *strategic direction* encompasses the processes and products by which the President, SECDEF, and CJCS provide *strategic guidance.*[1] Four key documents that provide that integration are the Unified Command Plan (UCP), GEF, the JSCP, and the GFMIG.

[1] *JP 5-0, Joint Operations Planning.*

2. National Defense Strategy (NDS): Ends, Ways, Means

a. The NDS describes the broad policy and strategic context within which the DOD operates and establishes key department-level planning parameters and priorities. See facing page for example of ends, ways, and means.[2]

b. Using the QDR[3] as its foundation, the DOD continually examines its approach – from objectives to capabilities and activities to resources – to ensure its best alignment for the nation, its allies and partners, and our uniformed military.

Example from 2008 NDS

The NDS identifies five major "ends," achieved by five "ways" that employ eight "means."[72]

1. Ends

Five ends guide defense policy and planning:

- Defend the Homeland.
- Win the Long War.
- Promote security.
- Deter conflict.
- Win our nation's wars.

2. Ways

Five ways achieve the NDS's ends:

- Shape the choices of key states.
- Prevent adversaries from acquiring or using WMD.
- Strengthen and expand alliances and partnerships.
- Secure U.S. strategic access and retain freedom of action.
- Integrate and unify our efforts: a new "Jointnesss."

3. Means

At the broadest level the DOD employs eight means to achieve its ends:

- The Total Force.
- Strategic communications.
- Intelligence and information.
- Organizational excellence.
- First-class technology and equipment.
- Alliances and partnerships.
- Security cooperation.
- Global posture.

b. These four documents are the principle sources of guidance for CCMDs steady-state campaign, deliberate and posture planning efforts. The GEF, guidance from the President and SECDEF, supersedes the Contingency Planning Guidance, Security Cooperation Guidance, and the Nuclear Weapons Planning Guidance and various policy memoranda related to GFM and Global Defense Posture (GDP). The GEF translates the NSS and the QDR/NDS into guidance that supports planning and efforts and complements the security goals outlined in the DOS Joint Strategic Plan (JSP). The intent is to link planning sub-systems and activities together to allow the DOD to provide CCDRs with realistic objectives, priorities, assumptions and resources for employing the force.

3. Guidance for Employment of the Force (GEF), Global Force Management Implementation Guidance (GFMIG) and Joint Strategic Capabilities Plan (JSCP)

a. Guidance for Employment of the Force. The GEF directs CCDRs to develop campaign plans designed to accomplish assigned strategic objectives. Campaign plans integrate steady-state security cooperation activities, "Phase 0" activities, and ongoing operations. The goal is to consolidate and integrate DOD planning guidance related to operations and other military activities into a single, overarching document. The GEF transitions the DOD's planning from a contingency-centric approach to a strategy-centric approach. Rather than initiating planning from the context of particular contingences, the strategy-centric approach requires commanders to begin planning from the perspective of achieving broad regional or functional objectives.

(1) The GEF is revised every two years with updates issued in interim years. USD(P), after SECDEF review and approval, will issue interim Strategic Guidance Statements (SGS), as required, to provide updated planning guidance.

(2) Under this approach, planning starts with the NDS, from which this document derives theater or functional strategic objectives prioritized appropriately for each CCDR.

(3) CCDRs are required to pursue these strategic objectives as they develop their theater or functional strategies, which they then translate into an integrated set of steady-state activities and operations by means of a campaign plan.

(4) Campaign plans provide the vehicle for linking steady-state shaping activities to current operations and contingency plans. They ensure that the various "Phase 0" components of CCMD contingency plans are integrated with each other and the command's broader security cooperation and shaping activities.

(5) Under this concept, contingency plans become "branches" to the campaign plan. Contingency plans are built to account for the possibility that steady-state shaping measures, security cooperation activities, and operations could fail to prevent aggression,

[2] National Defense Strategy

[3] The QDR statute stipulates that "each quadrennial defense review shall be conducted so as—(1) to delineate a national defense strategy consistent with the most recent NSS prescribed by the President pursuant to section 108 of the NSA of 1947 (50 U.S.C. 404a)." A statement of U.S. national defense strategy is not required as a separate document. The only statutory requirement is the mandate that the QDR "delineate a national defense strategy" and that the report on the QDR include "a comprehensive discussion of the national defense strategy of the United States." The first two QDR reports contained statements of defense strategy. In 2005, however, the DOD released a separate NDS document, which served as a basis for the 2006 QDR. DOD has since continued the practice, releasing a revised NDS statement in June 2008 in advance of the 2009 review process that concluded with the 2010 QDR report. The 2008 NDS articulated themes that were reflected in the strategic priorities of the 2009-2010 QDR, though with some significant changes in emphasis. (CRS,QDR 2010: Overview and Implications for National Security Planning. Stephen Daggett, May 2010).

preclude large-scale instability in a key state or region, or mitigate the effects of a major disaster. Contingency plans address scenarios that put one or more U.S. strategic objectives in jeopardy and leave the United States no other recourse than to address the problem at hand through military operations. Military operations can be in response to many scenarios, including armed aggression, regional instability, a humanitarian crisis, political bravado, or a natural disaster. Contingency plans should provide a range of military options coordinated with total USG response.

b. Global Force Management Implementation Guidance. The GFMIG integrates the complementary force assignment, apportionment, and allocation processes used to align U.S. forces, into a single document to improve the DOD's ability to manage forces from a global perspective. As such, the GFMIG provides force management implementation guidance and complements the GEF. The "Forces For" Unified Commands memorandum is incorporated within Section II of the GFMIG. This document is updated every two years or as required to reflect recommendations contained in forthcoming QDRs.

c. Joint Strategic Capabilities Plan. The JSCP normally will not repeat the guidance that is presented in the GEF. The two documents are complementary, not repetitive. The JSCP implements the strategic policy direction provided in the GEF and initiates the planning process for the development of campaign, campaign support, contingency, and posture plans. The JSCP does this by translating and consolidating GEF regional and functional guidance into specific campaign and deliberate planning requirements to CCDRs. The JSCP provides specific planning tasks that link to GEF strategic objectives, priorities, and security cooperation activities. The CCDR is expected to develop intermediate military objectives that contribute to the achievement of the GEF strategic objectives.

4. Strategic Guidance Focus

Strategic guidance will focus largely on solidifying guidance, agreeing on the framework assumptions and planning factors, establishing a common understanding of adversaries and their intentions, conducting initial interagency and/or coalition coordination (as authorized), and producing an approved CCDR mission statement. These outcomes form the foundation for continued planning. Strategic guidance is not always written, it can and will take other forms as well (i.e., State of the Union, presidential speeches, press releases, NSS, PD's, etc.).

5. Strategic Direction Focus

Strategic direction and supporting national-level activities, in concert with the efforts of CCDRs, ensure the following:

- National strategic objectives and termination criteria are clearly defined, understood, and achievable.
- AC is ready for combat and RC are appropriately manned, trained, and equipped in accordance with Title 10 USC responsibilities and prepared to become part of the total force upon mobilization.
- Intelligence, surveillance, and reconnaissance systems and efforts focus on the operational environment.
- Strategic guidance is current and timely;
- DOD, other intergovernmental organizations, allies, and coalition partners are fully integrated at the earliest time during planning and subsequent operations;
- All required support assets are ready.
- Multinational partners are available and integrated early in the planning process.
- Forces and associated sustaining capabilities deploy ready to support the commander's CONOPS.

Planning Initiation 11

Step #1	PLANNING INITIATION
Step #2	MISSION ANALYSIS
Step #3	COA DEVELOPMENT
Step #4	COA ANALYSIS AND WARGAMING
Step #5	COA COMPARISON
Step #6	COA APPROVAL
Step #7	PLAN OR ORDER DEVELOPMENT

1. Step 1 to JOPP — Planning Initiation

a. Linkage between Plan Initiation and National Strategic Endstate. The first step in the Joint Operation Planning Process (JOPP) is *Initiation*. Prior to joint operations, planning begins when an appropriate authority recognizes a potential for military capability to be employed in response to a contingency or crisis. JOPP begins when the President, SECDEF or CJCS decides on potential military options and directs CCDRs through guidance contained in the GEF, JSCP, Strategic Communication Guidance, and related strategic guidance statements (when applicable) to begin planning. However, CCDRs and other CDRs may initiate planning on their own authority when they identify a planning requirement not directed by higher authority. The CJCS may also issue a Warning Order in an actual crisis. Military options normally are developed in combination with other nonmilitary options so that the President can respond with all the appropriate instruments of national power.[1] For contingency plans and deliberate planning purposes, the JSCP serves as the primary guidance to begin planning and COA development. Planning is continuous once execution begins. However, planning initiation during execution is still relevant when there are significant changes to the current mission or the commander receives a mission for

[1] *JP 5-0, Joint Operations Planning.*

follow-on operations. The J-5 typically focuses on future planning of this nature while the J-3 focuses on current operations.

b. Strategic endstates and objectives are approved by the President with input from his closest advisors, staff, and administration officials. These strategic measures form the foundation that subordinate agencies, departments, and military planners use to develop strategic objectives that will support the overarching desired national endstate. A clear understanding of desired political goals and endstate is imperative at the strategic level to ensure that all elements of national power are applied effectively. For the military, clear delineation of strategic endstate is essential to ensuring that military force can be effectively and efficiently applied when necessary to support strategic success.

c. Current military doctrine recognizes three levels of war: strategic, operational and tactical. These three levels overlap. Planning and execution at each level is reliant on planning and execution at other levels. Clearly delineated strategic endstates and objectives forms the nucleus from which military plans at all levels evolve. Proper or improper identification of strategic endstates and objectives affect how military leaders plan and utilize military power to support attainment of those strategic measures. An incorrect interpretation of the strategic measure can lead to failure to accomplish the desired mission.

d. During planning initiation, deliberate planning tasks are transmitted, forces and resources are apportioned, and planning guidance is issued to the supported CCDR. During deliberate planning, CCDR's prepare contingency plans primarily in direct response to tasking in the JSCP.

e. When planning for crises, the CDR and staff will perform an assessment of the initiating directive to determine *time available* until mission execution, the current status of intelligence products and staff estimates, and other factors relevant to the specific planning situation. The CDR typically will provide **initial planning guidance,** which could specify time constraints, outline initial coordination requirements, authorize movement of key capabilities within the commander's authority, and direct other actions as necessary to provide the commanders current understanding of the operational environment (OE), the problem, and operational approach for the campaign or operation.

f. Strategic requirements or tasking for the planning of major contingencies may require the preparation of several alternative plans for the same requirement using different sets of forces and resources in order to preserve flexibility. For these reasons, contingency plans are based on reasonable assumptions (*hypothetical situation with reasonable expectation of future action*). See Chapter 7 (GFM) for detailed information on force planning.

Plans forecast, but do not predict.

A plan is a continuous, evolving framework of anticipated actions that guides subordinates through each phase of the joint operation. Any plan is a framework from which to adapt, not an exact blueprint. The measure of a good plan is not whether execution transpires as planned, but whether the plan facilitates effective action in the face of unforeseen events. Good plans foster initiative, account for uncertainty and friction, and mitigate risk.

Mission Analysis: Overview

12

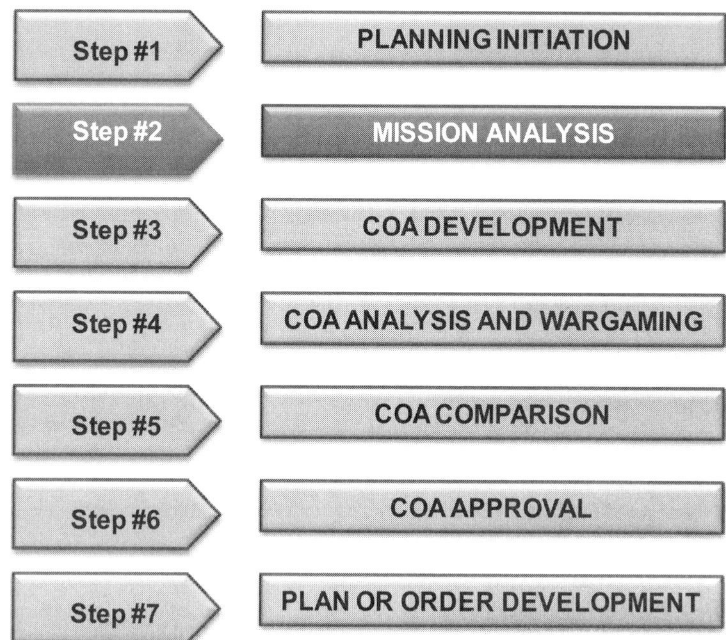

1. Step 2 to JOPP — Mission Analysis

The mission analysis process helps to build a common understanding of the problem to be solved and boundaries within which to solve it by key stakeholders. Mission analysis is used to study the assigned mission and to identify all tasks necessary to accomplish it. Mission analysis is critical because it provides direction to the commander and the staff, enabling them to focus effectively on the problem at hand.

> ***Primary products*** *of mission analysis are a restated mission statement, the initial intent statement, CDR's Critical Information Requirements (CCIR), and initial planning guidance (IPG).*

a. The CCDR is responsible for analyzing the mission and restating the mission for subordinate CDRs to begin their own estimate and planning efforts. Mission analysis is used to study the assigned mission and to identify all tasks necessary to accomplish it. Mission analysis is critical because it provides direction to the CCDR and the staff, enabling them to focus effectively on the problem at hand. *There is perhaps no step more critical to the JOPP and a succesful plan.*

One product of the mission analysis process is the mission statement. Your initial mission analysis as a staff will result in a "tentative" mission statement. This tentative mission statement is a recommendation for the commander based on mission analysis. This recommendation is presented to the commander for approval normally during the Mission Analysis Brief.

b. A primary consideration for a supported CCDR during mission analysis is the *national strategic endstate* and that set of national objectives and related guidance that define strategic success from the President's perspective. The endstate and national objectives will reflect the broadly expressed Political, Military, Economic, Social, Informational, Infrastructure (PMESII) and other circumstances that should exist after the conclusion of a campaign or operation. The CCDR also must consider multinational objectives associated with coalition or alliance operations.

c. The supported CCDR typically will specify a *theater strategic endstate*. While it will mirror many of the objectives of the national strategic endstate, the theater strategic endstate may contain other supporting objectives and conditions. This endstate normally will represent a point in time and/or circumstance beyond which the President does not require the military instrument of national power as the primary means to achieve remaining objectives of the national strategic endstate.

d. CCDRs include a discussion of the national strategic endstate and objectives in their initial planning guidance. This ensures that joint forces understand what the President wants the situation to look like at the conclusion of U.S. involvement. The CCDR and subordinate JFCs typically include the *military endstate* in their CDR's intent statement.[1]

e. During mission analysis, it is essential that the tasks (specified and essential task(s)) and their purposes are clearly stated to ensure planning encompasses all requirements; limitations (restraints-can't do, or constraints–must do) on actions that the CCDR or subordinate forces may take are understood; and the correlation between the CDRs' mission and intent, and those of higher, and other CDRs is understood.

f. The joint force's mission is the task or set of tasks, together with the purpose, that clearly indicates the action to be taken and the reason for doing so. The CCDR and staff can accomplish mission analysis through a number of logical tasks. Of these two, the *purpose* is preeminent. The CCDR can adjust his task to ensure he accomplishes the purpose. This is a critical aspect of *mission type orders* and the ability of subordinate CDRs to re-task themselves during rapidly changing circumstances and still fulfill the CDR's intent.

g. While all of these tasks will be addressed during the plan development process, it is critical to focus on the mission essential task(s) to ensure unity of effort and maximum use of limited resources. The mission essential task(s) defines success of the assigned mission.

Auftragstaktik (Mission Type Order): *Order issued to a lower unit that includes the accomplishment of the total mission assigned to the higher headquarters, or one that assigns a broad mission (as opposed to a detailed task), without specifying how it is to be accomplished.*

[1] *JP 5-0, Joint Operations Planning.*

13 Key-Steps to Mission Analysis

Key-Step — 1: Analyze Higher CDR's Mission and Intent

Key-Step — 2: Task Analysis, Determine Own Specified, Implied, and Essential Tasks

Key-Step — 3: Determine Known Facts, Assumptions, Current Status, or Conditions

Key-Step — 4: Determine Operational Limitations

- Constraints – "Must do"
- Restraints – "Can't do"

Key-Step — 5: Determine Own Military Endstate, Termination Criteria, Objectives and Initial Effects

Key-Step — 6: Determine Own and Enemy's Center(s) of Gravity, Critical Factors and Decisive Points

Key-Step — 7: Conduct Initial Force Structure Analysis (Apportioned Forces)

Key-Step — 8: Conduct Initial Risk Assessment

Key-Step — 9: Determine CDR's CCIR

- CFFI
- PIR

Key-Step — 10: Develop Tentative Mission Statement

Key-Step — 11: Develop Mission Analysis Brief

Key-Step — 12: Prepare Initial Staff Estimates

Key-Step — 13: Publish CDR's Planning Guidance and Initial Intent

**NOTE:* Previous JP 5-0, Joint Operations Planning, have listed numerous Mission Analysis Key-Steps. This document recognizes and addresses all Key-Steps of Joint Doctrine, and condenses them into the following 13 Key-Steps. This is done to allow a logical flow for planners to follow.*

h. Although some Key-Steps occur before others, mission analysis typically involves substantial parallel processing (spiral development) of information by the CDR and staff, particularly in a crisis situation. A primary example is the Joint Intelligence Preparation of the Operational Environment (JIPOE). JIPOE is a continuous process that includes defining the operational environment, describing the effects of the operational environment, evaluating the adversary, and determining and describing adversary potential and most dangerous COA(s). This planning process must begin at the earliest stage of campaign or operations planning and must be an integral part of, not an addition to, the overall planning effort. This is also true for logistics, medical, transportation, force and deployment planning to name just a few.

2. Planner Organization

Organizing the planning team and setting goals and objectives within a specific timeline can sometimes be as time consuming as the plan itself. If you enter the planning process with the following information/guidelines defined and understood, the actual planning process will go much smoother:

- Define clear organization planning responsibilities.
  - Who leads what efforts (topic, geographic, functions).
- Define the process for planning.
  - Planning organizations.
  - Product production/transition between organizations.
- Information flow.
  - Higher.
  - Lower.
  - Adjacent.
- Integrate other elements.
  - Coalition.
  - Interagency.
  - Host nation.
- Be sustainable in a 24/7/365 cycle.
  - Rapidly integrate augmentees.

A carelessly planned plan will take three times longer to complete than a carefully planned plan

Mission Analysis: JIPOE & IPIE

13

MISSION ANALYSIS and JOINT INTELLIGENCE PREPARATION of the OPERATIONAL ENVIRONMENT (JIPOE) and INTELLIGENCE PREPARATION of the INFORMATION ENVIRONMENT (IPIE)

> *"Nothing is more worthy of the attention of a good general than the endeavor to penetrate the designs of the enemy."*
>
> *Machiavelli, Discourses, 1517*

1. Overview

a. Joint Intelligence Preparation of the Operational Environment (JIPOE) is the analytical process used by joint intelligence organizations to produce intelligence assessments, estimates, and other intelligence products in support of the commander's decision-making process. It is a continuous process that involves four major steps: (1) defining the total operational environment; (2) describing the impact of the operational environment; (3) evaluating the adversary; and (4) determining and describing adversary potential courses of action (COAs), particularly the adversary's most likely COA and the COA most dangerous to friendly forces and mission accomplishment.

b. The process is used to analyze the physical domains (air, land, maritime and space); the information environment (which includes cyberspace), political, military, economic, social, information, and infrastructure (PMESII) systems; and all other relevant aspects of the operational environment, and to determine an adversary's capabilities to operate within that environment. JIPOE products are used by joint force, component, and supporting command staffs in preparing their estimates and are also applied during the analysis and selection of friendly COAs.[1]

2. JIPOE and the Intelligence Cycle

a. JIPOE is a dynamic process that both supports, and is supported by, each of the categories of intelligence operations that comprise the intelligence process, see Figure XIII-1(JP 2-01-3).

(1) JIPOE and Intelligence Planning and Direction. The JIPOE process provides the basic data and assumptions regarding the adversary and other relevant aspects of the operational environment that help the CDR and staff identify intelligence requirements, information requirements, and collection requirements. By identifying known adversary capabilities, and applying those against the impact of the operational environment, JIPOE provides the conceptual basis for the CDR to visualize and understand how the adversary might threaten the command or interfere with mission accomplishment. This analysis forms the basis for developing the commander's priority intelligence requirements (PIRs), which seek to answer those questions the CDR considers vital to the accomplishment of the assigned mission. Additionally, by identifying specific adversary COAs and Centers of

[1] *JP 2-01-3, Joint Intelligence Preparation of the Operational Environment.*

Gravity (COGs), JIPOE provides the basis for wargaming in which the staff "fights" each friendly and adversary COA. This wargaming process identifies decisions the CDR must make during execution and allows the J-2 to develop specific intelligence requirements to facilitate those decisions. JIPOE also identifies other critical information gaps regarding the adversary and other relevant aspects of the operational environment, which form the basis of a collection strategy that synchronizes and prioritizes collection needs and utilization of resources within the phases of the operation.[2]

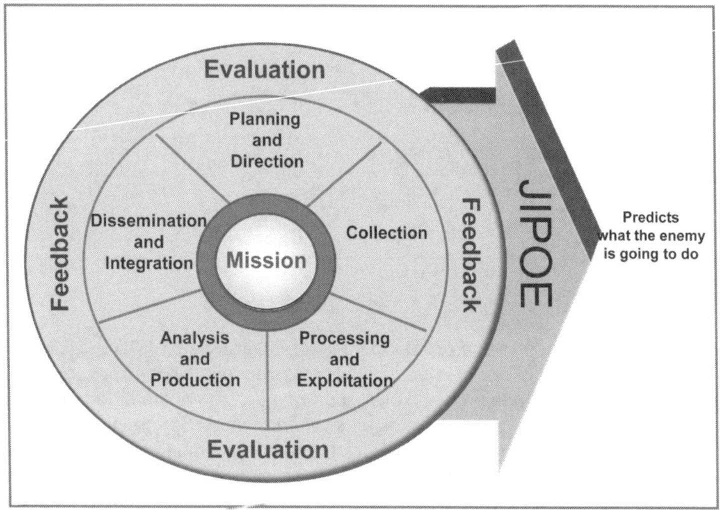

Figure XIII-1. JIPOE and the Intellegence Cycle.

(2) JIPOE and Intelligence Collection. JIPOE provides the foundation for the development of an optimal intelligence collection strategy by enabling analysts to identify the time, location, and type of anticipated adversary activity corresponding to each potential adversary COA. JIPOE products include several tools that facilitate the refinement of information requirements into specific collection requirements. JIPOE templates facilitate the analysis of all identified adversary COAs and identify named areas of interest (NAIs) where specified adversary activity, associated with each COA, may occur. JIPOE matrices are also produced that describe the indicators associated with each specified adversary activity. In addition to specifying the anticipated locations and type of adversary activity, JIPOE templates and matrices also forecast the times when such activity may occur, and can therefore facilitate the sequencing of intelligence collection requirements and the identification of the most effective methods of intelligence collection.

(3) JIPOE and Processing and Exploitation. The JIPOE process provides a disciplined yet dynamic time-phased methodology for optimizing the processing and exploiting of large amounts of data. The process enables JIPOE analysts to remain focused on the most critical aspects of the operational environment, especially the adversary. Incoming information and reports can be rapidly incorporated into existing JIPOE graphics, templates, and matrices. In this way, JIPOE products not only serve as excellent processing tools, but also provide a convenient medium for displaying the most up-to-date information, identifying critical information gaps, and supporting operational and campaign assessments.

[2] See JP 2-0, Joint Intelligence, for a more in-depth discussion of the relationship between intelligence requirements and information requirements. See JP 2-01, Joint and National Intelligence Support to Military Operations, for detailed guidance on the request for information (RFI) process.

(4) JIPOE and Analysis and Production. JIPOE products provide the foundation for the J-2's intelligence estimate. In fact, the JIPOE process parallels the paragraph sequence of the intelligence estimate format.[3]

(5) JIPOE and Dissemination and Integration. The J-2's intelligence estimate provides vital information that is required by the joint force staff to complete their estimates, and for subordinate commanders to continue concurrent planning activities. Timely dissemination of the intelligence estimate is therefore paramount to good operation planning. If time does not permit the preparation and dissemination of a written intelligence estimate, JIPOE templates, matrices, graphics, and other data sources can and should be disseminated to other joint force staff sections, and component and supporting commands, in order to facilitate their effective integration into operation planning. JIPOE geospatial perspectives should also be provided to systems supporting the common operational picture.

(6) JIPOE and Evaluation and Feedback. Consistent with the intelligence process, the J-2 staff continuously evaluates JIPOE products to ensure that they achieve and maintain the highest possible standards of intelligence excellence as discussed in JP 2-0, *Joint Intelligence.* These standards require that intelligence products anticipate the needs of the CDR and are timely, accurate, usable, complete, objective, and relevant. If JIPOE products fail to meet these standards, the J-2 should take immediate remedial action. The failure of the J-2 staff to achieve and maintain intelligence product excellence may contribute to the joint force's failing to accomplish its mission.

e. Roles and Responsibilities. Some critical billets of the JPIOE process are:

(1) CCDR. The CCDR is responsible for ensuring the standardization of JIPOE products within the command and subordinate joint forces, and for establishing theater procedures for collection management and the production and dissemination of intelligence products.

(2) J-2. The J-2 has the primary staff responsibility for planning, coordinating, and conducting the overall JIPOE analysis and production effort at the joint force level. Through the JIPOE process, the J-2 enhances the JFC's and other staff elements' ability to visualize all relevant aspects of the operational environment. The J-2 uses the JIPOE process to formulate and recommend PIRs for the CDR's approval, and develops information requirements that focus the intelligence effort (collection, processing, production, and dissemination) on questions crucial to joint force planning.

(3) CCDR Joint Intelligence Operations Center (JIOC). The JIOC is the focal point for the overall JIPOE analysis and production effort within the CCMD. It is responsible for managing collection requirements related to JIPOE and IPB efforts, and for producing intelligence products for the CCDR and subordinate commanders that support joint operation planning and ongoing operations. The JIOC ensures that the JIPOE production effort is accomplished in conjunction with all appropriate CCMD staff elements, particularly the Geospatial Intelligence (GEOINT), Meteorological and Oceanographic (METOC), and Information Operations (IO) staff officers. The JIOC also ensures that its JIPOE analysis is fully integrated with all IPB and JIPOE products produced by subordinate commands and other organizations. With the assistance of all appropriate joint force staff elements, the JIOC identifies information gaps in existing intelligence databases and formulates collection requirements and RFIs to address these shortfalls. Additionally, the CCMD JIOC may be requested to support another CCDR's federated intelligence requirements, to include JIPOE requirements. As a federated partner, the JIOC must be prepared to integrate into the overall federated intelligence architecture identified by the supported CCDR. All CCMD JIOCs are eligible to participate in federated intelligence support operations.

(4) Subordinate JFC/CDR. The subordinate CDRs clearly state their objectives, CONOPS, and operation planning guidance to their staffs and ensure that the staff fully understands their intent. Based on wargaming and the joint force staff's recommendation, the CDR selects a friendly COA and issues implementing orders. The CDR also approves the

[3] *See JP 2-01.3, JIPOE for greater details.*

list of intelligence requirements associated with that COA. The CDR then identifies those intelligence requirements most critical to the completion of the joint force's mission as PIRs.

(5) JTF Joint Intelligence Support Element (JISE) or JIOC. The intelligence organization at the JTF level is normally a JISE. However, the limited resources of a JISE will usually preclude a full JIPOE effort at the JTF level without substantial augmentation, reliance on reach back capability, and national-level assistance. To overcome this limitation, the CCDR may authorize the establishment of a JTF-level JIOC based on the scope, duration, and mission of the unit or JTF. A JTF JIOC is normally larger than a JISE and is responsible for complete air, space, ground, and maritime order of battle (OB) analysis; identification of adversary COGs; analysis of command and control (C2) and communications systems, targeting support; collection management; and maintenance of a 24-hour watch. Additionally, the JTF JIOC (if formed) serves as the focal point for planning, coordinating, and conducting JIPOE analysis and production at the subordinate joint force level. Most important, DIOCC forward element (DFE) personnel and liaison officers from DOD intelligence organizations provide the JTF JIOC with the means to obtain national support for the JIPOE effort. The JTF JIOC conducts its JIPOE analysis in conjunction with all other appropriate joint force and component command staff elements, particularly the Geospatial Information and Services (GI&S) and METOC staff officers.

(6) Subordinate Component Commands. The intelligence staffs of the subordinate component commands should ensure that appropriate IPB products are prepared for each domain in which the component command operates. Subordinate component commands should evaluate the specific factors in the operational environment that will affect friendly, neutral, and adversary COAs in and around their operational area and impact perceptions and support within their Area of Interest (AOI). More importantly, the analysis of the operational environment should better define those who are potentially friendly, potentially neutral, and potentially adversarial and the actions which would determine their orientation. These component command IPB products provide a level of detail and expertise that the J-2 should not attempt to duplicate, but must draw upon in order to form an integrated or "total" picture of an adversary's joint capabilities and probable COAs. Accordingly, the component commands should coordinate their IPB effort with the J-2 and with other component commands that have overlapping IPB responsibilities. This will ensure their IPB products are coordinated and disseminated in time to support the joint force's JIPOE effort.

(7) The Operations Directorate (J-3) and/or the Plans Directorate (J-5) Representative. The J-3 and/or J-5 ensure that all participants in the JIPOE effort are continuously updated on planning for both current and follow-on missions as well as on any anticipated change to the operational area. The J-3 and/or J-5 representative consolidates information on our own dispositions and provides the cell a clear understanding of friendly COGs, capabilities, and vulnerabilities. The J-3 and/or J-5 will conduct wargames that test friendly COAs against the complete set of adversary COAs developed during the JIPOE process. Based on the results of these wargames, the J-3 and/or J-5 will refine and determine the probability of success of each friendly COA against each adversary COA identified during the JIPOE process, and will make a recommendation to the CDR regarding which friendly COA best accomplishes the joint mission within the CDR's guidance and intent.[4]

> *"Know the enemy, know yourself -- your victory will never be endangered. Know the ground, know the weather -- your victory will then be total."*
>
> *Sun Tzu*
> *The Art of War, C. 500 B.C.*

[4] For more detailed guidance, see JP 2-01, Joint and National Intelligence Support to Military Operations.

3. Joint Intelligence Preparation of the Operational Environment (JIPOE)

a. JIPOE is a continuous process which enables CDRs and their staffs to visualize the full spectrum of adversary capabilities and potential COAs across all dimensions of the operational environment. JIPOE is a process that assists analysts to identify facts and assumptions about the operational environment and the adversary. This facilitates campaign planning and the development of friendly COAs by the joint force staff. JIPOE provides the basis for intelligence direction and synchronization that supports the COA selected by the CDR. JIPOE's main focus is on providing situational awareness and understanding of the operational environment and a predicative intelligence estimate designed to help the CDR discern the adversary's probable intent and most likely and most dangerous COA (Figure XIII-2 (JP 2-0)).

Purpose of Joint Intelliegence

- *Inform the commander*
- *Identify, define, and nominate objectives*
- *Support the planning and execution of operations*
- *Counter adversary deception and surprise*
- *Support friendly deception efforts*
- *Assess the effects of operations*

Figure XIII-2. Purpose of Joint Intelligence.

b. The JIPOE process assists CDRs and their staffs in achieving information superiority by identifying adversary COGs, focusing intelligence collection at the right time and place, and analyzing the impact of the operational environment on military operations. Understanding JIPOE is critical to mission success. Intelligence must be integrated with the overall plan from beginning to end utilizing the JOPP. JIPOE is a product of the intelligence staff estimate and is an integral part of the mission analysis process. JIPOE is the analytical process used by joint intelligence organizations to produce intelligence assessments, estimates, and other intelligence products in support of the joint force CDR's decision-making process. The primary purpose of the JIPOE is to support the CCDR's decision-making and planning by seeking to understand the operational environment and identifying, assessing, and estimating the enemy's COG, critical factors, capabilities, limitations, intentions, and enemy COAs (ECOA) that are most likely to be encountered based on the situation. Although JIPOE support to decision-making is both dynamic and continuous, it must also be "front loaded" in the sense that the majority of analysis must be completed early enough to be factored into the CDR's decision-making effort. JIPOE supports mission analysis by enabling the CDR and staff to visualize the full extent of the operational environment, to distinguish the known from the unknown, and to establish working assumptions regarding how adversary and friendly forces will interact within the operational environment. JIPOE also assists CDRs in formulating their planning guidance by identifying significant adversary capabilities and by pointing out critical operational environment factors, such as the locations of key geography, attitudes of indigenous populations, and potential land, air, and sea avenues of approach. JIPOE provides predictive intelligence designed to help the CDR discern the adversary's probable intent and most likely future COA. Simply stated, JIPOE helps the CDR to stay inside the adversary's decision-making cycle in order to react faster and make better decisions than the adversary.[5]

[5] *JP 2-01.3, Joint Intelligence Preparation of the Operational Environment.*

c. The intelligence directorates (J-2s) at all levels coordinate and supervise the JIPOE effort to support joint operation planning, enable commanders and other key personnel to visualize the full range of relevant aspects of the operational environment, identify adversary COGs, conduct assessment of friendly and enemy actions, and evaluate potential adversary and friendly COAs. The JIPOE effort must be fully coordinated, synchronized, and integrated with the separate Intelligence Preparation of the Battlespace (IPB) efforts of the component commands and Service intelligence centers. Additionally, JIPOE relies heavily on inputs from several related, specialized efforts, such as geospatial intelligence preparation of the environment (GPE) and medical intelligence preparation of the operational environment (MIPOE). All staff elements of the joint force and component commands fully participate in the JIPOE effort by providing information and data relative to their staff areas of expertise. However, CDRs and their subordinate commanders are the key players in planning and guiding the intelligence effort, and JIPOE plays a critical role in maximizing efficient intelligence operations, determining an acceptable COA, and developing a concept of operations (CONOPS). Therefore, commanders should integrate the JIPOE process and products into the joint force's planning, execution, and assessment efforts.

d. The analysts look at the operational environment from a systems perspective – looking at the operational environment major sub systems and then providing an assessment of the interrelationships between these systems. One approach is to examine the political, military, economic, social, informational, and infrastructure aspects of the operational environment, which are factors generally referred to as PMESII and desired effects. The PMESII construct offers a means to capture this information. Each PMESII factor is relevant and should be looked at critically when analyzing the operational environment. Understanding this environment has always included a perspective broader than just the adversary's military forces and other combat capabilities within the operational area. The planning, execution and assessment of joint operations require a holistic view of all *systems* (both military and non-military) that comprise the operational environment.

e. In addition to analyzing the conventional general military intelligence (GMI) products, JIPOE should analyze the environment from a systems perspective. Intelligence identifies and analyzes adversary and neutral systems and estimates how individual actions on one element of a system can effect other system components.

f. Using the JIPOE process, the Joint force J-2 manages the analysis and development of products that provide a systems understanding of the operational environment. This analysis identifies a number of nodes related to identified friendly objectives and effects—specific physical, functional, or behavioral systems, forces, information, and other components of the system. JIPOE analysts also identify links—the behavioral, physical, or functional relationship between nodes. Link and nodal analysis provide the basis for the identification of adversary COGs and decisive points for action to influence or change adversary system behavior. This methodology also provides the means by which intelligence personnel develop specific indicators of future adversary activity and support J3/5 COA development. It also enables analysts to understand how specific actions activities within the operational environment will influence other aspects of the operational environment.

g. JIPOE from a systems approach may require extensive resources (i.e., personnel with the proper expertise on the various aspects of the operational environment, and extensive collaboration). Although conceptually it is a sound practice, it may not always be possible to conduct a comprehensive JIPOE in this manner unless it is on a focused target set and operational environment, and sufficient time is allotted for this effort. It is critically important for intelligence personnel to understand the external resources available to support this effort. Like all intelligence collection and analysis, it is never complete, requiring continual update throughout planning and execution.

h. JIPOE Consists of Four Major Steps (see facing page, Figure XIII-3 (JP 2-01.3)):

Joint Intelligence Preparation of the Operational Environment (JIPOE)

JIPOE Consists of Four Major Steps:

- Define the operational environment.
- Describe the effects of the operational environment.
- Evaluate the adversary.
- Determine adversary COAs.

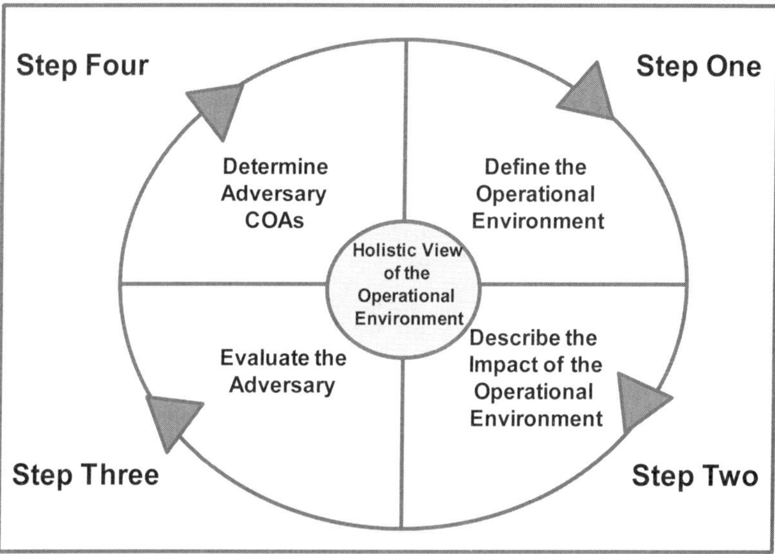

Figure XIII-3. JIPOE Process (JP 2-01.3)

Analysts use the JIPOE process to analyze, correlate, and fuse information pertaining to all aspects of the operational environment. The operational environment consists of the air, land, sea, space and associated adversary, friendly, and neutral systems. JIPOE is conducted both prior to and during joint force operations, as well as during planning for follow-on missions. The most current information available regarding the adversary situation and the operational environment is continuously integrated into the JIPOE process.

Joint Intelligence Preparation of the Operational Environment -Step One-

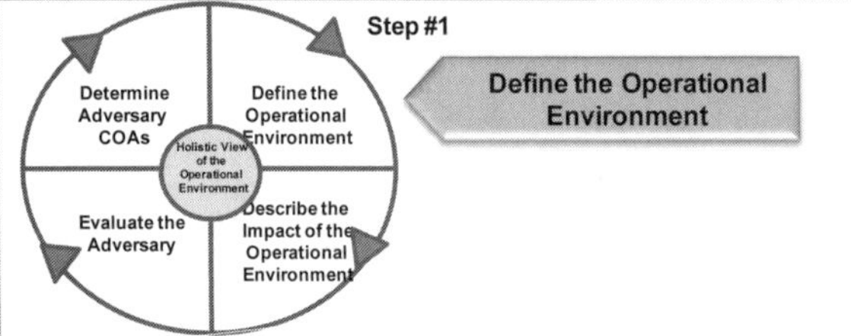

1. Identify the joint forces operational area
2. Analyze the missions and joint force commander's intent
3. Determine the significant characteristics of the operational environment
4. Establish the limits of the joint force's area of interest
5. Determine the level of detail required and feasible within the time available
6. Determine intelligence and information gaps, shortfalls and priorities
7. Collect material and submit requests for information to support further analysis

Figure XIII-4. Define the Operational Environment

i. The Staff Planners Role in JIPOE. The joint force J-2 has primary responsibility for planning, coordinating, and conducting the overall JIPOE analysis and production at the joint force level. However, JIPOE is a staff process – *not just* a J-2 process, and should be driven by the chief of staff. To ensure you're obtaining relevant and accurate intelligence support material for the CDR, and to ensure the most efficient and productive use of intelligence resources, the staff should take an active role in meeting with the J-2 and those analysts working on your production requirements. The staff provides information and data on the operational environment relative to their staff areas of expertise. They now also know you and understand intelligence in context with the operation you are planning or executing.

j. Step 1 — Define the Operational Environment. (Figure XIII-4, (JP 2-01.3)) During Step 1, the joint force staff assists the JFC and component CDRs in determining the dimensions of the joint force's operational environment by identifying the significant characteristics of the operational environment and gathering information relating to the operational environment and the adversary. Successfully defining the command's operational environment is critical to the outcome of the JIPOE process. The joint force J-2 staff works with other joint force and component command staff elements, including the IO planning staff, to formulate an initial survey of adversary, environmental, and other characteristics that may impact the friendly joint mission. Additionally, the joint force staff must also recognize that the operational environment extends beyond the geographic dimensions of land, air, sea, and space. It also includes nonphysical dimensions, such as the electromagnetic spectrum, automated information systems, and public opinion. These nonphysical dimensions may extend well beyond the joint force's designated operational areas, which will also impact determining the Area of Interest, or, according to the Joint Pub 1-02, *"that area of concern to the CDR, including the area of influence, areas adjacent thereto, and extending into enemy territory to the objectives of current or planned operations. This area also includes areas occupied by enemy forces that could jeopardize the accomplishment of the mission."* Understanding which characteristics are significant is done in context with the adversary, weather and terrain, neutral or benign population or elements, and most importantly with the CDR's intent and the mission, if specified. The significant characteristics, once identified, will provide focus and guide the remaining steps of JIPOE. Therefore, it is essential to conduct effective analysis of the operational environment to ensure the "right" characteris-

Joint Intelligence Preparation of the Operational Environment -Step Two-

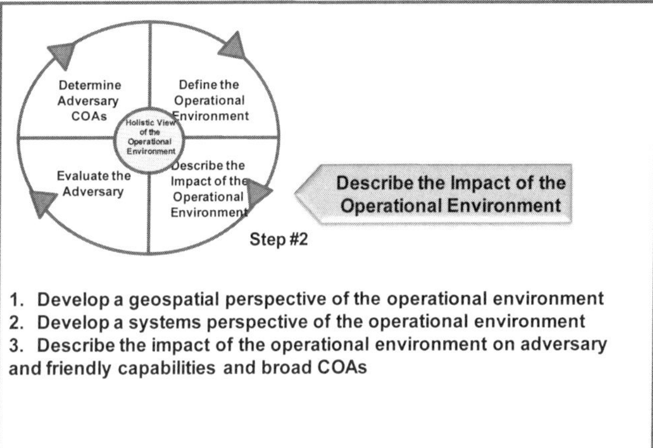

Figure XIII-5. Describe the Impact of the Operational Environment

tics were identified as significant. Identifying the wrong significant characteristics or simply not addressing them jeopardizes the integrity of the operation plan.

(1) The joint force J-2 staff evaluates the available intelligence data bases to determine if the necessary information is available to conduct the remainder of the JIPOE process. In nearly every situation, there will be gaps in the existing data bases. The gaps must be identified early in order for the joint force staff to initiate the appropriate intelligence collection requirements. The joint force J-2 will use the CDR's stated intent and initial PIR to establish priorities for intelligence collection, processing, production, and dissemination. The joint force J-2 staff initiates collection operations and issues RFIs to fill intelligence gaps to the level of detail required to conduct JIPOE. As additional information and intelligence is received, the J-2 staff updates all JIPOE products. If any assumptions are repudiated by new intelligence, the CDR, the J-3, and other appropriate staff elements should reexamine any evaluations and decisions that were based on those assumptions.

(2) Products from step one may include assessments of each significant characteristic, overlays of each, if applicable, and an understanding and graphical depiction of the operational area and possibly of the area of interests and entities therein which could affect our ability to accomplish our mission.

k. <u>Step 2 — Describe the Impact of the Operational Environment</u>. (Figure XIII-5, (JP 2-01.3)). Step 2, describing the operational environment impacts, *focuses on the environment.* The first action in describing operational environment effects is to analyze the military aspects of the terrain. The acronym that aids in addressing the various aspects of the operational environment is OCOKA - Observation and Fields of Fire, Concealment and Cover, Obstacles, Key Terrain, and Avenues of Approach. This analysis is followed by an evaluation of how these aspects of the operational environment will affect operations for both friendly and adversary forces.

(1) Products developed during this step might include overlays and matrices that depict the military effects of geography, meteorological (METOC) factors, demographics, and the electromagnetic and cyberspace environments. The primary product from JIPOE produced in step 2 is the *Modified Combined Operations Overlay* (MCOO) and is shown in Figure XIII-6, (JP 2-01.3). The MCOO is "a JIPOE product used to portray the effects of each battlespace dimension on military operations. It normally depicts militarily significant

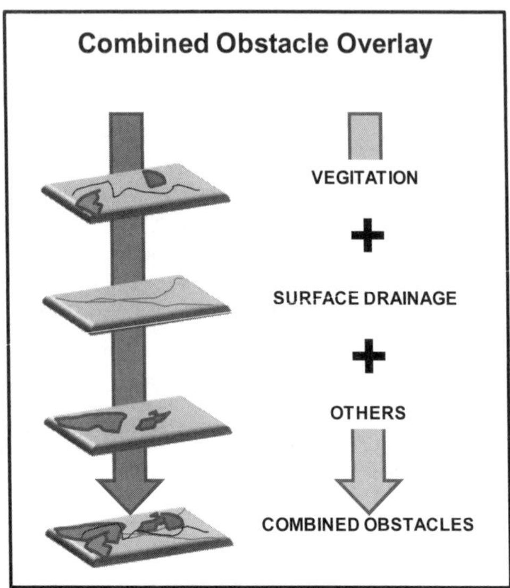

Figure XIII-6. Combined Obstacle Overlay.

aspects of the operational environment, such as obstacles restricting military movement, key geography, and military objectives."[6] Areas of the operational environment where the terrain predominantly favors one COA over others should be identified and graphically depicted. The most effective graphic technique is to construct a MCOO by depicting (in addition to the restricted and severely restricted areas already shown) such items as avenues of approach and mobility corridors, counter-mobility obstacle systems, defensible terrain, engagement areas, and key terrain.[7]

(2) A MCOO generally has standardized overlays associated with it. However, it is not a standardized product with respect to what it should portray simply because a CDR's requirements are based on his mission and intent – and they differ with each operation. Therefore, the MCOO should portray the relevant information necessary to support the CDR's understanding of the battlespace and decision-making process in context with his mission and intent. The results of terrain analysis should be disseminated to the joint force staff as soon as possible by way of the intelligence estimate (included in the order), documented analysis of the operational area, and the MCOO.

(3) Operational environments that you may be analyzing are broken down into dimensions, as follows:

- Land Dimension
- Maritime Dimension
- Air Dimension
- Space Dimension
- Electromagnetic Dimension
- Cyberspace Dimension
- Human Dimension
- Analysis of Weather and Effects
- Other Characteristics of the OE

[6] *Joint Pub 1-02.*

[7] *Refer to Joint Pub 2-01.3 JIPOE for more information concerning the types of MCOOs generated during step 2 of JIPOE.*

(a) *Land Dimension*. Analysis of the land dimension of the operational environment concentrates on terrain features such as transportation systems (road and bridge information), surface materials, ground water, natural obstacles such as large bodies of water and mountains, the types and distribution of vegetation, and the configuration of surface drainage and weather. Observation and fields of fire, concealment and cover, obstacles, key terrain, avenues of approach, and mobility corridors are examples of what is required to be evaluated to understand the terrain effects on your plan (Figure XIII-7, (JP 2-01.3)).

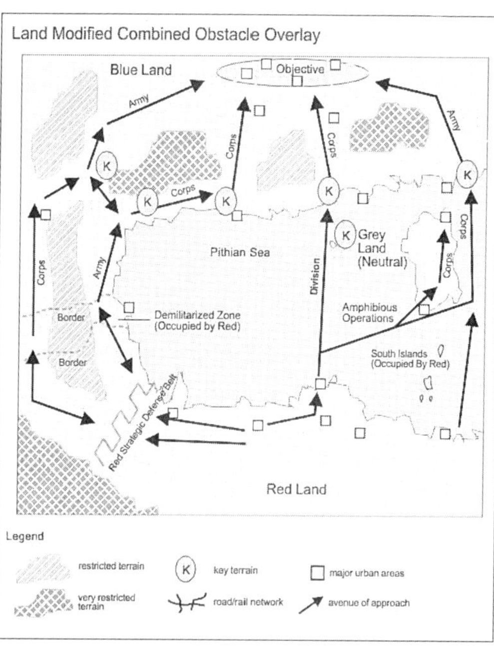

Figure XIII-7. Land MCOO

(b) *Maritime Dimension*. The maritime dimension of the operational environment is the sea and littoral environment in which all naval operations take place, including sea control, power projection, and amphibious operations. Key military aspects of the maritime environment can include maneuver space and chokepoints; natural harbors and anchorages; ports, airfields, and naval bases; sea lines of communications (SLOCs), and the hydrographic and topographic characteristics of the ocean floor and littoral land masses (Figure XIII-8, (JP 2-01.3)).

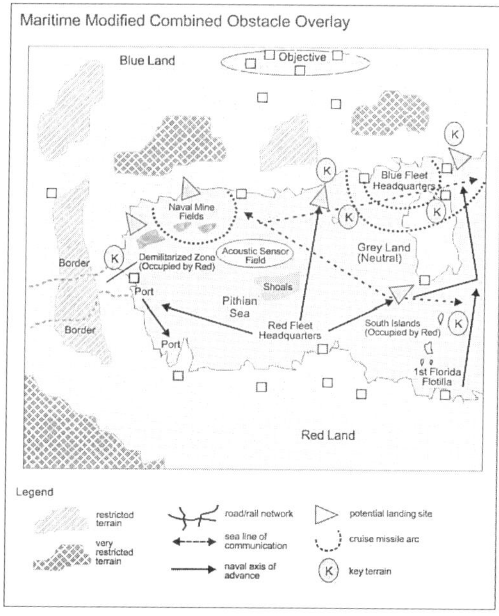

Figure XIII-8. Maritime MCOO

(c) _Air Dimension._ The air dimension of the operational environment is the environment in which military air and counter-air operations take place. It is the operating medium for fixed-wing and rotary-wing aircraft, air defense systems, unmanned aerial vehicles, cruise missiles, and some theater and anti-theater ballistic missile systems. The surface and air environments located between the target areas and air operations points of origin are susceptible to METOC conditions, surface and air borne missiles, lack of emergency airfields, restrictive air avenues of approach and operating altitude restrictions, to name a few (Figure XIII-9, (JP 2-01.3)).

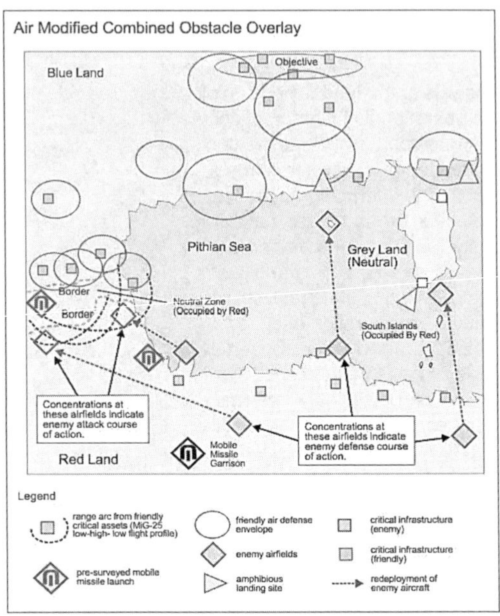

Figure XIII-9. Air MCOO

(d) _Space Dimension._ The space dimension of the operational environment begins at the lowest altitude at which a space object can maintain orbit around the earth (approximately 93 miles) and extends upward to approximately 22,300 miles (geosynchronous orbit). Forces that have access to this medium are afforded a wide array of options that can be used to leverage and enhance military capabilities. However, space systems are predictable in that they are placed into the orbits that maximize their mission capabilities. Once a satellite is tracked and its orbit determined, space operations and intelligence crews can usually predict its function and future position (assuming it does not maneuver). The path a satellite makes as it passes directly over portions of the earth can be predicted and displayed on a map as a satellite ground track (Figure XIII-10, (JP 2-01.3)).

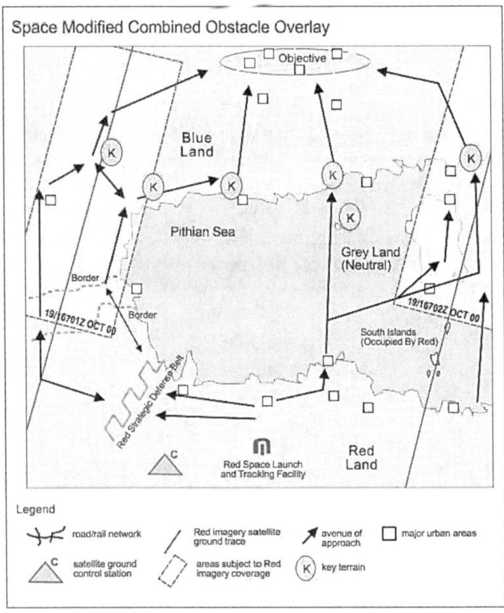

Figure XIII-10. Space MCOO

(e) Electromagnetic Dimension. The electromagnetic dimension of the operational environment includes all militarily significant portions of the electromagnetic spectrum, to include those frequencies associated with radio, radar, laser, electro-optic, and infrared equipment. It is a combination of the civil electromagnetic infrastructure; natural phenomena; and adversary, friendly, and neutral electromagnetic OB (Figure XIII-11, (JP 2-01.3)).

> *"Our information defines our decisions - Our decisions define our success."*
>
> *General James E. Cartwright VCJCS, April 2009*

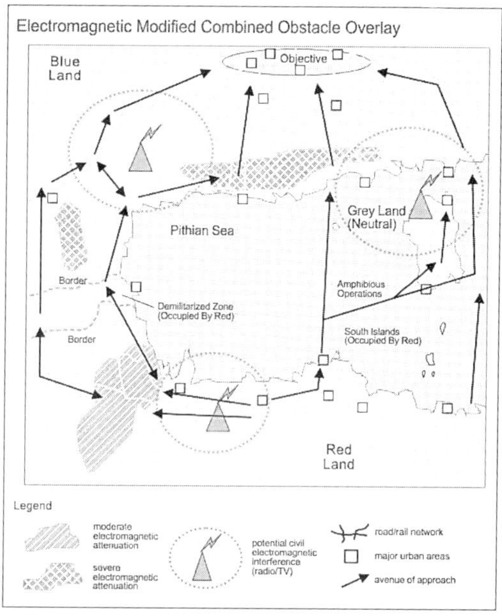

Figure XIII-11. Electromagnetic MCOO

(f) Human Dimension. The human dimension of the operational environment consists of various militarily significant sociological, cultural, demographic, and psychological characteristics of the friendly and adversary populace and leadership. It is the environment in which IO, such as psychological operations (PSYOP) and military deception are conducted. The analysis of the human dimension is a two-step process that: (1) identifies and assesses all human characteristics that may have an impact on the behavior of the populace as a whole, the military rank and file, and senior military and civil leaders; and (2) evaluates the effects of these human characteristics on military operations. Psychological profiles on military and political leaders may facilitate understanding an adversary's behavior, evaluating an adversary's vulnerability to deception, and assessing the relative probability of an adversary adopting various COAs.

(g) Analysis of Weather Effects. Weather is the state of the atmosphere regarding wind, temperature, precipitation, moisture, barometric pressure, and cloudiness. Climate is the composite or generally prevailing weather conditions of a region, averaged over a number of years. Initial studies of climatic effects may be prepared using available climatological data and/or seasonal outlooks requested from the DOD climate centers. These climate-based products are updated with outlooks and forecasts as more precise information is received concerning the actual weather conditions expected. METOC conditions affect the operational environment in several ways: the atmospheric and/or oceanographic environments can interact with, and thereby modify, the characteristics of each physical domain; or METOC can have a direct effect on military operations across all domains. The analysis of weather effects is a two-step process in which: (1) each military aspect of weather is analyzed; and (2) the effects of weather on military operations are evaluated. The joint force METOC officer is the source for weather information, and assists the joint force staff in determining the effects of METOC on adversary and friendly military operations. The overall effects of forecasted weather can be summarized in the form of a weather effects matrix (Figure XIII-12, (JP 2-01.3)).

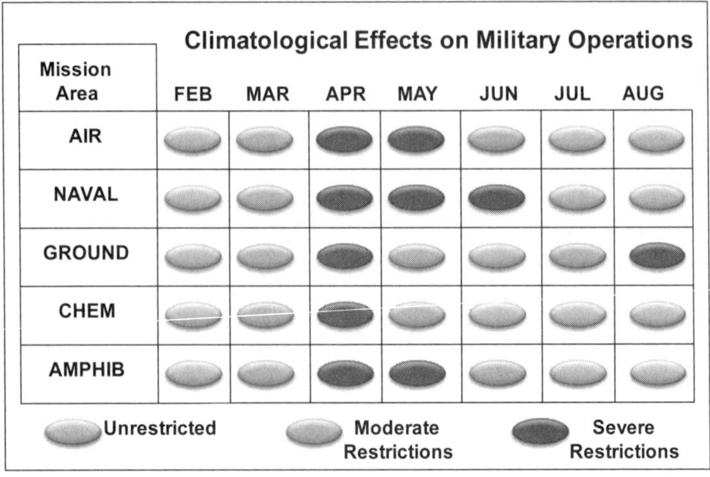

Figure XIII-12. Weather Effect Matrix.

(h) Others characteristics of the operational environment. Other characteristics include all those aspects of the operational environment that could affect friendly or adversary COAs that fall outside the parameters of the categories previously discussed. Because the relevant characteristics will depend upon the situation associated with each mission, there can be no definitive listing of characteristics appropriate under all circumstances. For example, the characteristics of the operational environment that may be relevant to a sustained humanitarian relief operation will be very different from those required for a joint combat operation against an adversary. Some examples to be addressed while evaluating the battlespace environment are time, political and military constraints, environmental and health hazards, infrastructure, industry, agriculture, economics, politics, and history. The country characteristics of an adversary nation should be developed through the analytic integration of all the social, economic, and political variables listed above. Country characteristics can also provide important clues as to where a nation may use military force and to what degree.

l. Step 3 — Evaluate the Adversary. (Figure XIII-13, (JP 2-01.3)) Step three of the JIPOE process, evaluating the adversary, identifies and evaluates the adversary's military and relevant civil COG, critical vulnerabilities (CVs), capabilities, limitations, and the doctrine and Tactics, Techniques and Procedures (TTPs) employed by adversary forces, absent any constraints that may be imposed by the operational environment described in step two. Failure to accurately evaluate the adversary may cause the command to be surprised by an unexpected adversary capability, or result in the unnecessary expenditure of limited resources against adversary force capabilities that do not exist.

(1) A COG can be viewed as the set of characteristics, capabilities, and sources of power from which a system derives its moral or physical strength, freedom of action, and will to act (more on COG in Chapter XIV, Mission Analysis Key-Step 6). The COG is always linked to the objective. If the objective changes, the center of gravity also could change. At the *strategic level,* a COG could be a military force, an alliance, a political or military leader, a set of critical capabilities or functions, or national will. At the *operational level* a COG often is associated with the adversary's military capabilities — such as a powerful element of the armed forces — but could include other capabilities in the operational environment. Since the adversary will protect the center of gravity, the COG invariably is found among strengths rather than among weaknesses or vulnerabilities. *CDRs consider not only the enemy* COGs, but also identify and protect their own COGs, which is a function of the J-3.

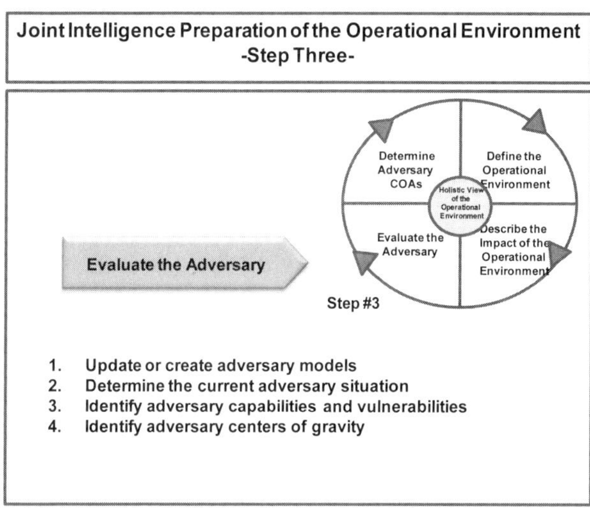

Joint Intelligence Preparation of the Operational Environment
-Step Three-
Determine Adversary COAs
Define the Operational Environment
Holistic View of the Operational Environment
Evaluate the Adversary
Describe the Impact of the Operational Environment
Evaluate the Adversary
Step #3
1. Update or create adversary models
2. Determine the current adversary situation
3. Identify adversary capabilities and vulnerabilities
4. Identify adversary centers of gravity

Figure XIII-13. Evaluate the Adversary

(2) The analysis of friendly and adversary COGs is a key step within the planning process. *Joint force intelligence analysts identify adversary COGs.* The analysis is conducted after gaining an understanding of the various systems in the operational environment. The analysis addresses political, military, economic, social, informational, and infrastructure systems of the operational environment, including the adversary's leadership, fielded forces, resources, population, transportation systems, and internal and external relationships. *The goal is to determine from which elements the adversary derives freedom of action, physical strength (means), and the will to fight.* The J-2 then attempts to determine if the tentative or candidate COGs truly are critical to the adversary's strategy. This analysis is a linchpin in the planning effort. After identifying friendly and adversary COGs, CDRs and their staffs must determine how to protect or attack them, respectively. An analysis of the identified COGs in terms of critical capabilities, requirements, and vulnerabilities is vital to this process.

(3) Understanding the relationship among the COGs not only permits, but also compels, greater precision in thought and expression in operational design. Planners should analyze COGs within a framework of three *critical factors — critical capabilities, requirements, and vulnerabilities* — to aid in this understanding. *Critical capabilities* are those that are considered crucial enablers for a center of gravity to function as such, and are essential to the accomplishment of the adversary's assumed objective(s). *Critical requirements* are essential conditions, resources, and means for a critical capability to be fully operational.

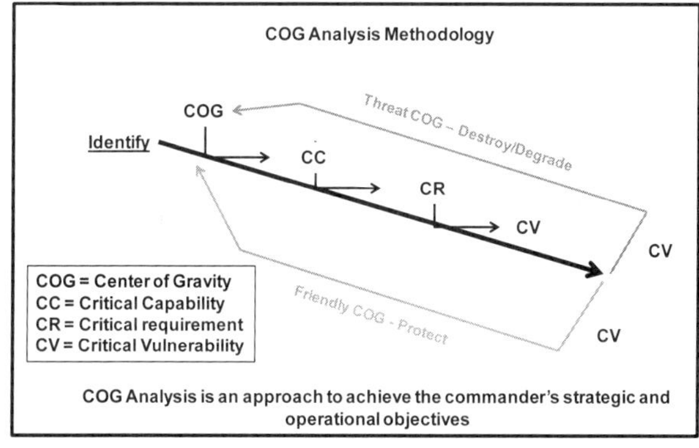

Critical vulnerabilities are those aspects or components of critical requirements that are deficient, or vulnerable to direct or indirect attack in a manner achieving decisive or significant results. Collectively, the terms above are referred to as *critical factors*. In general, a JFC must possess sufficient operational reach and combat power to take advantage of an adversary's critical vulnerabilities. Similarly, a supported CDR must protect friendly critical capabilities within the operational reach of an adversary. As a best practice, the J-2 will act as a "red cell" in helping to identify the friendly forces COG and conduct COG analysis to support an understanding of what must be protected.

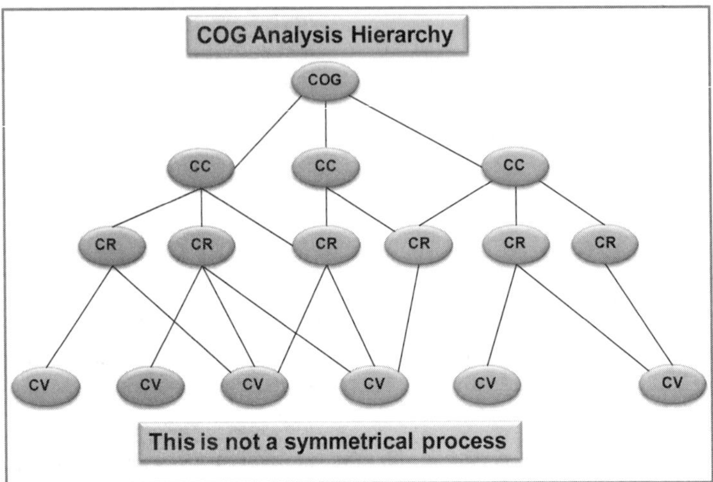

(4) In addition to the initial results of COG analysis, the primary products from JIPOE produced in JIPOE Step three are doctrinal templates, descriptions of the adversary's preferred tactics and options, and the identification of high-value targets (HVTs), which are targets that the enemy CDR requires for the successful completion of the mission. The loss of high-value targets would be expected to seriously degrade important enemy functions throughout the friendly CDR's area of interest.

(5) Adversary models depict how an opponent's military forces prefer to conduct operations under ideal conditions. They are based on a detailed study of the adversary's normal or "doctrinal" organization, equipment, and TTP. Adversary models are normally completed prior to deployment, and are continuously updated as required during military operations. The models consist of three major parts: graphical depictions of adversary doctrine or patterns of operations (doctrinal templates), descriptions of the adversary's preferred tactics and options, and the identification of high-value targets (HVTs).

(6) Doctrinal templates illustrate the employment patterns and dispositions preferred by an adversary when not constrained by the effects of the operational environment. They are usually scaled graphic depictions of adversary dispositions for specific types of military (conventional or unconventional) operations such as movements to contact, anti-surface warfare operations, insurgent attacks in urban areas, combat air patrols, and aerial ambushes. JIPOE utilizes single-Service doctrinal templates that portray adversary and, sea, air, special, or space operations, and produces joint doctrinal templates that portray the relationships between all the adversary's Service components when conducting joint operations.

(7) In addition to the graphic depiction of adversary operations portrayed on the doctrinal template, an adversary model must also include a *written description of an opponent's preferred tactics*. This description should address the types of activities and supporting operations that the various adversary units portrayed on the doctrinal template are expected to perform. It also contains a listing or description of the options (branches) available to the

adversary — should either the joint operation or any of the supporting operations fail — or subsequent operations (sequels) if they succeed.

(8) The adversary model must also include a list of HVTs. HVTs are those assets that the adversary CDR requires for the successful completion of the joint mission (and supporting missions) that are depicted and described on the joint doctrinal template. These targets are identified by combining operational judgment with an evaluation of the information contained in the joint doctrinal template and description. Assets are identified that are critical to the joint mission's success, that are key to each component's supporting operation, or that are crucial to the adoption of various branches or sequels to the joint operation. The joint targeting community collaborates in the identification of HVTs with the responsible producers for various intelligence product category codes.

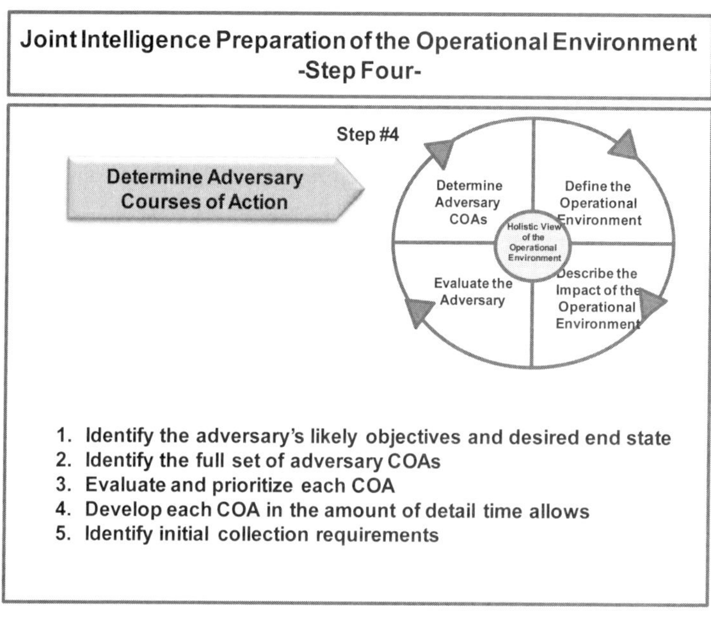

Joint Intelligence Preparation of the Operational Environment
-Step Four-
Step #4
Determine Adversary Courses of Action
Determine Adversary COAs
Define the Operational Environment
Holistic View of the Operational Environment
Evaluate the Adversary
Describe the Impact of the Operational Environment
1. Identify the adversary's likely objectives and desired end state
2. Identify the full set of adversary COAs
3. Evaluate and prioritize each COA
4. Develop each COA in the amount of detail time allows
5. Identify initial collection requirements

Figure XIII-14. Determine Adversary Courses of Action

m. Step 4 — Determine Adversary Courses of Action (COAs). (Figure XIII-14, (JP 2-01.3)) The first three steps of the JIPOE process help to satisfy the operational environment awareness requirements of the CDR and subordinate CDRs by analyzing the effects of the battlespace environment, assessing adversary doctrine and capabilities, and identifying adversary COGs. The fourth step of the JIPOE process seeks to go beyond operational environment *awareness* to help the CDR attain *knowledge* of the operational environment (i.e., a detailed understanding of the adversary's probable intent and future strategy). The process for Step four provides a disciplined methodology for analyzing the set of potential adversary COAs in order to identify the COA the adversary is most likely to adopt, and the COA that would be most dangerous to the friendly force or to mission accomplishment.

(1) The first activity in JIPOE Step four is to identify *the adversary's likely objectives and desired endstate* by analyzing the current adversary military and political situation, strategic and operational capabilities, and the country characteristics of the adversary nation, if applicable. The JIPOE analyst should begin by identifying the adversary's overall strategic objective, which will form the basis for identifying subordinate objectives and desired endstates.

(2) During this step, a consolidated list of all potential adversary COAs is constructed. At a minimum this list will include; (1) all COAs that the adversary's doctrine considers appropriate to the current situation and accomplishment of likely objectives, (2) all adversary COAs that could significantly influence the friendly mission, even if the adversary's doctrine considers them suboptimal under current conditions, and (3) all adversary COAs indicated by recent activities or events. Each COA is generated based on what we know of the adversary and how they operate (learned from Step three of JIPOE) to determine if the adversary can in fact accomplish the COA. If not, it is eliminated. J-2 analysts study how an adversary operates compared to the environment it must operate within, which we analyzed during Step two of JIPOE. Essentially, they superimpose the doctrinal adversary mode of operation on the environment. The result of this analysis is a full set of identified adversary COAs – time permitting. Adversary COAs that meet specific criteria are then completed. Much like friendly forces determine if their COAs meet specific criteria, J-2 personnel must also weigh the identified adversary COAs against certain criteria. The criteria generally includes: (1) *suitability*, (2) *feasibility*, (3) *acceptability*, (4) *uniqueness, and* (5) *consistency with their own doctrine.*

(3) Each COA should be developed in the amount of detail that time allows. Subject to the amount of time available for analysis, each adversary COA is developed in sufficient detail to describe, (1) the type of military operation, (2) the earliest time military action could commence, (3) the location of the sectors, zones of attack, avenues of approach, and objectives that make up the COA, (4) the OPLAN, to include scheme of maneuver and force dispositions, and (5) the objective or desired endstate. Each COA should be developed in the order of its probability of adoption, and should consist of a situation template, a description of the COA, and a listing of HVTs.

(4) A full set of identified adversary COAs are evaluated and ranked according to their likely order of adoption. The purpose of the prioritized list of adversary COAs is to provide a CDR and his staff with a starting point for the development of an OPLAN that takes into consideration the most likely adversary COA as well as the adversary COA most dangerous to the friendly force or mission accomplishment. The primary products produced in JIPOE Step four are the situation template and matrix, and the event template and matrix.[8]

4. Intelligence Preparation of the Information Environment (IPIE)

a. IPIE continues to evolve as we realize how critical a thorough understanding of the information environment is in modern warfare, and how valuable an accurate portrayal of the information environment is in facilitating effective planning and execution of information operations. However, to be valid, IPIE must be conducted as part of the J-2's JIPOE efforts. If conducted in isolation, IPIE will fail to provide a picture of the information environ-

> *"Nothing should be neglected in acquiring a knowledge of the geography and military statistics of their states, so as to know their material and moral capacity for attack and defense as well as the strategic advantages of the two parties."*
>
> *Jomini*
> *Precis de l'Art de la Guerre, 1838*

[8] *For more information on JIPOE, refer to JP 2-01.3, Intelligence Preparation of the Operational Environment. Also, see JP 5-0, Joint Operations Planning, JP 3-0, Joint Operations, and JFSC, Joint Information Operations Planning Handbook.*

Impact of the Information Environment on Military Operations

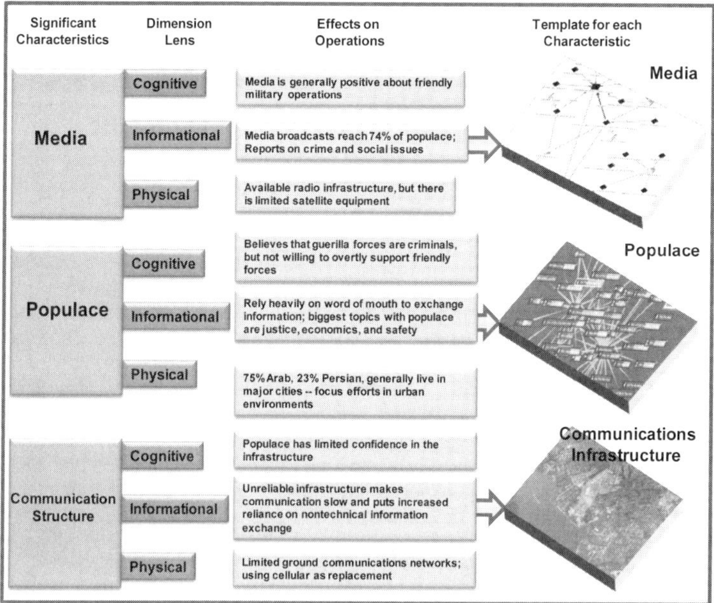

Significant Characteristics
Dimension Lens
Effects on Operations
Template for each Characteristic
Media
Cognitive
Media is generally positive about friendly military operations
Informational
Media broadcasts reach 74% of populace; Reports on crime and social issues
Physical
Available radio infrastructure, but there is limited satellite equipment
Media
Populace
Cognitive
Believes that guerilla forces are criminals, but not willing to overtly support friendly forces
Informational
Rely heavily on word of mouth to exchange information; biggest topics with populace are justice, economics, and safety
Physical
75% Arab, 23% Persian, generally live in major cities -- focus efforts in urban environments
Populace
Communication Structure
Cognitive
Populace has limited confidence in the infrastructure
Informational
Unreliable infrastructure makes communication slow and puts increased reliance on nontechnical information exchange
Physical
Limited ground communications networks; using cellular as replacement
Communications Infrastructure

Figure XIII-15. Impact of the Information Environment.

ment consistent with the other operating environments (i.e., land, sea, air, and space) and threat COAs generated by the J-2 staff.

b. The information environment is where humans observe, orient, decide, and act upon information, and is therefore the principal environment of decision making. This environment is pervasive to all activities worldwide, and is a common backdrop for the air, land, maritime, and space physical domains of the CDR's operational environment. The actors in the information environment include military and civilian leaders, decision makers, individuals, and organizations. Resources include the information itself and the materials and systems employed to collect, analyze, apply, disseminate, and display information and produce information-related products such as reports, orders, and leaflets. *Cyberspace* is a global domain within the information environment consisting of the interdependent network of information technology infrastructures, including the Internet, telecommunications networks, computer systems, and embedded processors and controllers. Within cyberspace, electronics and the electromagnetic spectrum are used to store, modify, and exchange data via networked systems. Significant characteristics of the information environment can be further evaluated within physical, informational, and cognitive dimensions (Figure XIII-15, (JP 2-01.3)).

c. There is only one reality and that exists within the physical dimension. Actions within the physical dimension are converted into selected data, information, and knowledge in the information dimension, which are interpreted in the mind of individuals to develop perceptions, awareness, understanding, beliefs, and values, etc., which friendly forces would like them to.

d. IPIE strives to understand the relationship between the three dimensions as it pertains to the operation being executed or planned and those factors that can affect successful mission accomplishment within the defined operational environment and outside of it. Understanding the relationship between the dimensions enables one to understand the first, second and third order effects of an action that takes place in the physical domain. Thus, IO planners can more effectively plan IO initiatives to achieve desired effects, contribute to ensuring the CDR avoids scenarios that could achieve undesired effects, and also act as advisors to those who must understand the second and third order effects of actions planned to take place or that have taken place within the battlespace. Keep in mind that the operational environment affected may extend beyond that of the geographic area within which friendly forces operate. Needless to say, the foundation for effectively interpreting actions within the information environment is sound intelligence.

e. From an IO perspective, *the operational environment is the conceptual volume in which the CDR seeks to dominate the enemy.* The operational environment expands and contracts in relation to the CDR's ability to acquire and engage the enemy, or can change as the CDR's vision of the operational environment changes. It encompasses all three dimensions and is influenced by the operational dimensions of time, tempo, depth, and synchronization. The operational environment is not assigned by a higher CDR nor is it constrained by assigned boundaries. A command's operational environment is determined by the range of direct fire weapons, artillery, aviation, and electronic warfare (EW). The operational environment extends beyond the area of operation and may include the home station of a deploying friendly force. For IO, the operational environment is the volume of space in which friendly forces can influence the information environment. The information environment potentially expands the command's operational environment as the effects of IO elements like psychological operations and public affairs can extend well beyond the range of conventional weapon systems.

Support to Joint Operations Planning

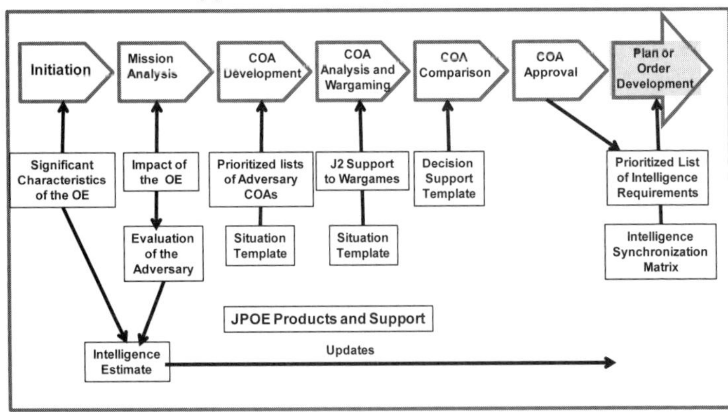

Figure XIII-16. Support to JOPP.

5. JIPOE Support to JOPP

JIPOE supports joint operation planning by identifying significant facts and assumptions about the operational environment. This information includes details regarding adversary critical vulnerabilities, capabilities, decisive points, limitations, COGs, and potential COAs. JIPOE products are used by the CDR to produce the commander's estimate of the situation and CONOPS, and by the joint force staff to produce their respective staff estimates. Various intelligence products such as DIA produced dynamic threat assessments (DTAs), baseline JIPOE products, and other locally produced assessments, will contribute to developing and enhancing comprehensive intelligence estimates. JIPOE products also help to provide the framework used by the joint force staff to develop, wargame, and compare friendly COAs and provide a foundation for the CDR's decision regarding which friendly COA to adopt. JIPOE support is crucial throughout the steps of the JOPP (see Figure XIII-16 (JP 2-01.3)).

The Information Environment (Dimensions)

The Physical Dimension

The physical dimension is composed of the command and control (C2) systems, and supporting infrastructures that enable individuals and organizations to conduct operations across the air, land, sea, and space domains. It is also the dimension where physical platforms and the communications networks that connect them reside. This includes the means of transmission, infrastructure, technologies, groups, and populations. The physical dimension extends beyond the operational area to encompass those theater and national capabilities (such as systems, databases, centers of excellence, subject-matter experts) that support the CDR's C2 and decision-making requirements. Cyberspace encompasses many physical dimension capabilities, but others exist outside cyberspace. Examples include important information infrastructures such as television, radio, and newsprint, even though these access cyberspace to support their products. Likewise, individuals are non-cyberspace transmitters and receivers of information.

The Information Dimension

The informational dimension links the physical and cognitive dimensions. The joint force uses cyberspace capabilities, cyberspace operations, and non-cyberspace ways and means to collect, process, store, disseminate, display, and protect information and related products. The informational dimension focuses on the content and flow of information, and it is in this dimension that the commander communicates intent and commands and controls military forces. The relative vulnerability of various aspects of the informational dimension (whether due to poor physical security, improper operator training, or lack of safeguards) combined with the level of sophistication of an adversary's CNA capabilities, may help analysts determine an adversary's potential method of attack.

The Cognitive Dimension

The cognitive dimension encompasses the minds of those who transmit, receive, and respond to or act on information. In this dimension, people think, perceive, visualize, understand, and decide. These activities may be affected by a commander's psychological characteristics, personal motivations, and training. Factors such as leadership, morale, unit cohesion, emotion, state of mind, level of training, experience, situational awareness, as well as public opinion, perceptions, media, public information, and rumors may also affect cognition. Particularly in operations characteristic of irregular warfare (such as counterinsurgency), operations can succeed or fail based on how adept the commander and staff are at understanding and operating in the cognitive dimension with respect to the target population. The analysis of the cognitive dimension is a two-step process that: (1) identifies and assesses all human characteristics that may have an impact on the behavior of the populace as a whole, the military rank and file, and senior military and civil leaders; and (2) evaluates the influence these human characteristics have on military operations.

a. The JIPOE effort should facilitate parallel planning by all strategic, operational, and tactical units involved in the operation. JIPOE products developed to support strategic-level planning should also be simultaneously disseminated to all appropriate operational and tactical headquarters. This is especially true during initial planning periods when headquarters at intermediate echelons may tend to filter information as it travels down to tactical units.

b. The integration of Service component IPB products with the CDRs' JIPOE effort creates a synergy in which an adversary's COAs may provide indicators as to the adversary's overall capabilities, intentions, desired endstate, and strategy. Specifically, JIPOE products facilitate operation planning by determining the following:

(1) The idiosyncrasies and decision-making patterns of the adversary strategic leadership and field commanders.

(2) The adversary's strategy, intention, or strategic concept of operations, which should include the adversary's desired endstate, perception of friendly vulnerabilities, and adversary intentions regarding those vulnerabilities.

(3) The composition, dispositions, movements, strengths, doctrine, tactics, training, and combat effectiveness of major adversary forces that can influence friendly actions in the theater and operational areas.

(4) The adversary's principal strategic and operational objectives and lines of operation.

(5) The adversary's strategic and operational sustainment capabilities.

(6) COGs and decisive points throughout the adversary's operational and strategic depths.

(7) The adversary's ability to conduct IO and use or access data from all systems.

(8) The adversary's regional strategic vulnerabilities.

(9) The adversary's capability to conduct asymmetric attacks against friendly global critical support nodes (e.g., electric power grids, oil and gas pipelines, prepositioned supply depots).

(10) The adversary's relationship with possible allies and the ability to enlist their support.

(11) The adversary's defensive and offensive vulnerabilities in depth.

(12) The adversary's capability to operate advanced warfighting systems (e.g., smart weapons and sensors) in adverse METOC conditions.

(13) Key nodes, links, and exploitable vulnerabilities within an adversary system.

b. Plan Initiation. A preliminary or abbreviated JIPOE analysis pertaining to potential contingencies and significant characteristics of the operational environment should precede and inform the initiation phase of joint operation planning. During the initiation phase, DIA produces a DTA for each top priority plan identified in the GEF and continuously updates each DTA as relevant aspects of the operational environment change. CCMD intelligence analysts accelerate JIPOE Step one activities by continuously monitoring the situation, alerting the CDR and staff to developments that may impact the operation planning effort, updating existing JIPOE products, and initiating new intelligence collection or production requirements. Additionally, the CDR may decide to form a JIPOE coordination cell to coordinate support and help analyze the initiating directive to determine time available until mission execution, the current status of JIPOE products and related staff estimates, and other factors relevant to the specific planning situation.

c. Mission Analysis. In order for the joint force staff to identify potential COAs, the CDR must formulate planning guidance based on an analysis of the friendly mission. This analysis helps to identify specified, implied, and essential tasks, any constraints on the application of military force, the CDR's task and purpose (restated mission), and possible follow-on missions. JIPOE supports mission analysis by enabling the CDR and joint force staff to visualize the full extent of the operational environment, to distinguish the known from the unknown, and to establish working assumptions regarding how adversary and friendly forces will interact within the constraints of the operational environment. JIPOE assists CDRs in formulating their planning guidance by identifying significant adversary capabilities and by pointing out critical factors, such as the locations of key geography, attitudes of indigenous populations, and potential land, air, and sea avenues of approach. Mission analysis and CDR guidance form the basis for the subsequent development of friendly COAs by the joint force staff. It is therefore imperative that an initial version of the impact of the operational environment, evaluation of the adversary and adversary COAs be briefed to the CDR at the mission analysis briefing. This is critical to enabling the CDR to provide sufficient guidance for friendly COA development.

d. COA Development. The J-2 ensures that all adversary COAs are identified, evaluated, and prioritized (JIPOE Step four) in sufficient time to be integrated into the friendly COA development effort. Additionally, the evaluation of the adversary (JIPOE Step three) is used by the J-3 and J-5 to estimate force ratios. The process of estimating force ratios may be complicated due to wide disparities between friendly and adversary unit organization, equipment capabilities, training, and morale. In such situations, the J-2, J-3, and J-5 may choose to develop local techniques and procedures for evaluating adversary units and equipment in terms of friendly force equivalents. The J-3 also depends heavily on JIPOE products prepared during the analysis of the adversary situation and the evaluation of other relevant aspects of the operational environment in order to formulate initial friendly force dispositions and schemes of maneuver.

Additionally, the JIPOE analysis of HVTs is used by the J-3 and J-5 to identify targets whose loss to the adversary would significantly contribute to the success of a friendly COA. These targets are refined through wargaming and are designated as HPTs. JIPOE also provides significant input to the formulation of deception plans by analyzing adversary intelligence collection capabilities and the perceptual biases of adversary decision makers.

e. COA Analysis and Wargaming. Assumptions regarding the operational environment and adversary must be realistic. Avoid constructing assumptions that are deliberately designed to support premature conclusions or conceptual bias that favors one COA over another. For example, the joint force staff must guard against seizing upon one adversary COA as a "given" simply because it fits preconceived notions or is a "convenient" match for an already-favored friendly COA. Rather, the staff should plan to counter *all* adversary COAs identified during the JIPOE process. It is imperative that CDRs and their staffs recognize that the least likely adversary COA may be the one actually adopted precisely because it is the least likely, and therefore may be intended to maximize surprise. The wargame should follow a sequence of "action — reaction — counteraction" in which the J-2, JIOC, or red team personnel play the roles of adversary commanders.

(1) Decision Support Template.
The decision support template (see Figure XIII-17, (JP 2-01.3)) is essentially a combined intelligence estimate and operations estimate in graphic form. It relates the detail contained on the event template (prepared during JIPOE step four) to the times and locations of critical areas, events, and activities that would necessitate a command decision, such as shifting the location of the main effort or redeploying forces. Although the decision support template does not dictate decisions to the CDR, it is a useful tool for indicating points in time and space (decision points) where action by the CDR may be required. The decision support template is constructed by combining the event template with data developed during the wargame. The J-2, J-3, J-4, J-5, and J-6 collaborate in the production of the decision support template, which is fully coordinated with all joint force staff elements. The decision support template displays TAIs, avenues of approach, objectives, and time phase lines derived from the JIPOE event template.

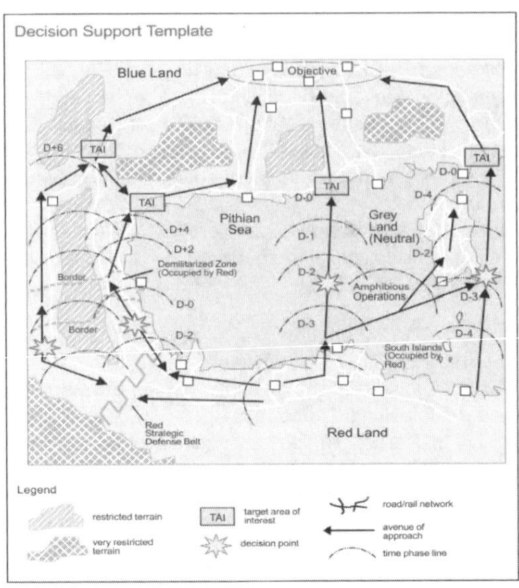

Figure XIII-17. Decision Support Template

f. COA Comparison. Following wargaming, the staff compares friendly COAs to identify the one that has the highest probability of success against the full set of adversary COAs as depicted on the decision support template. Each joint force staff section uses different criteria for comparing friendly COAs, according to their own staff area of expertise. For example, the J-3 and J-5 compare friendly COAs based on the friendly force's ability to defeat each adversary COA, whereas the J-2 assesses the overall capabilities of intelligence collection and production to support each friendly COA. Additionally, each staff section must ensure that they have fully considered the CDR's initial planning guidance for COA selection.

g. COA Approval. After comparing friendly COAs, each joint force staff element presents its findings to the remainder of the staff. Together they determine which friendly COA they will recommend to the CDR. The J-3 then briefs the COAs to the CDR using graphic aids, such as the decision support template and matrix. The CDR decides upon a COA and announces the CONOPS.

h. Plan or Order Development. Using the results of wargaming associated with the selected COA, the joint force staff prepares plans and orders that implement the CDR's decision. The J-2 prioritizes intelligence requirements and synchronizes intelligence collection requirements to support the COA selected by the CDR.

See JP 2-01.3, JIPOE, for a more in-depth discussion of the relationship between the JOPP and, JIPOE.

> *"When I took a decision, or adopted an alternative, it was after studying every relevant — and many an irrelevant — factor. Geography, tribal structure, religion, social customs, language, appetites, standards — all were at my finger-ends. The enemy I knew almost like my own side. I risked myself among them a hundred times, to learn."*
>
> *Colonel T. E. Lawrence*
> *Letter to Liddell Hart, 26 June 1933*

Mission Analysis: Key Steps

14

I. Step 2 to JOPP - Mission Analysis: Key Steps

> ***Upon receipt of strategic tasking:***
>
> *• Planners analyze the tasking to determine where the use of military power and expertise would be appropriate.*
>
> *• Analyze each strategic objective to determine if we can accomplish it with military power or support its accomplishment.*

1. Key-Step — 1: Analyze Higher Commander's Mission and Intent — "Receipt of mission"

The decision-making process begins with the receipt or anticipation of a new mission. This can either come from an order issued by higher headquarters, or derived from an ongoing operation.

a. As soon as a new mission is received, a warning order is issued to the staff alerting them of the pending planning process. The staff prepares for the mission analysis immediately on receipt of a warning order by gathering the tools needed to do mission analysis. Once the new mission is received, the commander and the staff must do a quick initial assessment. It is designed to optimize the commander's use of time while preserving time for subordinate commanders to plan and complete combat preparations. This assessment:

- Determines the time available from mission receipt to mission execution.
- The staff assesses the scope of the assigned mission, intent, endstate, objectives, and other guidance from the next higher command, JSCP, etc. (purpose, method, endstate).
- Determines the time needed to plan, prepare for, and execute the mission for own and subordinate units.
- Determines the JIPOE.
- Determines the staff estimates already available to assist planning.
- Determine whether the mission can be accomplished in a single operation, or will likely require a campaign due to its complexity and likely duration and intensity.

b. At the CCDR level, this would be strategic guidance issued by the President, SECDEF or CJCS. The commander must draw broad conclusions as to the character of the forthcoming military action. The commander should not make assumptions about issues not addressed by the higher commander and if the higher headquarters' directive is unclear, ambiguous, or confusing, the commander should seek clarification.

c. A main concern for a commander during mission analysis is to study not only the mission, but also the intent of the higher commander. Within the breadth and depth of today's operational environment, effective decentralized control cannot occur without a shared vision. Without a commander's intent that expresses that common vision, unity of effort is difficult to achieve. In order to turn information into decisions, and decisions into actions that are "about right," commanders must understand the higher commander's "intent."

While the commander's intent has previously been considered inherent in the mission and concept of operations, most often you will see it explicitly detailed in the plan/order. Successfully communicating the more enduring intent allows the force to continue the mission even though circumstances have changed and the previously developed plan/concept of operations is no longer valid.

d. The higher commander's intent is normally found in Paragraph 3, Execution, of the higher commander's guidance. The intent statement of the higher echelon commander should then be repeated in paragraph 1, Situation, of your own OPLAN or OPORD to ensure that the staff and supporting commanders understand it. Each subordinate commander's intent must be framed and embedded within the context of the higher commander's intent, and they must be nested both horizontally and vertically to achieve a common military endstate.

e. Staff officers must update their staff estimates and other critical information constantly. This information allows them to develop assumptions that are necessary to the planning process. Staff officers must be aggressive in obtaining this information. Reporting of this information must be a *push system* versus a pull system. Subordinate units must rapidly update their reports as the situation changes.

f. The critical product of this assessment is an initial allocation of available time, especially in a crisis. The commander and the staff must balance the desire for detailed planning against the need for immediate action. The commander must provide guidance to subordinate units as early as possible to allow subordinates the maximum time for their own planning and preparation for operations. This, in turn, requires aggressive coordination, deconfliction, integration, and assessment of plans at all levels, both vertically and horizontally.

g. As a general rule, the commander allocates a minimum of two-thirds of available time for subordinate units to conduct their planning and preparation. This leaves one-third of the time for the commander and his staff to do their *planning*. They use the other two-thirds for their own preparation.

h. Time, more than any other factor, determines the detail with which the staff can plan. Once time allocation is made, the commander must determine how much time will be dedicated to each step within the JOPP, or to abbreviate the process.

OIF Lessons Learned at the CCMD Level:

- *Planning organization is critical*
- *Planning is evolutionary; embrace change as a constant*
- *Must explain military operations in simple terms*
- *Risk is different at each level*

2. Key-Step — 2: Determine Own Specified, Implied, and Essential Tasks

Any mission consists of two elements: the *task(s)* to be accomplished by one's forces and their *purpose*. If a mission has multiple tasks, then the priority of each task should be clearly expressed. Usually this is done by the sequence in which the tasks are presented. There might be a situation in which a commander has been given such broad guidance that all or part of the mission would need to be deduced. Deduction should be based on an appreciation of the general situation and an understanding of the superior's objective. Consequently, deduced tasks must have a reasonable chance of accomplishment and should secure results that support the superior commander's objective.

> ***Tasks***
>
> • *Determine what we've been told to do (**specified tasks**)*
>
> • *Determine what else we need to do to accomplish our mission (**implied tasks**)*
>
> • *Deterimine which tasks are **essential** to accomplishment of the mission*
>
> ***Task Analysis***
>
> • ***State the task(s)***. *The task is the job or function assigned to a subordinate unit or command by a higher authority. A mission can contain a single task, but it often contains two or more tasks. If there are multiple tasks, they normally will all be related to a single purpose.*

a. Determine specified, implied, and essential tasks, by reviewing strategic communication guidance and other documents used during Function I, Strategic Guidance and Initiation, in order to develop a concise mission statement. Specified and implied strategic tasks are derived from specific Presidential, SECDEF guidance, national (or multinational) planning guidance documents such as the GEF, JSCP, the UCP, or from CCDR initiatives. The national military objectives form the basis of the campaign's mission statement.

(1) Specified Task — A task that is specifically assigned to an organization by its higher headquarters. Tasks *listed* in the mission received from higher headquarters are specified or stated (assigned) tasks. They are what the higher commander wants accomplished. The commander's specified tasks may be found in paragraph 3b, (Execution-Tasks) section of the order, but could also be contained elsewhere, for example in the JSCP or other national guidance.

> ***Specified Tasks (examples)***
>
> • *Assure allies, coalition partners and friends in the African Region are committed to uphold treaty obligations (USAFRICOM Security Cooperation Strategy)*
>
> • *Deter Mass Migration (USSOUTHCOM TCP, Haiti)*
>
> • *Prepare CONPLAN w/TPFDD for the defense of Country Blue in case of external aggression and in support of U.S. interests (JSCP)*

(2) Implied Task — After identifying the specified tasks, the commander identifies additional major tasks necessary to accomplish the assigned mission. Though not facts, these additional major tasks are implied tasks, which are sometimes *deduced* from detailed analysis of the order of the higher commander, known enemy situation, and the commander's knowledge of the physical environment. *Therefore, the implied tasks subsequently included in the commander's restated mission should be limited to those considered critical to the accomplishment of the assigned mission.* Implied tasks do not include routine or standing operating procedures (SOPs) that are performed to accomplish any type of mission by friendly forces. Hence, tasks that are inherent responsibilities of the commander (providing protection of the flank of his own unit, reconnaissance, deception, etc.) are not considered implied tasks. The exceptions are only those routine tasks that cannot successfully be carried out without support or coordination of other friendly commanders. An example of an implied task is: if the JTF Commander was given a specified task to seize a seaport facility, the implied task might be the requirement to establish maritime superiority within the area of operations before the assault.

(3) Essential Tasks — Essential tasks are determined from the list of both specified and implied tasks. They are those tasks that must be executed to achieve the conditions that define mission success. Depending on the scope of the mission's purpose, some of the specified and implied tasks might need to be synthesized and re-written as an essential task. *Only essential tasks should be included in the mission statement.*

Implied Tasks

- *Additional major tasks necessary to accomplish the assigned mission*
- *Not routine or SOPs, except if the routine task cannot successfully be carried out without support or coordination with another CDR*
- *Example:*

- Specified Task: Seize the Port of Red

- Implied Task: Establish maritime superiority within the AO prior to D-day

Implied Tasks (example)

- *Develop/Expand the FID program*
- *Secure lines of communication in region to ensure unfettered flow of forces and equipment*
- *Build capacity to conduct stability operations, HA/DR, CN*

Essential Tasks (examples)

- *Defend Country Blue.*
- *Secure international support for the conduct of military operations.*
- *Maintain/Restore regional stability.*
- *Build usable, relevant and enduring capabilities to improve GOH's ability to provide for its own security needs. (Operation Unified Response).*

3. Key-Step — 3: Determine Known Facts, Assumptions, Current Status, or Conditions

The staff assembles both facts and assumptions to support the planning process and initial planning guidance. What does the organization know about the current situation and status?

a. Fact. A *fact* is a statement of information known to be true (such as verified locations of friendly and adversary force dispositions).[1]

Facts (examples)

- *Government of Haiti (GoH) does not have adequate emergency management capabilities.*
- *GoH has little or no control over its borders.*
- *The United States has an embassy in Red.*
- *Red has a mutual defense pact with Country Orange.*
- *Blue airfields would require extensive improvements to support modern operations.*

[1] *JP 5-0, Joint Operations Planning.*

b. <u>Assumptions</u>. An assumption is used in the absence of facts that the commander *needs* to continue planning. It is a supposition on the current situation or a presupposition on the future course of events, either or both assumed to be true in the absence of positive proof, necessary to enable the commander in the process of planning to complete an estimate of the situation and make a decision on the COA. An assumption encompasses the issues over which a commander normally does not have control, both friendly and adversary.

- If you make an assumption, you must direct resources towards turning it into a fact (intelligence collection, RFI etc.) and/or develop a branch plan.
- Assumptions that address gaps in knowledge are critical for the planning process to continue.
- Subordinate commanders must treat assumptions given by the higher headquarters as *facts*. If the commander or staff does not concur with the higher commander's planning assumptions, they should be challenged before continuing with the planning process. All assumptions should be continually reviewed to ensure validity.
- When dealing with an assumption, changes to the plan may need to be developed should the assumption prove to be incorrect.
- Because of their influence on planning, the fewest possible assumptions are included in a plan.
- A *valid assumption* has three characteristics: it is *logical, realistic, and essential* for the planning to continue. Assumptions should be continually re-validated.
- Assumptions are made for both friendly and adversary situations. The planner should assume that the adversary would use every capability at his disposal (i.e., nuclear, biological, and chemical (NBC), asymmetric approach, etc.) and operate in the most efficient manner possible.
- Planners should never assume an adversary has less capability than anticipated, nor assume that key friendly forces have more capability than has been demonstrated.

> ***Assumptions (examples)***
>
> *• LOCs outside the theater will remain open.*
>
> *• Annual pre-planned U.S. force movements (annual and temporarily deployed) continue at historic levels.*
>
> *• The USG will make significant, long-term resource commitment to reconstruction in Haiti.*
>
> *• NATO countries will provide basing and over-flight.*
>
> *• Red supported terrorists will conduct operations in Blue in attempt to destabilize the government.*

(1) Assumptions are used in the planning process at each command echelon. Usually, commanders and their staffs should make assumptions that fall within the scope of their operational environment. We often see that the higher the command echelon, the more assumptions will be made. Assumptions enable the commander and the staff to continue planning despite a lack of concrete information. They are artificial devices to fill gaps in actual knowledge, but they play a crucial role in planning. A poor assumption may partially or completely invalidate the entire plan—to account for a possible wrong assumption, planners should consider developing branches to the basic plan. Assumptions should be kept at a minimum.

> *"Extended DOD participation in Haitian recovery will impact Global Force Posture/Capability." (Operation Unified Response)*

(2) Assumptions are not rigid. Their validation will influence intelligence collection. They must be continuously checked, revalidated, and adjusted until they are proven as facts or are overcome by events. Ask yourself three simple questions while considering an assumption:

- Is it *logical*?
- Is it *realistic*?
- Is it *essential for planning*?

(3) Probably one of the most important considerations of an assumption is that if you cannot validate the assumption during the planning process by either rejecting it or turning it into a fact, then you *MUST* consider a branch plan to cover the possible repercussions of an invalid assumption during execution.

4. Key-Step — 4: Determine Operational Limitations (Limiting Factors): Constraints/Restraints

Operational limitations are actions required or prohibited by higher authority and other restrictions that limit the commander's freedom of action, such as diplomatic agreements, political and economic conditions in affected countries, host nation issues and support agreements.

a. A *constraint* is a requirement placed on the command by a higher command that *dictates an action*, thus restricting freedom of action (*must do*). The superior's directive normally indicates circumstances and limitations under which one's own forces will initiate and/or continue their actions. Therefore, the higher commander may impose some constraints on the commander's freedom of action with respect to the actions to be conducted. These constraints will affect the selection of COAs and the planning process. Examples include tasks by the higher command that specify: "Be prepared to . . . "; "Not earlier than . . . "; "Not later than . . ."; "Use coalition forces . . ." Time is often a constraint, because it affects the time available for planning or execution of certain tasks.

> ***Constraint example:*** *General Eisenhower was required to liberate Paris instead of bypassing it during the 1944 campaign in France.*

> ***Constraint example:*** *Simultaneous Humanitarian Assistance Operation beginning the same day as the air campaign on the kick-off of Operation Enduring Freedom.*

b. A ***restraint*** is a requirement placed on the command by a higher command that *prohibits an action*, thus restricting freedom of action (can't do).

> ***Restraint example:*** *General MacArthur was prohibited from striking Chinese targets north of the Yalu River during the Korean War.*

> ***Restraint example:*** *President Musharraf's request that the American plan for Afghanistan not involve the Indian government nor their military.*

c. Some operational limitations are commonly expressed as *Rules of Engagement* (ROE). Operational limitations may restrict or bind COA selection or may even impede implementation of the chosen COA. These ROE or operational limitations become more complex in multinational or coalition operations. Commanders must examine the operational limitations imposed on them, understand their impacts, and develop options that minimize these impacts in order to promote maximum freedom of action during execution.[2]

Note: Constraints and restraints collectively comprise "operational limitations" on the commander's freedom of action. Remember restraints and constraints do not include doctrinal considerations. Do not include self-imposed limitations during this portion of the process

5. Key-Step Five: Determine Termination Criteria, Own Military Endstates, Objectives and Initial Effects

> *"One should not take the first step in war without considering the last."*
>
> *Carl von Clausewitz, On War, edited and translated by Michael Howard*

a. Termination of Joint Operations.

(1) Because the very nature of termination criteria begins shaping the futures of contesting nations or groups, it is imperative that we fundamentally understand what termination criteria are. What *military conditions* must be produced in the theater to achieve the strategic goal and how those military conditions serve to leverage the transition from war to peace is a fundamental aspect of conflict termination. It is important to recognize that these conditions defined during mission analysis may change as the operation planning, and then the operation, unfolds. Nonetheless, the process of explicitly and clearly defining terminal conditions is an important one since it requires careful dialogue between civilian (strategic) and military (operational) leadership which may, in turn, offer some greater assurance that the defined endstate is both politically acceptable and militarily attainable.

> ***Termination Criteria***
>
> - *In many cases, military power can be applied to establish a specified condition that serves as a transition point for the application of other instruments to augment or replace military efforts.*
> - *Other means of power will provide the driving force for achievement of the objectives.*
> - *The planner must determine what that point is and define the conditions that exist at that point in time and space.*
> - *Those conditions are event-driven rather than time-driven to define the point at which the President no longer requires the military instrument of power.*
> - *This list of conditions comprises the termination criteria.*
> - *We re-write this list into paragraph format that contains the commander's vision.*
> - *We call this the military endstate.*

(2) Based on the President's strategic objectives that comprise a desired national strategic endstate, the supported JFC can develop and propose *termination criteria* which are the specified standards approved by the President or the SECDEF that must be met before a joint operation can be concluded. Military operations seek to end war on favorable terms. Knowing when to end a war and how to preserve the objectives achieved are vital components of campaign design and relate to theater-strategic planning. Before hostilities begin, the theater commander must have a clear sense of what the endstate is. He also needs to know whether (and how) ending the conflict at the point he has recommended

[2] *JP 5-0, Joint Operations Planning.*

will contribute to the overall strategic goals,[3] so conflict termination can be considered the link or leverage between endstate and post hostilities.[4] The design and implementation of leverage and the ability to know how and when to terminate operations are part of the overall implementation of operational design. Not only should termination of operations be considered from the outset of planning, but it should also be a coordinated effort with appropriate other government agencies (OGA), international governmental organizations (IGO), non-governmental organizations (NGO), and multi-national partners.

Termination Criteria (example)

- *Blue borders are secure.*
- *A stable security environment exists in Blue, Red, and Orange.*
- *Red no longer poses a threat to regional countries.*
- *Non-DOD agencies and/or international agencies effectively lead and conduct reconstruction and humanitarian assistance operations.*
- *U.S. military forces return to shaping and security cooperation activities.*

(3) If the termination criteria have been properly set and met, the necessary leverage should exist to prevent the enemy from renewing hostilities and to dissuade other adversaries from interfering. Moreover, the national strategic endstate for which the United States fought should be secured by the leverage that U.S. and multinational forces have gained and can maintain.

(4) As discussed, in order to have acceptable termination criteria, it is of utmost importance to have an achievable national strategic endstate based on clear national strategic objectives.[5] One of the first considerations in Operational Design that a commander must address is:

"What conditions are required to achieve the objectives?" ***(Ends)***

(5) This consideration requires clarity first on strategic objectives (President, SECDEF) before operational military objectives (CCMD, JFC) are defined. This consideration also requires a measure of clarity on those "conditions" which are required to achieve the strategic objectives. Those conditions are the conflict termination criteria which is the bridge over which armed conflict transitions into a post-conflict phase and between the political objectives and the desired strategic endstate.[6]

(6) The commander will be in a position to provide the President, SECDEF and CJCS with critical information on enemy intent, objectives, strategy, and chances of success. Commanders consider the nature and type of conflict, the objectives of military force, the plans and operations that will most affect the enemy's judgment of cost and risk, and the impact of military operations on alliance and coalition warfare.[7] Military strategic advice to political authorities regarding termination criteria should be reviewed for military feasibility, adequacy, and acceptability as well as estimates of time, costs, and the military forces required to reach the criteria.[8]

[3] *Joint Publication 1, Doctrine for the Armed Forces of the United States, Joint Publication 3-0, Joint Operations.*

[4] *Should Deterrence Fail: War Termination in Campaign Planning, James W. Reed, Parameters, Summer 1993.*

[5] *Joint Publication 1, Doctrine for the Armed Forces of the United States.*

[6] *Joint Operational Warfare, Dr. Milan Vego, 20 September 2007, IX-180.*

[7] *FM 100-5, Operations, FM 3-0, FM 3-0, Operations.*

[8] *Joint Publication 1, Doctrine for the Armed Forces of the United States.*

(a) The commander will then formulate termination criteria that he reasons will set the enduring conditions to achieving the military and strategic endstates. Those properly conceived termination criteria are key to ensuring that the achieved military objectives endure.

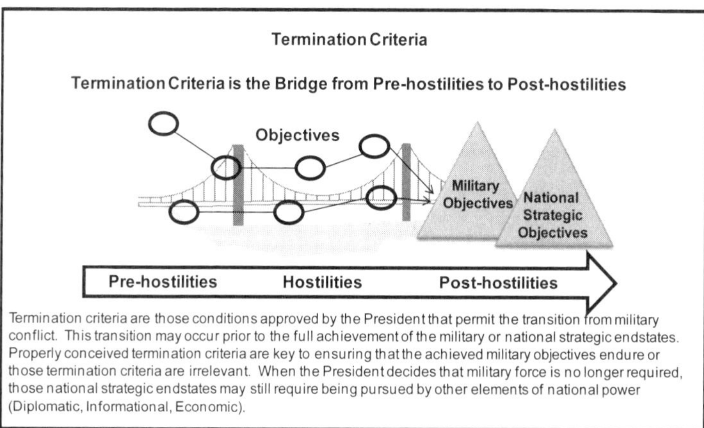

Termination Criteria
Termination Criteria is the Bridge from Pre-hostilities to Post-hostilities
Objectives
Military Objectives
National Strategic Objectives
Pre-hostilities
Hostilities
Post-hostilities
Termination criteria are those conditions approved by the President that permit the transition from military conflict. This transition may occur prior to the full achievement of the military or national strategic endstates. Properly conceived termination criteria are key to ensuring that the achieved military objectives endure or those termination criteria are irrelevant. When the President decides that military force is no longer required, those national strategic endstates may still require being pursued by other elements of national power (Diplomatic, Informational, Economic).

Figure XIV-1. Termination Criteria are set during Pre-hostilities and represent the bridge from Hostilities to Post-Hostilities.

(7) To facilitate development of effective termination criteria, it must be understood that U.S. forces must follow through in not only the "dominate" phase, but also the "stabilize" and "enable civil authority" phases to achieve the leverage sufficient to impose a lasting solution.[9]

(8) Further, development of a military endstate is complementary to and supports attaining the specified termination criteria and national strategic endstate. As stated, the termination of operations must be considered from the outset of planning in a coordinated effort with relevant agencies, organizations, and multinational partners. This ability to understand how and when to terminate operations is instrumental to operational design.[10]

(9) There are three approaches to obtaining national strategic objectives by military force:[11]

(a) Imposed settlement: destroy critical functions and assets and/or adversary military capabilities.

(b) Negotiated settlement: political, diplomatic and military. Negotiating power springs from two sources: military success and military potential. In the past 200 years, more than half of the wars have ended in negotiated settlements. However, some wars may not have a formal ending, because the defeated side is unwilling to negotiate or because there is no one with whom to negotiate. Such conflicts are often transformed into insurgency and counterinsurgency, for example the US experience in Iraq and the French experience in Indochina.

[9] *JP 5-0, Joint Operation Planning, JP 3-0, Joint Operations.*

[10] *Ibid*

[11] *Joint Operational Warfare, Dr. Milan Vego, 20 September 2007, p. IX-178.*

(c) Indirect Approach: erode adversary's power, influence, and will; undermine the credibility and legitimacy of their political authority; and undermine their influence and control over, and support by, the indigenous population. This is in relation to the irregular challenges posed by non-state actors and is irregular warfare.[12]

(10) In its strategic context, military victory is measured in the attainment of the national strategic endstate and associated termination criteria. Termination criteria for a negotiated settlement will differ significantly than those of an imposed settlement.

> *"The success of Desert Storm militarily was astounding, but Saddam Hussein reminded the world for the next twelve years that the political victory was incomplete. In April 2003, Operation Iraqi Freedom (OIF) completed the task of ousting Saddam Hussein, but a political victory in Iraq remains unsecured today."*
>
> *The Theater Commander: Planning for Conflict Termination; Colonel John W. Guthrie, U.S. Army War College, Carlisle Barracks, Carlisle, PA.*

(11) At the operational level, then, the military contribution should serve to increase (or at least not decrease) the leverage available to national decision-makers during the terminal phases of a conflict.

(12) One could view conflict termination not as the end of hostilities, but as the transition to a new post-conflict phase characterized by both civil and military problems. This consideration implies an especially important role for various civil affairs functions. It also implies a requirement to plan the interagency transfer of certain responsibilities to national, international, or nongovernmental agencies. Effective battle hand-off requires planning and coordination before the fact.[13]

> *"Those who can win a war well can rarely make a good peace, and those who could make a good peace would never have won the war."*
>
> *Winston Churchill, My Early Life: A Roving Commission*

(13) Termination Summary. Knowing when to terminate military operations and how to preserve achieved advantages is essential to achieving the national strategic endstate. When and under what circumstances to *suspend* or *terminate* military operations is a *political* decision. Even so, it is essential that the CJCS and the supported JFC advise the President and SECDEF during the decision-making process. The supported JFC should ensure that political leaders understand the implications, both immediate and long term, of a suspension of hostilities at any point in the conflict. Once established, the national strategic objectives enable the supported commander to develop the military endstate, recommended termination criteria, and supporting military objectives.

(a) Planners must plan for conflict termination from the outset of the planning process and update these plans as the campaign evolves. To maintain the proper perspective, they must know what constitutes an acceptable political-military endstate (i.e., what military conditions must exist to justify a cessation of combat operations?). In examining the proposed national strategic endstate, the CCDR and the staff must consider whether it has reasonable assurance of ending the fundamental problem or underlying conditions that instigated the conflict in the first place.

(b) When addressing conflict termination, campaign planners must consider a wide variety of operational issues to include disengagement, force protection, transition to post-

[12] Joint Publication 1, Doctrine for the Armed Forces of the United States.

[13] Should Deterrence Fail: War Termination in Campaign Planning, James W. Reed, Parameters, Summer 1993.

conflict operations, and reconstitution and redeployment. Planners must also anticipate the nature of post-conflict operations, where the focus will likely shift to Stability, Security, Transition and Reconstruction (SSTR); for example, peace operations, foreign humanitarian assistance, or enforcement of exclusion zones.

(c) In formulating the theater campaign plan, the CCDR and staff should consider the following:

1 Conflict termination, end of the joint operation, is a key aspect of the campaign planning process.

2 Emphasize backward planning; decision makers should not take the first step toward hostilities or war without considering the last step.

3 Define the conditions of the termination phase. The military objectives must support the political aims — the campaign's conflict termination process is a part of a larger *implicit bargaining process*, even while hostilities continue. The military contribution can significantly affect the political leverage available to influence that process.

4 Consider how efforts to eliminate or degrade an opponent's command and control (C2) may affect, positively or negatively, efforts to achieve the termination objectives. Will opponents be able to affect a cease-fire or otherwise control the actions of their forces?

5 Interagency coordination plays a major role in the termination phase. View conflict termination not just as the end of a joint operation and disengagement by joint forces, but as the transition to a new post-hostilities phase characterized by both civil and military problems.[14]

(d) Since war is fought for political aims, it is only successful when such aims are ultimately achieved. Success on the battlefield does not always lead to a successful strategic endstate. Making sure that it does requires the close collaboration of political and military leaders. As noted, a period of post-hostilities activities may exist in the period from the immediate end of the hostilities to the accomplishment of the national strategic goals and objectives, and it is imperative that this be recognized and planned for.

> *"The object of war is a better state of peace — even if only from your own point of view. Hence it is essential to conduct war with constant regard to the peace you devise."*
>
> *B.H. Liddell Hart, Strategy (2nd revised edition) New York: Penguin Group, 1991*

b. Endstate. Once the termination criteria are understood, operational design continues with the determination of the *strategic and military endstates and objectives*. The endstate gets to "why" we are developing a campaign plan and seeks to answer the question: "How does the United States strategic leadership want the OE (i.e., the region and/or potential adversary) to behave at the conclusion of the campaign?" Objectives normally answer the question of "what" needs to be done to achieve the endstate. As you might expect, the distinction between endstates and objectives can be very vague. In designing and planning, we must recognize and define two mutually supporting endstates in a single campaign – a national strategic endstate and a theater-strategic/military endstate.

[14] *JP 5-0, Joint Operation Planning.*

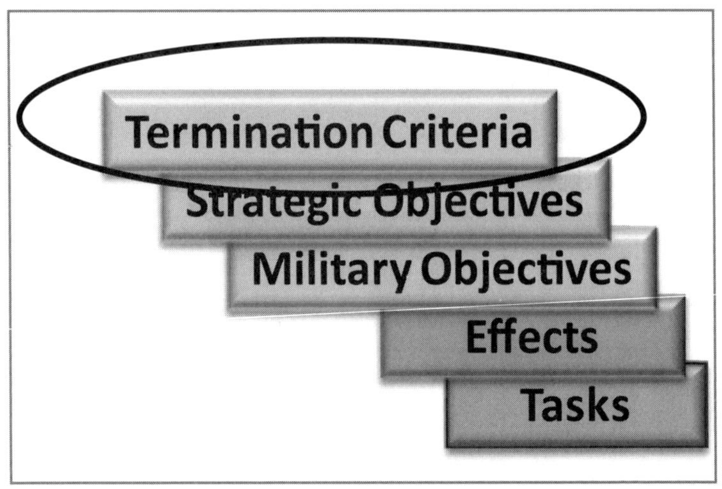

(1) The *national strategic endstate* describes the President's political, informational, economic, and military vision for the region or theater when operations conclude. National strategic endstates are derived from President/SECDEF guidance that is often vague. More often than not, senior military leaders will assist the President/SECDEF in developing that endstate. Below is an example of a national strategic endstate:

> *"Secure, stable, democratic, inclusive, non-hostile and prosperous Haiti." (USSOUTHCOM TCP)*

> *"An economically viable and stable Country X, without the capability to coerce its neighbors."*

(2) The *theater strategic or military endstate* is a subset of the national strategic endstate discussed above and generally describes the military conditions that must be met to satisfy the objectives of the national strategic endstate. (We will develop our military endstate after analyzing the tasks required by strategic direction.) Strategic objectives clarify and expand upon endstate by *clearly defining* the decisive goals that must be achieved in order to ensure we achieve U.S. policy, for example:

- *Loss of life and human suffering of Haitians minimized.*
- *Scope of crisis no longer exceeds capability of HN, UN and international efforts.*
- *US military personnel are redeployed.*

- *Country X deterred from coercing its neighbors.*
- *X ceases support to regional terrorism.*
- *X's WMD inventory, production and delivery means reduced.*

(a) Getting the answers to the above decisive goals is what makes mission analysis different at this level when compared to the tactical level – you will not find the clear and definitive guidance in one location that you may be used to. There is no "higher order" to cut and paste from. Instead the NSS, NDS, NMS, PPD, SECDEF and Presidential speeches, and verbal guidance all provide input to help define an endstate and corresponding objectives. With so many sources of guidance, consistency is normally an issue to overcome. Though not directive in nature, guidance contained in various United States interagency and even international directives, such as United Nations Security Council Resolutions (UNSCRs), will also impact campaign endstates and objectives.

(3) This endstate normally will represent a point in time or circumstance beyond which the President does not require the military instrument of national power to achieve remaining objectives of the national strategic endstate. While the military endstate typically will mirror many of the conditions of the national strategic endstate, it may contain other contributory or supporting conditions. Aside from its obvious role in accomplishing both the national and military strategic objectives, clearly defining the military endstate conditions promotes unified action, facilitates synchronization, and helps clarify (and may reduce) the risk associated with the joint campaign or operation. Commanders should include the military endstate in their planning guidance and commander's intent statement.[5]

(4) The CJCS or the supported CCDR may recommend a military endstate, but the President or SECDEF should formally approve it. A clearly defined military endstate complements and *supports attaining the specified termination criteria and objectives associated with other instruments of national power.* The military endstate helps affected CCDRs modify their theater strategic estimates and begin mission analysis even without a pre-existing OPLAN. The CCDR must work closely with the civilian leadership to ensure a clearly defined military endstate is established and that the military objectives support the political ones, and will lead to the desired endstate. This backward planning cycle ensures a military victory with lasting political success.

(5) The CCDR also should anticipate that military capability likely would be required in some capacity in support of other instruments of national power, potentially before, during, and after any required large-scale combat. Commanders and their staffs must understand that many factors can affect national strategic objectives, possibly causing the endstate to change even as military operations unfold. A clearly defined endstate is just as necessary for situations across the range of military operations that might not require large-scale combat. While there may not be an armed adversary to confront in some situations, the JFC still must think in terms of ends, ways, and means that will lead to success and endstate attainment.

(6) Remember that a primary consideration for a supported CCDR during mission analysis is the *national strategic endstate* — that set of national objectives and related guidance that define strategic success from the President's perspective. This national strategic endstate will reflect the "broadly" expressed political, military, economic, social, informational, and other circumstances that should exist after the conclusion of a campaign or operation. Below is an example of a broad national strategic endstate directly from the *Chairman's National Military Strategic Plan for the War on Terrorism:*

> *"Violent extremist ideology and terrorist attacks eliminated as a threat to the way of life of free and open societies. A global environment that is inhospitable to violent extremism, wherein countries have the capacity to govern their own territories, including both the physical and virtual domains of their jurisdictions. Partner countries have in place laws, information sharing, and other arrangements, that allow them to defeat terrorists as they emerge, at the local and regional levels."*
>
> *Chairman's National Military Strategic Plan for the War on Terrorism*

[5] *JP 5-0, Joint Operations Planning.*

(7) Often, the military endstate is achieved before the national strategic endstate. While it will mirror many of the objectives of the national strategic endstate, the theater strategic or military endstate may contain other supporting objectives and conditions. An example of a theater strategic or military endstate:

> *"A defeated Country X where WMD delivery, production, and storage, as well as conventional force projection capabilities, are destroyed, and its remaining military is reorganized to adequately defend its borders."*

(8) CDRs include a discussion of the national strategic endstate in their initial planning guidance. This ensures that joint forces understand what the President wants the situation to look like at the conclusion of U.S. involvement. The CCDR and subordinate JFCs typically include the *military endstate* in their intent statement.

> *"My intent is to persuade country X through a show of coalition force to stop intimidating its neighbors and cooperate with diplomatic efforts to abandon its WMD programs. If X continues its belligerence and expansion of WMD programs, we will use force to reduce X's ability to threaten its neighbors, and restore the regional military balance of power. Before U.S. and coalition forces redeploy, X's military will be reduced by half, its modern equipment destroyed, its capability to project force across its borders eliminated, and its WMD stores, production capacity, and delivery systems eliminated."*

c. Objectives. *An objective is a clearly defined, decisive, and attainable goal toward which every military operation is directed.* Objectives and their supporting effects provide the basis for identifying tasks to be accomplished.

(1) Objectives Prescribe Friendly Goals. They constitute the aim of military operations and are necessarily linked to political objectives (simply defined as: what we want to accomplish). Military objectives are one of the most important considerations in campaign and operational design. They specify what must be accomplished and provide the basis for describing campaign effects.

(2) A clear and concise endstate allows planners to examine objectives that support the endstate. Objectives describe what must be achieved to reach the endstate. These are usually expressed in military, political, economic, and informational terms and help define and clarify what military planners must do to support the achievement of the national strategic endstate. Objectives developed at the national-strategic and theater-strategic levels are the defined, decisive, and attainable goals towards which all operations-not just military operations-and activities are directed within the JOA.

National Strategic Objectives from the QDR, February 2010:

- *Prevail in Today's Wars.*
- *Prevent and Deter Conflict.*
- *Prepare to defeat adversaries and succeed in a Wide Range of Contingencies.*
- *Preserve and enhance the All-Volunteer Force.*

(3) Objective statements do not suggest or infer the ways and means for accomplishment. Passive voice is a convention that can assist a commander in distinguishing an objective from an effect and state objectives without inferring potential ways and means, thereby broadening the range of possible actions to achieve the objectives.

(4) An objective statement should have the following attributes:

- Establishes a single goal: a desired result, providing one concise "end" toward which operations are directed.
- Is specific: identifies the key system, node or link to be affected.
- Link to higher-level objectives directly or indirectly.
- Unambiguous as possible.
- Uses passive voice.
- Prescriptive in nature.
- Does not infer causality: no words (nouns or verbs) that suggest ways and/or means.

"Provide direct, urgent humanitarian relief support to the government and people of Haiti." (Operation Unified Response)

(5) Objectives are written as concise descriptive statements: *Country Red is no longer a threat to regional peace. Country Blue's IRBM capability is eliminated.*

Determine the Valid Objective

- *Eliminate terrorists in Brown.*
  - *Uses active voice, re-write as follows: "Terrorists are eliminated in Brown."*
- *Pre-invasion borders are restored after offensive operations.*
  - *Uses a causality to explain how it will achieve the effect, re-write as follows: "Restoration of pre-invasion boarders."*
- *A free and democratic Blue under the duly elected government.*
  - *This objective is correct.*

(6) At the outset of Operation Iraqi Freedom, the SECDEF set eight mission objectives for the operation. The eight mission objectives for *Operation Iraqi Freedom* were:

- End the regime of Saddam Hussein.
- Eliminate Iraq's weapons of mass destruction.
- Capture or drive out terrorists.
- Collect intelligence on terrorist networks.
- Collect intelligence on Iraq's weapons of mass destruction activity.
- Secure Iraq's oil fields.
- Deliver humanitarian relief and end sanctions.
- Help Iraq achieve representative self-government and ensure its territorial integrity.

(7) The stated U.S. military objectives for *Operation Enduring Freedom* were:

- The destruction of terrorist training camps and infrastructure within Afghanistan.
- The capture of al Qaeda leaders.
- The cessation of terrorist activities in Afghanistan.
- Make clear to Taliban leaders that the harboring of terrorists is unacceptable.
- Acquire intelligence on al Qaeda and Taliban resources.
- Develop relations with groups opposed to the Taliban.
- Prevent the use of Afghanistan as a safe haven for terrorists.
- Destroy the Taliban military allowing opposition forces to succeed in their struggle.
- Military force would help facilitate the delivering of humanitarian supplies to the Afghan people.

(8) Strategic military objectives define the role of military forces in the larger context of national strategic objectives. This focus on strategic military objectives is one of the most important considerations in operational design. The nature of the political aim, taken in balance with the sources of national strength and vulnerabilities, must be compared with the strengths and vulnerabilities of the adversary and/or other factors in the operational environment to arrive at reasonably attainable strategic military objectives. Strategic objectives must dominate the planning process at every juncture.[16]

> *"During Desert Storm the U.S. Central Command's sweeping envelopment maneuver was brilliantly effective, not only because it neutralized the Republican Guard forces, the Iraqi army's center of gravity, it also placed a significant allied force in position to threaten Baghdad, thus creating added incentive for Iraq to agree to an early cease-fire. An operational decision had affected an opponent's strategic calculus by creating additional allied leverage."*
>
> *Should Deterrence Fail: War Termination in Campaign Planning,James W. Reed, Parameters, Summer 1993*

(9) Objectives may change over time. You may see a change in operational level objectives that still support strategic or national level objectives. You may even see changes in strategic objectives that of course will affect both operational and tactical level objectives. Figure XIV-2[17] on the following pages is an historical example of how objectives can change throughout a conflict due to operational necessity and strategic guidance.

[16] *JP 5-0, Joint Operations Planning.*

[17] *Warfare Studies Institute, Joint Air Operations planning Course, Joint Air Estimate Planning Handbook, Version 5, 30 January 2005, pg 15-16.*

CHANGING OBJECTIVES

The Korean War—A case of changing political and national objectives while engaged in combat.

The Korean War clearly demonstrates the linkage between the political and military objectives. The political objectives changed three times during the conflict, mandating major revisions of and limitations to the campaign plans.

OBJECTIVE: FREE THE REPUBLIC OF KOREA

The Democratic People's Republic of Korea invaded the Republic of Korea on 24 June 1950. President Harry S. Truman heeded the request of the United Nations Security Council that all members "furnish such assistance to the Republic of Korea as may be necessary to repel armed attack and restore international peace and security in the area." [18] *This translated into guidance to United Nations Command Far East Command, then commanded by Gen Douglas MacArthur, "to drive forward to the 38th parallel, thus clearing the Republic of Korea of invasion forces."* [19] *MacArthur accomplished this by first heavily reinforcing the remaining pocket of South Korean resistance around Pusan with United States military forces. Using these forces he pushed northward and executed the highly successful amphibious landing of two divisions behind enemy lines at Inchon. Airpower was used to wage a comprehensive interdiction campaign against the enemy's overextended supply routes. United Nations forces achieved the original objective by October 1950.*

OBJECTIVE: CHANGE 1, FREE ALL OF KOREA

In view of the success at Inchon and the rapid progress of United Nations forces northward, the original objective was expanded. "We regarded," said Secretary of Defense Marshall, "that there was no . . . legal prohibition against passing the 38th parallel." [20] *The feeling was that the safety of the Republic of Korea would remain in jeopardy as long as remnants of the North Korean Army survived in North Korea.*[21] *This was expressed in a UN resolution on 7 October 1950 requiring "all necessary steps be taken to ensure conditions of stability throughout Korea."*[22] *MacArthur then extended the counteroffensive into North Korea. However, the enemy's logistic tail extended northward into the People's Republic of China. Because the United Nations and United States did not want to draw China into the war, targets in China were off-limits. For this reason, use of airpower was limited largely to close air support. United Nations forces advanced to near the Chinese border. The second objective was achieved, temporarily at least, by November 1950.*

[18] *Robert Frank Futrell, Ideas, Concepts, Doctrine, Vol. 1 (Maxwell AFB, Ala.: Air University Press, 1989), 293.*

[19] *Ibid., 297.*

[20] *Ibid.*

[21] *Ibid.*

[22] *Ibid.*

OBJECTIVE: CHANGE 2, SEEK CEASE-FIRE, RESOLVE BY NEGOTIATION

On 26 November 1950, the Chinese Communists launched a massive counterattack that shattered the United Nations forces, forcing a retreat from North Korea. MacArthur realized he was in a no-win situation and requested permission to attack targets in China. The Joint Chiefs of Staff's (JCS) guidance was neither to win nor to quit; they could only order him to hold. They vaguely explained that, if necessary, he should defend himself in successive lines and that successful resistance at some point in Korea would be "highly desirable," but that Asia was "not the place to fight a major war." [23]*On 14 December, at the request of the United States, the United Nations adopted a resolution proposing immediate steps be taken to end the fighting in Korea and to settle existing issues by peaceful means. "On 9 January 1951, the Joint Chiefs of Staff informed MacArthur that while the war would be limited to Korea, he should inflict as much damage upon the enemy as possible."* [24] *Limiting the conflict to Korea negated our ability to use naval and airpower to strategically strike enemy centers of gravity located within China. On 11 April 1951, Truman explained the military objective of Korea was to "repel attack. . . to restore peace. . . to avoid the spread of the conflict."* [25] *The political objectives and the military reality placed MacArthur in a difficult situation. MacArthur proved unwilling to accept these limited objectives and was openly critical of the Truman administration. Truman relieved him of command.*

The massive Chinese attacks mounted in January and April of 1951 failed because of poor logistical support. United Nations forces sought to exact heavy casualties upon the enemy rather than to defend specific geographical objectives. As the Chinese and North Koreans pressed forward, their lines of communication were extended and came under heavy air interdiction attack. By May 1951, United Nations forces had driven forward on all fronts. With communist forces becoming exhausted, negotiations for a cease-fire began on 10 July 1951. The quest for the third objective finally ended on 27 July 1953 with implementation of a cease-fire that is still in force.

Figure XIV-2. Changing Objectives.

"Sometimes military and political objectives become decisive points in the operational design because they are essential not only to reach the endstate, but critical to affecting the enemy's center(s) of gravity or protecting friendly center(s) of gravity. Above all, keep in mind that while objectives can be rephrased as decisive points, decisive points are not synonymous with objectives. Objectives always refer to endstate. Decisive points always refer to operational level center(s) of gravity."

Operational Design: A Methodology for Planners,
Dr. Keith Dickson, Professor of Military Studies, JFSC

[23] *William Manchester, American Caesar (New York: Dell Publishing, 1983), 617.*

[24] *Robert Frank Futrell, Ideas, Concepts, Doctrine, Vol. 1 (Maxwell AFB, Ala.: Air University Press, 1989), 302.*

[25] *Ibid.*

d. Effects. One of the most confusing aspects encountered in recent joint doctrinal changes has been the infusion of *effects* language and processes into the planning process. The issue is made even more difficult in view of the conceptual overlap between *effects theory* with the established principles of operational art. There have been numerous effects-related models promulgated by a number of agencies. These include Effects-Based Operations (EBO) and Effects-Based Planning (EBP). Neither of these terms are codified in U.S. joint doctrine, though one will often see them used in some staffs. Presently, Joint doctrine simply uses the term "effects."

(1) Effects help CDR's and their staffs understand and measure conditions for achieving objectives. An effect is a physical and/or behavioral state of a system that results from an action, a set of actions, or another effect. Just as termination conditions support/define achieving the endstate, a *desired effect* can also be thought of as a condition that can support achieving an associated objective, while an *undesired effect* is a condition that can inhibit progress toward an objective.

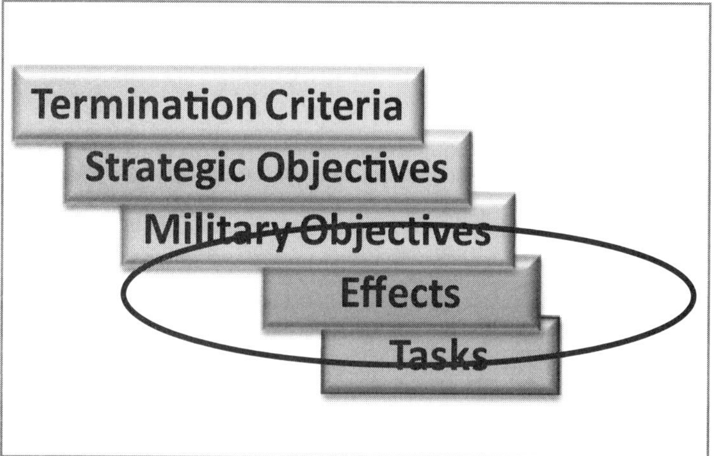

Effects are derived from objectives. *They help bridge the gap between objectives and tasks by describing the conditions that need to be established or avoided within the JOA to achieve the desired endstate.*

(2) The use of effects planning is not new; good CDRs and staffs have always thought and planned this way. Effects should not be over-engineered into a list of equations, data bases and checklists. The use of effects during planning is reflected in the steps of JOPP as a way to clarify the relationship between objectives and tasks and help the CDR and staffs determine conditions for achieving objectives.

(3) Before one can appreciate the commonality between operational art and effects-based concepts, it is important to understand the fundamental tenets of operational art. This document cannot detail all aspects of operational art, and those unfamiliar with operational art must first gain an awareness of this theoretical framework before launching into operational-level planning. This lack of knowledge of operational art is often the source of much of the current disconnect found in many effects-based concepts. The most critical of these is the nature of the objective and its relationship to the *center of gravity* (COG). Please review the discussion on the COG under Key-Step 6.

(a) Operational design is defined by joint doctrine as, "The conception and construction of the framework that underpins a campaign or major operation plan and its subsequent execution." Central to operational design is for the planning staff to understand the *desired* endstate and requisite objectives necessary to achieve the desired endstate.

(b) An objective is a clearly defined, decisive, and attainable goal toward which every military operation is directed. Objectives and their supporting effects provide the basis for identifying tasks to be accomplished. Joint operation planning integrates military actions and capabilities with those of other instruments of national power in time, space, and purpose in unified action to achieve the JFC's objectives.

(4) As observed Key-Step 6, an enemy's COG is inextricably linked to its objective (just as the friendly COG is linked to its own objectives). In fact, this is a critical aspect to determining an enemy's COG, a fact that is generally unclear in effects–based concepts. An enemy's interconnected PMESII system is meaningless if one does not first (correctly) assess an enemy's probable endstate and objectives. One must remember that the *raison d'être* for a COG is to accomplish an objective.

(5) Destroy or defeat an enemy's COG and you have severely inhibited an enemy's ability to achieve his objective (unless another source of power assumes the role of COG). In a theoretical construct, one should see an enemy COG as something that stands between the friendly COG and the friendly objective(s). Thus, an operation is invariably focused upon an enemy's COG in order to achieve *friendly* objectives.

(6) Effects vs. Objectives.

(a) The next point that often serves to confuse joint planners is the role of effects in the COG-Objective construct. Joint doctrine defines an effect as:

> ***Effect**: (1) The physical or behavioral state of a system that results from an action, a set of actions, or another effect. (2) The result, outcome, or consequence of an action. (3) A change to a condition, behavior, or degree of freedom. JP 3-0, 22 March 2010*

(b) Thus, an effect is an outcome from an action, and in its simplest terms it is the condition that one hopes for (if a desired effect) upon the accomplishment of a task or action assigned to a subordinate command. In operational art terms, an effect may be considered as the consequence from attacking/controlling a decisive point or an enemy's critical vulnerability/requirement and/or capability enroute to the defeat of an enemy's COG. A desired effect should become apparent as the planning staff completes its COG analysis and is often found in the commander's planning guidance and/or intent. *An effect, however, is not necessarily an end onto itself.* Rather, an effect's purpose should be to ensure a friendly COG's progress to its respective objective (s). Figure XVI-3 offers a graphic depiction of the concept.

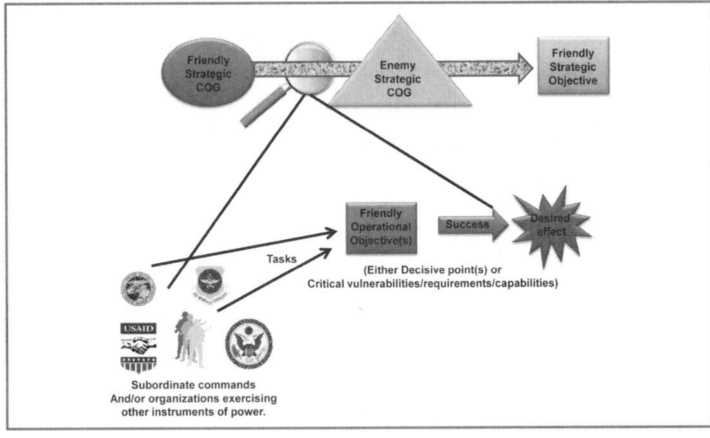

Figure XIV-3. Effects vs. Objectives (NWC 4111H).

(c) Figure XIV-3 illustrates the relationship between objectives, effects, and COGs. Upon determination of one's own objectives and COG, the planning staff assesses an enemy's probable objective(s) and its related COG. The friendly operations are then focused upon defeating the enemy COG enroute to the friendly objective. Since a direct attack upon the enemy COG is frequently impractical (or too costly), friendly COAs will often focus on attacking through a combination of Critical Vulnerabilities (CV), Critical Requirements (CR), Critical Capabilities (CC), and decisive points (DP) that are integral to the enemy COG. If, as depicted in Figure XIV-3, the enemy COG being attacked is at the strategic level of war, these attacks/operations become tasks for subordinate commands, functions and/or agencies against the applicable CVs, CRs, CCs, and DPs. The successful accomplishment of these subordinate operations produced desired effects (either facilitating the friendly COG and/or degrading the enemy COG).

These effects, coupled most likely with other friendly-induced effects from other complementary operations, are intended to combine/to defeat/degrade the enemy COG and allow for the accomplishment of the friendly strategic objective.

> *"Every attempt to make war easy and safe will result in humiliation and disaster."*
>
> *General William T. Sherman*

(7) The language used to craft effects is as critical as the language used to describe objectives and tasks. Effects' planning focuses all elements of national and international power to achieving national/coalition and theater strategic objectives.

(a) Although joint doctrine does not prescribe a specific convention for writing a desired effect statement, *there are four primary considerations.*

- Each desired effect should link directly to one or more objectives.
- The effect should be measurable.
- The statement should not specify ways and means for accomplishment.
- The effect should be distinguishable from the objective it supports as a condition for success, not as another objective or a task.

(b) The effect should be distinguishable from the objective it supports as a condition for success, not as another objective or a task. The same considerations apply to writing an undesired effect statement.

(8) During mission analysis, the commander considers how to achieve national and theater-strategic objectives, knowing that these likely will involve the efforts of other U.S. agencies and multinational partners.

(9) At the operational and tactical levels, subordinate commanders use the effects and broad tasks prescribed by the JFCs as the context in which to develop their supporting mission plans. Normally, the effects established by the JFCs have been developed in collaboration with subordinate commands and supporting agencies. At a minimum, the development of effects should have the full participation by the JFCs, as well as empowered representatives from interagency organizations.

(10) The use of effects in planning helps commanders and staffs use other elements of operational design more effectively by clarifying the relationships between Centers of Gravity (COG), Lines of Operation (LOO), Lines of Effect (LOE), Decisive Points, and Termination Criteria. The JFC and planners continue to develop and refine effects throughout JOPP planning steps. Monitoring progress toward attaining desired effects and avoiding undesired effects continues throughout execution.[26]

[26] *JP 5-0, Joint Operations Planning.*

(11) With a common set of desired and undesired effects, the commander can issue guidance and intent to his staff and components, and work with other stakeholders to accomplish fused, synchronized, and appropriate actions on PMESII systems within the operational environment (beyond military on military) to attain the desired effects and achieve objectives.

Operational Design and Effects

- *Developed by CCMD based on an assessment of the operational environment and objectives.*
- *Integrate, to extent possible, all "stakeholder" input – JTF, Embassy (s) and other non-military stakeholders.*
- *Achieved by the cumulative outcome of the DIME tasks/activities executed over time - multiple phases.*
- *A JTF CDR may develop specific effects within his JOA to facilitate attainment of the campaign effects and objectives. These must be nested within those developed by the higher command.*

(12) Key to the effects is full participation of all of the players – military and other elements of national power – in a fully inclusive process of assessing, planning and eventually directing actions.

Why Use Effects

- *Establishes clear links between what we want to do (objectives), to what has to be done to achieve objectives (effects) and the actions (tasks) required to achieve desired effects.*
- *Recognizes that today's adaptive adversary operates within a complex, interconnected operational environment. Focus on creating effects on its major sub systems (PMESII) – desired and undesired.*
- *Provides a format for better integration of interagency by focusing operational level planning on effects, not tactical tasks.*
- *Uses military actions to set the conditions for and exploits the affects of interagency actions – military may not be the decisive national (DIME) capability.*
- *Provides a means to measure progress in achieving objectives.*

e. Effects in the Planning Process-The Bottom Line. While the commander and unit SOP will dictate how the staff will apply effects-based processes, for the purposes of this document, one may find the following considerations of use:

(1) Operational Design.

(a) Apply the constructs of operational art. Ensure clarity between objectives and desired effects.

(b) When assessing possible enemy COGs, ensure the JIPOE is inclusive of the full PMESII (as appropriate to the level of operation) in its analysis. Recognize the duplication of the concept of nodes and links with the terminology of COG deconstruction. To avoid confusion, do not use both sets of terminology in the same analysis.

(2) Commander's Intent. If the commander has expressed specific desired effects and success criteria, ensure they are captured in the commander's intent and/or planning guidance.

(3) Assessment. Select and assess meaningful criteria as appropriate, though continue to maintain a broader perspective that cannot be conveyed by color-coded indicators.

f. Tasks. Theater tasks are those actions that the CCDR must execute in order to achieve the strategic effects which support overarching strategic objectives. These tasks are designed to be the broad theater strategic tasks from which subordinates derive their own operational and tactical tasks.

Task: *An action or activity (derived from an analysis of the mission and concept of operations) assigned to an individual or organization to provide a capability.*

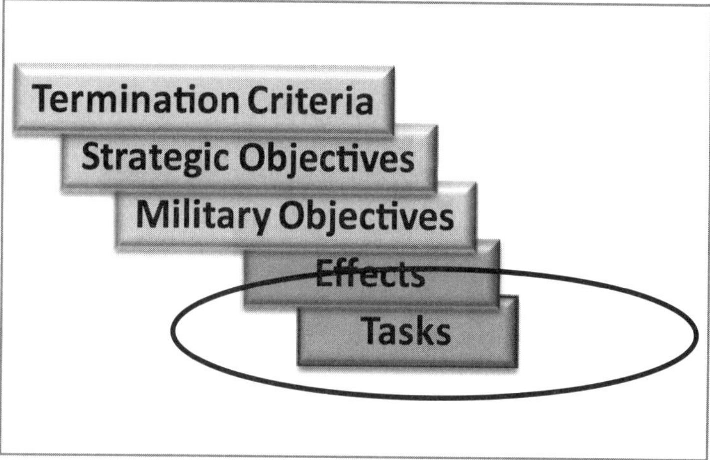

(1) Once the CDR and staff understand the objectives and effects that define the operation, they then match appropriate tasks to desired effects. Task determination begins during mission analysis, extends through COA development and selection, and provides the basis for the tasks eventually assigned to subordinate and supporting commands in the operation plan or order.

Objectives *prescribe friendly goals.*

Effects *describe the state of system behavior in the operational environment.*

Tasks *direct friendly action.*

(2) The CDR emphasizes the development of effects-related tasks early in the planning process because of the obvious importance of these tasks to objective accomplishment. Each of these tasks aligns to one or more effects, reflects action on a specific system or node, is written in active voice, and can be assigned to an organization in the operation plan or order. Support tasks such as those related to logistics and communications also are identified during mission analysis.

(3) *Task Assessment*. *Mission success criteria* describe the standards for determining mission accomplishment. The CDR may include these criteria in the planning guidance so that the staff and components better understand what constitutes mission success. When these criteria are related to the termination criteria, which typically apply to the end of a joint operation and disengagement by joint forces, this often signals the end of the use of the military instrument of national power. Mission success criteria can also apply to any joint operation, subordinate phase, and joint force component operation. These criteria help the CDR determine if and when to move to the next major operation or phase.

• The initial set of these criteria determined during mission analysis becomes the basis for *assessment*. Developing an assessment plan begins during planning and is continuous.

• Ensure your staff gather tools and assessment data, understand current and desired condition, develop assessment measures and potential indicators, develop a collection plan, assign responsibilities for generating recommendations and identify any feedback mechanisms.

• Assessment uses indicators, *measures of performance (MOPs)* and *measures of effectiveness (MOEs)* to indicate progress toward achieving objectives. If the mission is unambiguous and limited in time and scope, mission success criteria could be readily identifiable and linked directly to the mission statement. For example, if a JFC's mission is to evacuate U.S. personnel from the U.S. Embassy in Grayland, then the mission analysis could identify two primary success criteria: (1) all U.S. personnel are evacuated and (2) evacuation is completed before D-Day. However, more complex operations may require MOEs and MOPs for each task, effect, and phase of the operation. For example, if a JFC's specified tasks are to ensure friendly transit through the Strait of Gray, eject Redland forces from Grayland, and restore stability along the Grayland-Redland border, then mission analysis should indicate many potential success criteria — measured by MOEs and MOPs — some for each desired effect and task.

• Measuring the status of tasks, effects, and objectives becomes the basis for reports to senior commanders and civilian leaders on the progress of the operation. The CCDR can then advise the President and SECDEF accordingly and adjust operations as required. Whether in a supported or supporting role, JFCs at all levels should develop their mission success criteria with a clear understanding of termination criteria established by the CJCS and SECDEF.

(4) *Assessment Process and Measures.* The assessment process uses indicators to shape a staff's collection effort, it uses MOPs to evaluate task performance at all levels of war and MOEs to determine progress of operations toward achieving objectives.

MOEs help answer questions like: "Are we doing the right things? Are our actions producing the desired effects? Are alternative actions required? MOPs are closely associated with task accomplishment. MOPs help answer questions like: "Was the action taken? Were the tasks completed to standard? How much effort was involved? Well-devised measures can help the commanders and staffs understand the causal relationship between specific tasks and desired effects.

MOE	MOP	INDICATOR
Answers the question: Are we doing the right things?	Answers the question: Are we doing things right?	Answers the question: What is the status of this MOE or MOP?
Measures purpose accomplishment	Measures task completion	Measures raw data inputs to inform MOEs and MOPs
Measures *why* in the mission statement	Measures *what* in the mission Statement	Information used to make measuring what or why possible
No hierarchical relationship to MOPs	No hierarchical relationship to MOEs	Subordinate to MOEs and MOPs
Often formally tracked in formal assessment plans	Often formally tracked in execution Matrixes	Often formally tracked in formal assessment plans
Typically challenging to choose the correct ones	Typically simple to choose the correct ones	Typically as challenging to select correctly as the supporting MOE or MOP

Assessment Terms

•A MOE is a criterion used to assess changes in system behavior, capability, or operational environment that is tied to measuring the attainment of an end state, achievement of an objective, or creation of an effect (JP 3-0).

•A MOP is a criterion used to assess friendly actions that is tied to Measuring task accomplishment (JP 3-0).

•In the context of assessment, an indicator is an item of information that provides insight into a measure of effectiveness or measure of performance.

• Indicators take the form of reports from subordinates, surveys and polls, and information requirements. Indicators help to answer the question *"What is the current status of this MOE or MOP?"* A single indicator can inform multiple MOPs and MOEs. Examples of indicators for the MOE "Decrease in insurgent activity" are – number of hostile actions per Province each week, number of munitions caches found per Province each week.

• MOEs assess changes in system behavior, capability, or operational environment. *Are we doing things right?* They are the operational measures of changes in conditions, both positive and negative. They measure the attainment of an endstate, achievement of an objective, or creation of an effect; they do not measure task performance. These measures typically are more subjective than MOPs, and can be crafted as either qualitative or quantitative. MOEs can be based on quantitative measures to reflect a trend and show progress toward a measurable threshold. Examples of MOEs for the objective to "Provide a safe and secure environment" may include – decrease in insurgent activity or increase in population trust of host-nation security forces.

• MOPs measure task performance. MOPs help answer questions such as *"Was the action taken?"* or *"Were the tasks completed to standard?"* They generally focus on the friendly force and are generally quantitative, but also can apply qualitative attributes to task accomplishment. A MOP confirms or denies that a task has been properly performed. MOPs are commonly found and tracked at all levels in execution matrixes. They are measures that characterize physical or functional attributes relating to the [operation], system operation measured or estimated under specific conditions. MOPs are used in most aspects of combat assessment, since the latter typically seeks specific, quantitative data or a direct observation of an event to determine accomplishment of tactical tasks. But, MOPs have relevance for noncombat operations as well (e.g., tons of relief supplies delivered or noncombatants evacuated). MOPs also can be used to measure operational and strategic tasks, but the type of measurement may not be as precise or as easy to observe. Evaluating task accomplishment using MOPs is relatively straightforward and often results in a yes or no answer; for example, Route XX cleared, $150,000 spent on dam construction, Al Jabber airfield secured.

(5) The assessment process and related measures should be *relevant, measurable, responsive*, and *resourced* so there is no false impression of accomplishment. Quantitative measures can be helpful in this regard.

• *Relevant*. MOPs and MOEs should be relevant to the task, effect, operation, the operational environment, the endstate, and the commander's decisions. This criterion helps avoid collecting and analyzing information that is of no value to a specific operation. It also helps ensure efficiency by eliminating redundant efforts.

• *Measurable*. Assessment measures should have qualitative or quantitative standards they can be measured against. To effectively measure change, a baseline measurement should be established prior to execution to facilitate accurate assessment throughout the operation.

**Note: Both MOPs and MOEs can be quantitative or qualitative in nature, but meaningful quantitative measures are preferred because they are less susceptible to subjective interpretation.*

• *Responsive*. Assessment processes should detect situation changes quickly enough to enable effective response by the staff and timely decisions by the commander. The JFC and staff should consider the time required for an action or actions to produce desired results within the operational environment and develop indicators that can respond accordingly. Many actions directed by the JFC require time to implement and may take even longer to produce a measurable result.

• *Resourced*. To be effective, assessment must be adequately resourced. Staffs should ensure resource requirements for data collection efforts and analysis are built into plans and monitored. Effective assessment can avoid duplication of tasks and unnecessary actions, which in turn can help preserve combat power.

**Note: The planning staff may find the Universal Joint Tasks in the Universal.*

Joint Task List (UJTL CJCSM 3500.04D) as a helpful starting point for crafting assessment criteria for an operation. For example, using the same Grayland Noncombatant Evacuation Operation (NEO) scenario mentioned earlier, the UJTL offers the Task descriptions and potential measures. Example follows:

UJTL Task

ST 8.4.3 Coordinate Evacuation and Repatriation of Noncombatants from Theater
Task Description: third-country resources to evacuate US dependents, USG civilian employees, and private citizens (US and third-country) from the theater and support the repatriation of appropriate personnel to the US. Such operations are conducted in support of the Department of State. Theater organizations at various echelons provide support (for example, medical, transportation, and security) to noncombatants (JP 3-0, 3-07, 3-07.5, 3-08v2, 3-10, 3-57, CJCSI 3110.14, CJCSM 3122.03).

Potential Measures offered by the UJTL

M1	Days	To organize and deploy fully operational JTF.
M2	Hours	To evacuate noncombatants (once CCDR directed to conduct evacuation).
M3	Hours	To evaluate situation and present recommendations to decision maker(s).
M4	Percent	Of US citizens and designated foreign nationals accounted for by name during evacuation.
M5	Percent	Of US citizens and designated foreign nationals accounted for.
M6	Percent	Of US citizens and designated foreign nationals evacuated.
M7	Percent	Of US citizens desiring, evacuated.
M8	Percent	Of evacuees available and desiring evacuation, moved (IAW OPLAN)
M9	Yes/No	NEO plans include actions in the event of NBC attack.

THINKING CAP: *Operations have always required assessment, this is nothing new. One should, however, be alert to the potential pitfalls encountered by a staff that becomes so enamored with color-coded MOE/MOP indicators that it fails to remember the nature of warfare. History is replete with these examples. The operational environment is constantly changing as enumerable human and environmental factors exert their influences. The enemy also has a vote, and is fighting to win! Friction has not disapperaed from the operational environment and friendly intelligence collection and analysis may be flawed. And, finally, as noted above, not everything is quantifiable – commanders and their staffs will also have to rely on their experience and intuition.*

Where applicable, the theater tasks are linked to appropriate universal joint tasks in the UJTL. This association further develops the scope of these tasks while providing further linkage to Joint Capability Areas (JCA) association, capability gap identification, and CCDR resource/Planning Programming, Budgeting, and Execution (PPBE) planning.

6. Key-Step — 6: Determine Own and Enemy's Center(s) of Gravity (COG), Critical Factors and Decisive Points

> *"The adversary's COG... would be attacked carefully, with measured means, because its indiscriminate destruction, while useful in defeating military forces, might bring undesirable consequences in rebuilding that same society."*
>
> *Creating a New Center of Gravity: A New Model for Campaign Planning, Watson, Bryan C.*

a. Clausewitzian concepts such as friction, fog and culminating points, to name a few, abound in our military vernacular. But arguably, none has been discussed, debated nor written on more than the Clausewitzian concept of COG or main point. For the United States military, the origins of the COG concept are rooted in the Cold War. The COG concept matured in the American mindset largely during an era when the United States military was focused heavily (and almost exclusively) on producing doctrine that would win wars decisively against a conventional traditional military force and nation state, "especially in such places as the Fulda Gap."[27]

(1) The COG concept has served us for years as a giant lens for focusing strategic and operational efforts to achieve decisive results. However, for the current generation of military professionals, the conflicts in Iraq and Afghanistan have evoked a disquieting epiphany: battlefield victory is useless without an ensuing political victory.[28] The ongoing military efforts in these countries find the U.S. military engaged in prolonged insurgencies and postwar reconstruction operations far removed from decisive battle. Furthermore, the strategic landscape suggests that the future for the United States military will be rife with other such "ambiguous and uncomfortable wars—and their aftermath."[29] This has evoked a corresponding renaissance in American doctrinal thinking and with it, not surprisingly, a number of proposals to redefine the COG.[30]

(2) As accepted definitions for COG are discussed in this document, planners must also strive to understand today's ambiguous environment and take the learned elements and acclimate and adjust ahead of an adaptable adversary.

(3) To understand the concept of COG we must begin with a discussion on the COG as a focal point for identifying critical factors; sources of strength as well as weaknesses and systemic vulnerabilities. Joint doctrine defines *center of gravity* as "the set of characteristics, capabilities, and sources of power from which a system derives its moral or physical strength, freedom of action, and will to act."[31] This definition states in modern terms the classic description offered by Clausewitz, "the hub of all power and movement, on which everything depends," the point at which all our energies should be directed."

[27] Rudolph M. Janiczek, A Concept At The Crossroads: Rethinking The Center Of Gravity, October 2007.

[28] This enduring dictum also comes to us courtesy of Clausewitz: "The political object is the goal, war is the means of reaching it, and the means can never be considered in isolation from their purpose." On War, p. 87.

[29] Max G. Manwaring, The Inescapable Global Security Arena, Carlisle Barracks, PA: Strategic Studies Institute, U.S. Army War College, 2002, p. 2.

[30] Rudolph M. Janiczek, A Concept At The Crossroads: Rethinking The Center Of Gravity, October 2007.

[31] JP 1-02,DOD Dictionary of Military and Associated Term1, JP 5-0, Joint Operation Planning.

(4) COG can be categorized as either physical or moral, and they are not limited in scope to military forces. A *physical* COG, such as a capital city or a military force, is typically easier to identify, target, and assess. Physical COGs can often be influenced solely by military means. However, the U.S. experiences in Iraq and Afghanistan exemplify the notion that active conflict can outlast the neutralization of a perceived COG. Neither the demise of the Taliban nor the removal of Saddam Hussein's regime brought an end to violence in either theater. *As with the Lernaean Hydra from Greek methology, for each head removed, two more grew back.* The U.S. finds itself engaged with elements of the former regimes as well as a multitude of other groups with varying interests and motivations. At a minimum, the nature of the COG has changed in each case.[32] This change, or evolution of the COG, may be exploited by understanding the civil dimension or *moral* COG. The moral centers of gravity are dynamic and inherently related to human factors: civilian populations, a charismatic or key leader, powerful ruling elite or strong-willed populace. Influencing a moral COG is far more difficult and typically cannot be accomplished by military means alone. The intangible and complex nature of moral COG and their related vulnerabilities necessitates the collective, integrated efforts of the instruments of power. As an example, it is a common understanding that access to, and influence over, civilian populations is a source of strength for insurgent movements and arguably terrorist networks.[32] As noted in FM 3-24/MCWP 3-33.5 "the ability to generate and sustain popular support, or at least acquiescence and tolerance, often has the greatest impact on the insurgency's long-term effectiveness. This ability is usually the insurgency's center of gravity." Engaging the civil COG and/or vulnerabilities utilizing a focused interagency approach during the planning process, the commander may ultimately shape, mitigate a threat, or gain awareness over it, degrading the insurgency's long-term effectiveness.

> *"The source of strength, or civil COG, for those responsible for the genocide in Rwanda in 1994 was the civilian population. Planners of the genocide recruited Hutu Burundi refugees and militias from the lower economic classes. A combination of anti-Tutsi propaganda and physical threats fueled their massive participation in the slaughter. According to interviews with survivors of the massacres, most of the 50,000 recruited killers were peasants just like their victims. Any organization tasked to stop the killing would have had to influence the civil COG—the peasant population."*
>
> *Sele, Richard K., Engaging Civil Centers of Gravity and Vulnerabilities.*

(5) At the strategic level, a COG might be an alliance, a political or military leader, a set of critical capabilities or functions, or national will. At the operational level a COG often is associated with the adversary's military capabilities — such as a powerful element of the armed forces — but could include other factors in the operational environment associated with the adversary's civil, political, economic, social, information, and infrastructure systems.

> *During the 1990-91 Persian Gulf War the coalition itself was identified as a friendly strategic COG, and the CCDR took measures to protect it, to include deployment of theater missile defense systems.*

(6) There is no certainty that a single COG will emerge at the strategic and operational levels. It is possible that no COG will emerge below the strategic level. At the tactical level, the COG concept has no utility; for us to speak of a tactical COG, the tactical level of war would have to exist independent of the operational and strategic level. It is commonly accepted that the tactical equivalent of the COG is the objective. Modern writings and

[32] *Sele, Richard K., Engaging Civil Centers of Gravity and Vulnerabilities.*

[33] *Ibid.*

understanding of the COG has evolved beyond the term's pre-industrial roots to include the possibility of multiple COGs existing at the strategic and operational levels.

(7) At all levels, COGs are interrelated. Strategic COGs have associated decisive points that may be vulnerable at the operational level, just as operational COGs may be vulnerable to tactical-level actions. Therefore, analysis of friendly and enemy COGs is a continuous and related process that begins during planning and continues throughout a major operation or campaign. Figure XIV-4 shows a number of characteristics that can be associated with a COG.

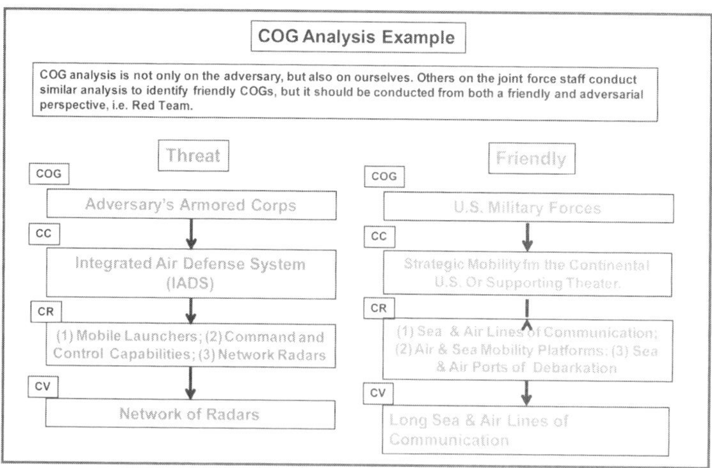
COG Analysis Example
COG analysis is not only on the adversary, but also on ourselves. Others on the joint force staff conduct similar analysis to identify friendly COGs, but it should be conducted from both a friendly and adversarial perspective, i.e. Red Team.
Threat
COG
Adversary's Armored Corps
CC
Integrated Air Defense System (IADS)
CR
(1) Mobile Launchers; (2) Command and Control Capabilities; (3) Network Radars
CV
Network of Radars
Friendly
COG
U.S. Military Forces
CC
Strategic Mobility fm the Continental U.S. Or Supporting Theater.
CR
(1) Sea & Air Lines of Communication; (2) Air & Sea Mobility Platforms; (3) Sea & Air Ports of Debarkation
CV
Long Sea & Air Lines of Communication

Figure XIV-4. Threat and Friendly Physical COG Analysis Example.[34]

(8) The essence of "operational art" lies in being able to produce the right combination of effects in time, space, and purpose relative to a COG to neutralize, weaken, destroy (consistent with desired endstate/commanders intent), or otherwise exploit it in a manner that best helps achieve military objectives and attain the military endstate. *In theory*, this is the most direct path to mission accomplishment. While doing this the commander must also plan for protecting friendly potential COGs, such as agreements with neutral and friendly nations for transit of forces, information and networks, coalition relationships, and U.S. and international public opinion.

(9) COGs *are not vulnerabilities, however,* within every COG lies inherent vulnerabilities that when attacked can render those COGs weaker and even more susceptible to direct attack and eventual destruction. This process cannot be taken lightly, since a faulty conclusion resulting from a poor or hasty analysis can have very serious consequences, such as the inability to achieve strategic and operational objectives at an acceptable cost. Planners must continually analyze and refine COGs.

(10) An additional insight on the COG is provided by Antulio Echevarria in his Strategic Studies Institute paper, *Clausewitz's Center of Gravity; It's Not What We Thought.* Echevarria maintains that the COG needs to be redefined as a "focal point," not as a strength or weakness or a source of strength. A COG is more than a critical capability; it is the point where a certain centripetal force seems to exist, something that holds everything else together. As an example he offers the following:

[34] *Joint Information Operations Planning Handbook.*

" ...al-Qa'ida cells might operate globally, but they are united by their hatred of apostasy.[35] This hatred, not Osama bin Laden, is their CoG. They apparently perceive the United States and its Western values as the enemy CoG (though they do not use the term) in their war against "apostate" Muslim leaders. Decisively defeating al-Qa'ida will involve neutralizing its CoG, but this will require the use of diplomatic and informational initiatives more than military action.

Commanders and their staffs need to identify where the connections—and the gaps—exist in the enemy's system as a whole before deciding whether a center of gravity exists. The CoG concept does not apply if enemy elements are not connected sufficiently. In other words, successful antiterrorist operations in Afghanistan may not cause al-Qa'ida cells in Europe or Singapore to collapse." [36]

b. The adversarial context pertinent to COG analysis takes place within the broader operational environment context. A system's perspective of the operational environment assists in understanding the adversary's COGs. In combat operations, this involves knowledge of how an adversary organizes, fights, and makes decisions, and of their physical and psychological strengths and weaknesses. Moreover, the CCDR and staff must understand other operational environment systems and their interaction with the military system (Figure XIV -5). This holistic understanding helps commanders and their staffs identify COGs, *critical factors*, and *decisive points* to formulate LOO[37] (LOO discussed in detail in Chapter 15).

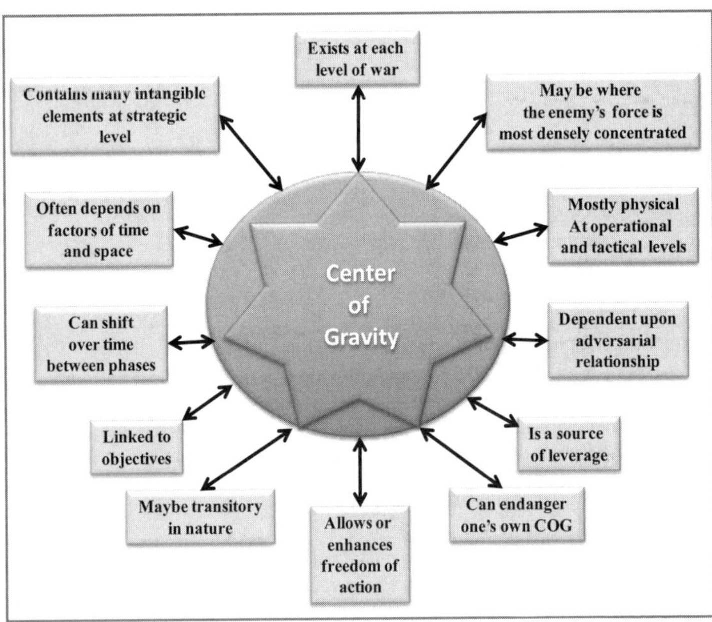

Figure XIV-5. Characteristics of Center Gravity.

[35] *Al-Qa'ida (the Base)," International Policy Institute for Counter-Terrorism, on the World Wide Web at www.ict.org.il/inter_ter (accessed 3 April 2002).*

[36] *Echevarria, Antulio J. II, LtCol U.S. Army. Clausewitz's Center of Gravity: It's Not What We Thought, Naval War College Review, Winter 2003, Vol. LVI, No.1.*

[37] *JP 3-0, Joint Operations.*

> *"All too often, the center of gravity is understood as being one of the enemy's vulnerabilities. On the contrary, a center of gravity is found invariably among critical strengths-never critical weaknesses or critical vulnerabilities."*
>
> *Operational Warfare, Milan N. Vego, 2000*

(1) Critical Factors. All COGs have inherent *"critical capabilities"* — those means that are considered crucial enablers for the adversary's COG to function and essential to the accomplishment of the adversary's assumed objective(s). These critical capabilities permit an adversary's COG to resist the military endstate. In turn, all critical capabilities have essential *"critical requirements"* — those essential conditions, resources, and means for a critical capability to be fully operational. *Critical vulnerabilities* are those aspects or components of the adversary's critical requirements which are deficient or vulnerable to direct or indirect attack that will create decisive or significant effects disproportionate to the military resources applied. Collectively, these are referred to as *"critical factors."*

(a) Critical factors, including *capabilities* and *vulnerabilities*, become decisive points as they are identified and friendly forces can operate against them. They become apparent only after thoroughly analyzing the opponent's centers of gravity. Some of the critical factors may lie beyond the reach or ability of the military forces to affect them.

For example, the continuing sympathy of a broad ethnic group toward an insurgent group may be a critical capability for the insurgents. However, affecting that affinity lies beyond the capability of the military force in the JOA, and therefore it does not become a decisive point.

> *"Another error is to confuse a decisive point with the center of gravity. Although closely related, decisive points are not necessarily sources of strength; they could be found among critical weaknesses. These weaknesses are only relevant if they are open to attack and facilitate an attack on the enemy center of gravity. Once the center of gravity is determined, decisive points are identified and targeted."*
>
> *Operational Warfare, Milan N. Vego, NWC 2000 p.30;, Bruce L. Kidder, Center of Gravity: Dispelling the Myths (Carlisle Barracks, PA. US Army War College, 1996), p.12*

1 *Critical capabilities* are the essential conditions, resources, and means required for a center of gravity to be fully operational. They are the attributes of a center of gravity that can affect friendly force operations within the context of a given scenario, situation, or mission. Critical capabilities may vary between phases of an operation, function at various echelons, or change in character as an operation progresses. At the operational and tactical levels, critical capabilities represent a potential for action in that they possess an ability to inflict destruction, seize objectives or create effects, or prevent friendly forces from achieving mission success. For example, critical requirements can include the resources necessary to deploy a force into combat, the popular support a national leader needs to remain in power, or the perceived legitimacy required to maintain popular support. Critical capabilities are also the primary means through which to isolate, dislocate, disintegrate, or destroy a COG.

2 *Critical vulnerabilities* are aspects or components of those critical requirements that are deficient or vulnerable to direct or indirect attack in a manner that will create decisive or significant effects. Commanders may neutralize, weaken, or destroy an adversary's COG by attacking through critical vulnerabilities. For example, an enemy armored Corps could represent an operational-level COG. Deficient mobile air defenses protecting the enemy corps could be a critical vulnerability that friendly forces could exploit.

Critical capabilities and vulnerabilities are interrelated. The loss of one critical capability may expose vulnerabilities in other critical capabilities; the loss of a critical capability may initiate a cascading effect that accelerates the eventual collapse of a COG. The analysis of a COG and its critical factors will reveal these systemic relationships and their inherent vulnerabilities.

(b) Critical factors are the keys to protecting or neutralizing COGs during a campaign or major operation. To achieve success, the force must possess sufficient combat power and operational reach to affect critical factors. Similarly, commanders must protect those friendly critical capabilities within the enemy's operational reach.

1 Operational Approach. *Direct versus Indirect. Operational approach* is the manner in which a commander attacks the enemy COG. The essence of COG analysis lies in determining an approach to the COG that can neutralize, weaken, or destroy it.

a The *direct approach* is to apply combat power directly against the enemy's COG or principal strength. In most cases, however, the enemy COG is well-protected and not vulnerable to a direct attack.

b The *indirect approach* is to attack the enemy COG by applying combat power against a series of decisive points that avoid enemy strengths.

In theory, direct attacks against enemy COGs resulting in their neutralization or destruction is the most direct path to victory — if it can be done in a prudent manner (as defined by the military and political dynamics of the moment). Where direct attacks against enemy COGs mean attacking into an opponent's strength, CCDRs should seek an indirect approach until conditions are established that permit successful direct attacks. In this manner, the enemy's critical vulnerabilities can offer indirect pathways to gain leverage over its COGs. For example, if the operational COG is a large enemy force, the joint force may attack it indirectly by isolating it from its C2, severing its LOCs, and defeating or degrading its protection capabilities. In this way, CCDRs employ a synchronized and integrated combination of operations to weaken enemy COGs indirectly by attacking critical requirements, which are sufficiently vulnerable.

CENTER OF GRAVITY AND CRITICAL VULNERABILITIES

"During the Battle of Britain in 1940, an operational center of gravity for Britain was the Royal Air Force Fighter Command. A critical capability for Fighter Command was the ability to meet Luftwaffe attacks in a timely manner. The critical requirement linked to that specific critical capability was advance warning regarding the timing, strength and direction of Luftwaffe attacks. The critical vulnerability linked to that specific critical requirement was the fragility and vulnerability of the British radar system that provided the advance warning. However, the Germans did not realize the importance of the radar system and did not follow up their early attacks against it."

Joe Strange, Marine Corps University, Perspectives on Warfighting, Number 4, 1996

(2) COG *analysis* is thorough and detailed; faulty conclusions drawn from hasty or abbreviated analyses can adversely affect operations, waste critical resources, and expose forces to undue risk (Figure XIV-6). A thorough, systemic understanding of the operational environment facilitates identifying and targeting enemy COGs. This understanding requires detailed knowledge of how adversaries and enemies organize, fight, and make decisions. Knowledge of their physical and psychological strengths and weaknesses is also needed. In addition, commanders should understand how military forces interact with the other systems (political, military, economic, social, information, and infrastructure) within the

operational environment. This understanding helps planners identify decisive points and COGs, and formulate lines of operation. As each center of gravity is identified, analysis is further refined by understanding the operational approach to each center of gravity and the decisive points associated with them.[38]

(a) Decisive Points. Decisive points emerge from an analysis of endstate, objectives, and center(s) of gravity. A *decisive point* is a place, key event, or enabling system that allows commanders to gain a marked advantage over an enemy and greatly influence the outcome of an operation. Decisive points are not COGs; they are keys to attacking or protecting their critical requirements. When it is not feasible to attack a COG directly, commanders focus operations to weaken or neutralize the critical requirements (critical vulnerabilities) upon which it depends. These critical vulnerabilities are decisive points, providing the indirect means to weaken or collapse the COG. Operational-level decisive points provide the greatest leverage on COGs.[39]

(b) Some decisive points are geographic, such as port facilities, transportation networks and nodes, and bases of operations. Other physical decisive points may include elements of an enemy force. Events, such as commitment of the enemy operational reserve or reopening a major oil refinery, may also be decisive points. A common characteristic of decisive points is their relative importance to the COG; the nature of a decisive point compels the enemy to commit significant effort to defend or marginalize it. The loss of a decisive point weakens a COG and may expose more vulnerabilities as it begins to collapse.

(c) Decisive points assume a different character during stability or civil support operations. They may be less tangible and more closely associated with critical events and conditions. For example, they may include repairing a vital water treatment facility, establishing a training academy for national security forces, securing a major election site, or quantifiably reducing crime. In Hurricane Andrew relief operations in 1992, for example, reopening schools became a decisive point. None of these decisive points are physical, yet all are vital to establishing the conditions for a transition to a legitimate civil authority. In an operation predominated by stability tasks, this transition is typically a supporting condition of the desired endstate.

(d) A situation typically presents more decisive points than the force can control, destroy, or neutralize with available resources. Through critical factors analysis, commanders identify the decisive points that offer the greatest leverage on COGs. They designate the most important decisive points as objectives and allocate enough resources to achieve the desired results on them. Decisive points that enable commanders to seize, retain or exploit the initiative are crucial. Controlling these decisive points during operations helps commanders retain and exploit the initiative. If the enemy maintains control of a decisive point, it may exhaust friendly momentum, force early culmination, or facilitate a counterattack. Decisive points shape the design of operations. They help commanders select objectives that are clearly decisive relative to the endstate; they ensure that vital resources are focused only on those objectives that are clearly defined, attainable, and directly contribute to establishing the conditions that comprise that endstate.

> *"The first task… in planning for war is to identify the enemy's centers of gravity, and if possible, trace them back to a single one."*
>
> *Carl von Clausewitz, On War, 1832*

[38] *FM 3-0, Operations.*

[39] *FM 3-0, Operations.*

(e) Decisive points are always oriented on the key vulnerabilities that can only be identified through the COG or another method of systems analysis. Generally, commanders attack adversary vulnerabilities at decisive points so that the results they achieve are disproportional to the military and other resources applied. Consequently, commanders and their staffs must analyze the operational environment and determine which systems' nodes or links or key events offer the best opportunity to affect the enemy's COGs or to gain or maintain the initiative. The commander then designates them as decisive points, incorporates them in the LOO, and *allocates sufficient resources* to produce the desired effects against them.

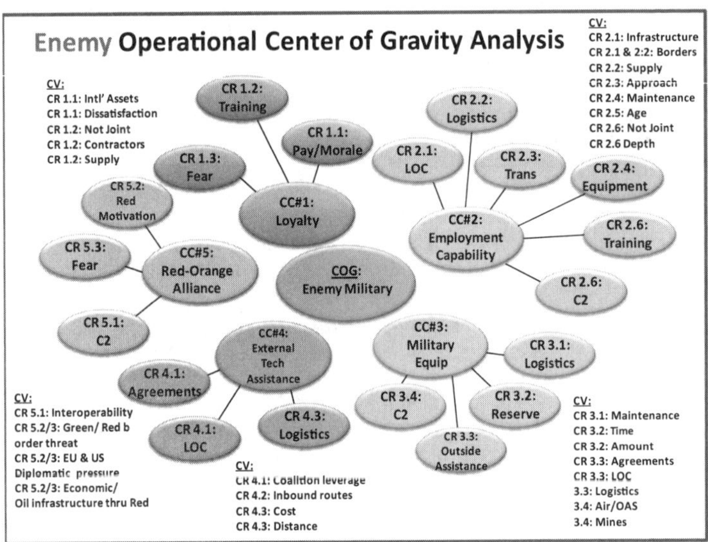

Figure XIV-6. Example of Enemy Operational Center of Gravity and Critical Factors.

> *"However absorbed a commander may be in the elaboration of his own thoughts, it is sometimes necessary to take the enemy into consideration."*
>
> *Winston Churchill, The World Crisis, 1911-1918;1923*

c. No COG discussion is complete until we look at the whole operational environment and take a comprehensive look at all the systems in this environment relevant to the mission and operation at hand. A system is a functionally related group of elements forming a complex whole. A system's view to understanding the operational environment considers more than just an adversary's military capabilities, order of battle, and tactics. Instead, it strives to provide a perspective of interrelated systems PMESII, and others, that comprise the operational environment relevant to a specific operation (Figure XIV-7 (JP 5-0)). A system's perspective provides the CCDR and staff with a common frame of reference for collaborative planning with other government agencies' (OGA) counterparts to determine and coordinate necessary actions that are beyond the CCDR's command authority.

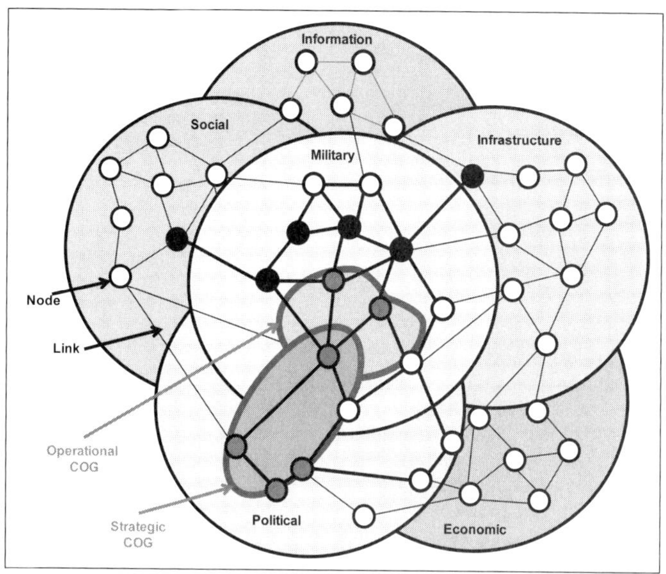

Figure XIV-7. Identifying Centers of Gravity.

(1) Understanding PMESII systems, their interaction with each other, and how system relationships will change over time will increase the JFC's knowledge of how actions within a system can affect other system components. Among other benefits, this perspective helps intelligence analysts identify potential sources from which to gain indications and warning, and facilitates understanding the continuous and complex interaction of friendly, adversary, and neutral systems. A systems understanding also supports operational design by enhancing elements such as centers of gravity, LOOs, and decisive points. This helps commanders and their staffs visualize and design a broad approach to mission accomplishment early in the planning process, which makes detailed planning more efficient. For example, Figure XIV-7A (JP 5-0) depicts the critical capabilities, critical requirements, and critical vulnerabilities associated with two of the adversary's strategic and operational COGs.

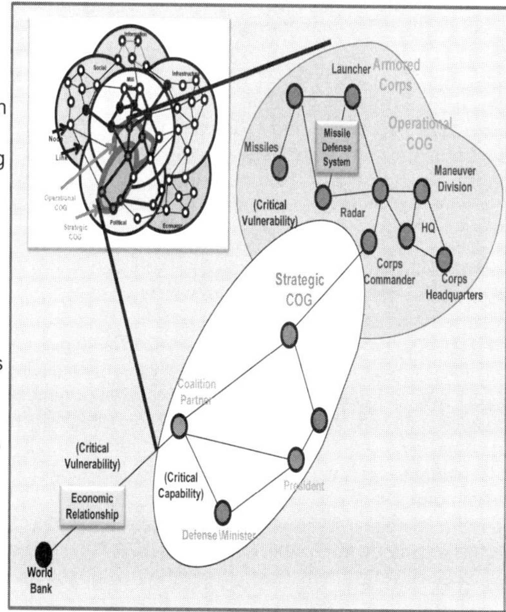

Figure XIV-7A. Analyzing Critical Factors

> *"The traditional military-centric single center of gravity focus that worked so well in the cold war doesn't allow us to accurately analyze, describe, and visualize today's emerging networked, adaptable, asymmetric adversary. This adversary has no single identifiable 'source of all power.' Rather, because of globalization, the information revolution, and, in some cases, the non-state characteristic of our adversary, this form of adversary can only be described (and holistically attacked) as a system of systems."*
>
> *Insights on Joint Operations: The Art and Science, Gen (Ret) Gary Luck, September 2005.*

d. Center of Gravity Determination.[40] (The discussion which frames the COG analysis process is provided by the Naval War College.) While primarily a strategic and operational level concern, the identification of both the enemy and friendly COG is an essential element of any plan. If the staff gets this part wrong, the operation will at best be inefficient and, at worst, end in failure. The commander and staff should be deeply involved in a dialogue with the higher joint force headquarters planning staff during this critical analysis. While tactical-level organizations may not be party to the formulation of a COG analysis, they most certainly will be participants in the execution of the resulting tactical objectives and tasks that are derived from the analysis. Therefore, even tactical commanders and their planning staffs should be familiar with the process and reasoning used for the COGs analysis in order to place their own operations in the proper context.

e. The purpose of this section is to provide the planner with a brief review of each of the information requirements displayed earlier in this chapter. This section is not intended to replace the extensive study of the nuances of COG analysis that all planners should strive to master; rather, it is intended to identify information requirements and to offer some considerations in the application of the collected data. The reader will note that the JOPP has the staff collecting information for both the enemy and friendly COGs. Neither can be identified nor considered in a vacuum—a common staff planning mistake. The struggle between opposing forces employing their unique means and ways to achieve their respective ends (objectives) is a dynamic that can only be appreciated if they are viewed collectively. While the explanations and examples provided below are for enemy COGs analysis, the process is the same for determining and analyzing friendly COGs. The only differences are in the planning actions taken once the analysis is completed. Planners develop courses of action that focus on defeating the enemy's COG while at the same time mitigating risks to their own COG.

f. The Center of Gravity Flow Chart, Figure XIV-8, illustrates the flow used to identify a COG and to determine the ways in which it can be attacked. Each step of the process, as it corresponds to the numbers in Figure XIV-8, is described below. Later in this section an example, Desert Storm Enemy COG Analysis, is provided.

(1) Step 1: Identifying the Objective(s). Identifying the objective is a critical first step. Before one can determine a COG, the objective(s) must be identified. If this portion of the analysis is flawed, then the error infects the remainder of the process. The planner should first determine the ultimate (strategic or operational) objectives and then the supporting intermediate (operational or major tactical) objectives. The operational objectives should show a direct relationship to the strategic objectives. If this linkage between strategic and operational objectives cannot be established, the objectives are suspect. Objectives, and particularly strategic objectives, usually have requirements/tasks that fall primarily into the responsibility of instruments of power other than the military. These are still important to identify since the military may have a supporting role in their accomplishment.

[40] Sweeny, Dr. Patrick C., Naval War College, NWC 4111H, Joint Operations Planning Process Workbook and NWP 5-01, Navy Planning.

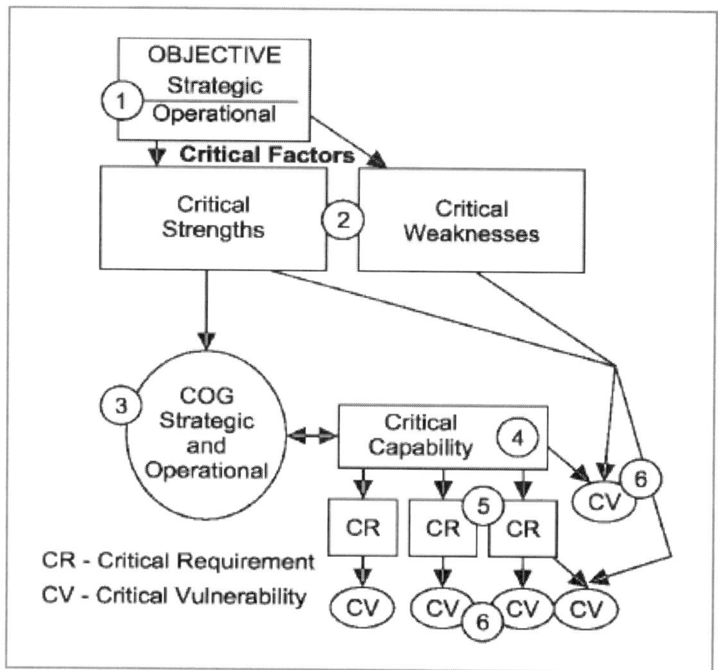

Figure XIV-8. Center of Gravity Flow Chart.

(2) Step 2: Identify Critical Factors. Critical factors are those attributes considered crucial for the accomplishment of the objective. These factors that in effect describe the environment (in relationship to the objective) must be identified and classified as either sufficient (critical strength) or insufficient (critical weakness). Critical factors are a cumulative term for critical strengths and critical weaknesses of a military or nonmilitary source of power; they can be quantifiable (tangible) or unquantifiable (intangible). Critical factors are present at each level of war; they require constant attention because they are relative and subject to changes resulting from the actions of one's forces or of the enemy's actions. It is important while conducting the analysis for this step that planners maintain a sharp eye on the objectives identified in the first step—each level of war has critical factors that are unique to that level.

(a) The questions that should be asked when determining critical factors for the enemy are:

- "What are the attributes, both tangible and intangible, that the enemy has and must use in order to attain his strategic (operational) objective?" *These are critical strengths. The second question is,*
- "What are the attributes, both tangible and intangible, that the enemy has and must use in order to achieve his strategic (operational) objective, but which are weak and may impede the enemy while attempting to attain his objective?" *These are critical weaknesses.*

(b) The answers to these two questions will produce a range of critical strengths and critical weaknesses associated with specific levels of war. One should note that, like the close relationship expected to be found between strategic and operational objectives, there will undoubtedly be some critical strengths and critical weaknesses that have a similar close relationship between the corresponding critical factors. For example, a strategic critical weakness, such as a strategic leader having a tenuous communications link to his fielded forces, may also create an operational critical weakness for fielded forces unable to reliably communicate with their higher command.

(3) Step 3: Identify the Centers of Gravity. Joint doctrine defines a COG as "The source of power that provides moral or physical strength, freedom of action, or will to act."[41]While the definition is helpful for assisting in the identification of the operational COG, when considering the strategic COG, a planner should be alert to the fact that the definition is not focused upon only the military aspects of the analysis. In view of the discussion in the first step, when strategic objectives are being identified planners should consider the broader application of the definition, remembering that the role of instruments of power other than the military may prevail.

The COGs at each level of war should be found among the listed critical strengths identified within the critical factors of Step Two. While all of the identified strengths are critical, the planner must deduce which among those capabilities identified rise(s) above all others in importance in accomplishing the objective (that is, those tangible and intangible elements of combat power that would accomplish the assigned objectives). This critical strength is the COG. This does not diminish the importance of the other critical strengths; however, it forces the planner to examine closely the relationships of the various critical strengths to one another and the objective. This close examination of interrelationships could be improved by using a systems perspective of the operational environment. Such a study may well offer the planner an enhanced understanding of an adversary's COG and its interdependencies. See JP 5-0 for more information on the systems approach to COG refinement. This analysis of these relationships will prove important in the next step.

(4) Step 4: Identify Critical Capabilities. Joint doctrine defines a critical capability as: *"a means that is considered a crucial enabler for a COG to function as such and is essential to the accomplishment of the specified or assumed objective(s)."*[42]

(a) If the COG is a physical force (often the case at the operational level), the commander and staff may wish to begin their examination of critical capabilities by reviewing the integration, support, and protection elements of the enemy's combat power as they apply to the COG.

(b) Many of these elements are often found in the joint functions as described in the Universal Joint Task List (C2, intelligence, sustainment, protection, fires, and movement and maneuver). Moreover, these capabilities often are located within the critical strengths and weaknesses identified in Step Two.

(c) The planner should be alert for two major considerations. First, although a capability is a critical strength, if it bears no relationship to the identified COG, it cannot be considered a critical capability. The second consideration is that although some capability may be perceived as a critical weakness, if it is an essential enabler for the enemy COG, then it is a critical capability, albeit weak in nature.

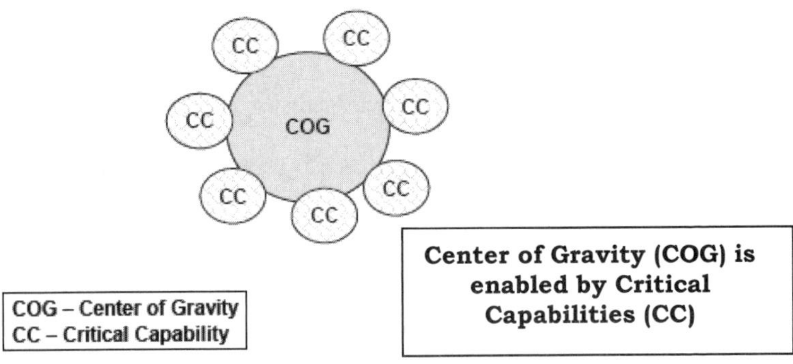

[41] *JP 1-02, DOD Dictionary of Military and Associated Terms.*

[42] *Ibid.*

(5) Step 5: Identify Critical Requirements. Once a COG's critical capabilities are identified, the next step is for the staff to identify those essential conditions, resources, and means for a critical capability to be fully operational.

(a) These are the critical requirements that support each of the critical capabilities. This is essentially a detailed view of what comprises a critical capability.

> ***Example**: a critical capability for an operational COG to accomplish its mission might be its ability to exert C2—its ability to receive direction as well as communicate directives to subordinates. The critical requirements might include **tangible** requirements such as: communication nodes, antennas, frequency bands, individual command posts, spare parts, bandwidth, specific satellites, and so forth. It may also include **intangibles** such as commander's perceptions and morale.*

**Note: Planners should be cautious at this point. One is often presented with a wealth of potential targets or tasks as each critical capability is peeled back and the numerous supporting critical requirements are identified. There is often a temptation to stop at this point of the analysis and begin constructing target lists. Such an action could result in a waste of resources and may not be sufficient to achieve the desired effects. The planner should find the sixth step as a more effective way to achieve the defeat of a COG*

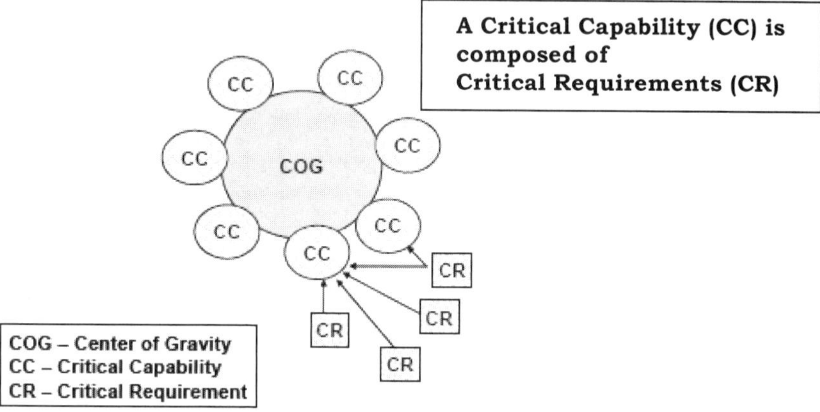

(6) Step 6: Identify Critical Vulnerabilities. Joint doctrine defines a critical vulnerability as "an aspect of a critical requirement which is deficient or vulnerable to direct or indirect attack that will create decisive or significant effects."[43]

The planner should contemplate those critical capabilities and their supporting critical requirements in this regard, keeping in mind that these weaknesses must bear a direct relationship to a COG and its supporting critical capabilities for it to be assessed as a critical vulnerability. Striking a weakness that bears no such relationship is simply a measure taken to harvest "low hanging fruit" that offers no decisive benefit. The planner should also take this opportunity to consider the previously assembled lists of critical strengths and critical weaknesses from Step Two to determine if there are any critical factors with a close relationship to the COG that were not captured in the previous critical capability/critical requirement steps (steps four and five).

[43] *JP 3-0, Joint Planning.*

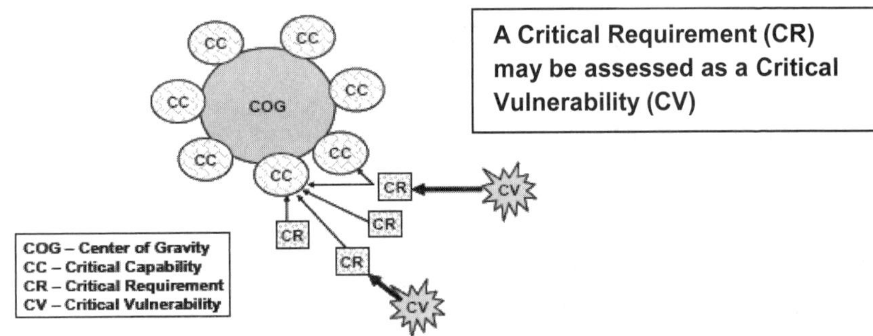

**Note: While the planner first seeks critical weaknesses within the critical capabilities and supporting critical requirements as implied by the definition, there might be opportunities found in critical strengths that provide decisive or significant results disproportionate to the military resources applied. An example might be the integrated air defense (IAD) that is protecting an operational COG. While this critical capability might be assessed as strength, its neutralization and the subsequent opening of the COG to direct attack may be assessed by the commander as more favorable in regard to the amount of resources and time expended to achieve the desired effects.*

(7) Step 7: Identify Decisive Points. Joint doctrine defines decisive points as "a geographic place, specific key event, critical factor, or function that, when acted upon, allows commanders to gain a marked advantage over an adversary or contribute materially to achieving success."[44] As with all previous steps, the value of a DP is directly related to its relationship to a COG and its objective:

(a) In the example shown in Figure XIV-9, from a friendly COG perspective, DPs 1 and 4, which provide access to the friendly COG, must be protected from attacks by the enemy COG. Decisive Points 2 and 3, which provide decisive access to the enemy COG, become friendly objectives or tasks. If there is no relationship, it is not a DP. *A DP is neutral in nature; that is, it is by definition as important to both the enemy and friendly commanders.*

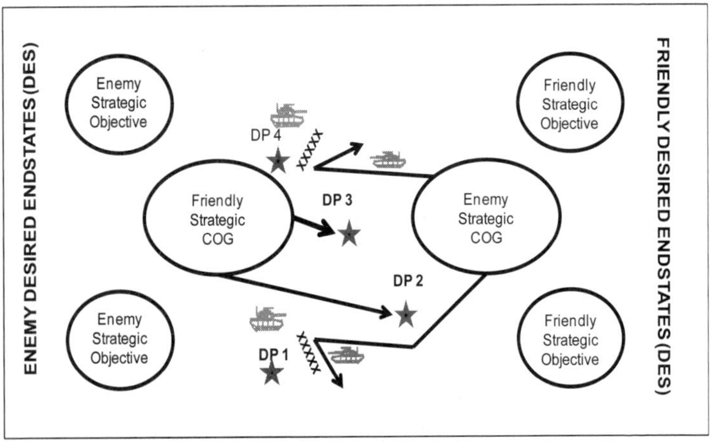

Figure XIV-9. Theoretical Relationship of Two Opposing COGs and Their Decisive Points.

[44] *JP 3-0, Joint Planning.*

(b) If, for example, an APOD/SPOD complex is a DP for a friendly commander, enabling that commander to project the COG through it on the way to the objective, then the enemy commander will also assess the complex as a threat to the enemy COG and should attempt to deny the friendly force commander control of the DP. In both cases, this DP, if within the capability of the force, will undoubtedly become an objective or task assigned to both enemy and friendly subordinate commands. Using this APOD/SPOD DP example, one might find the friendly JFC assigning the JFMCC the tactical task of "Seize Redland SPOD NLT D+2 in order to support the flow of JTF Blue Sword forces into Redland."

**Note: The planner must remember that this is a dynamic process. Any changes in the information considered in the first two steps of this process require the staff to revalidate its conclusions and subsequent supporting operations. As objectives change, the sources of power required to achieve the desired endstate might also change. As new sources of strength appear in the operational environment, how do they interact?*

(c) The table below provides an example enemy COG analysis (note that the same must be done for the friendly COG to ensure measures are taken to protect one's own COG). This example is not intended to be exhaustive and serves only as an illustrative example, exploring only a single critical capability and its associated critical requirements, and offering simply a selection of DPs.

Enemy Center of Gravity Determination
(Desert Storm Enemy COG Analysis)

Identify

1a. Strategic Objectives (s)
• Retain Kuwait as 19th Province
• Enhance Saddam Hussein's hold on power
• Increase Iraq's political and military influence in the Arab world
• Increase Iraq's power and influence within OPEC
1b. Operational Objective(s)
• Defeat or neutralize a coalition attack to liberate Kuwait
• Prevent coalition forces from obtaining air superiority
• Prevent coalition forces from obtaining sea control in the northern part of the Persian Gulf
2a. Critical Strengths Integrated Air Defense (IAD)
• WMD
• Land-based ballistic missiles (Scuds)
• Republican Guards in the Kuwait Theater of Operations (KTO)
• Forces are in defensive positions
• Saddam and his strategic C2
• Combat experienced units and commanders
• Missile-armed surface combatants
• Sea Mine inventories and delivery platforms
2b. Critical Weaknesses
• World Opinion: Arab world outrage
• Long and exposed Land
• LOCs from Iraq to KTO
• Combat skills and readiness of the Air Force
• Numerical and qualitative inferiority of naval forces
• Low morale and poor discipline of regular forces
• Class IX for weapon systems
• Inadequate forces to protect the Iraq-Iranian border
3a. Strategic COG
Saddam and his inner circle security apparatus
3b. Operational COG
Republican Guards in the Kuwait Theater of Operations (KTO)
Note: For the sake of brevity, this example will only examine the Operational COG

> *"How can one man say what he should do himself, if he is ignorant of what his adversary is about?"*
>
> *Lt Gen Antoine-Henri, Baron de Jomini, 1838*

4. Critical Capabilities

- Sustain Rep Guard forces in KTO (Log)
- Receive strategic direction and provide directives to subordinate units (C3)
- Protect forces from coalition airpower (Integrated Air Defense –IAD)
- Employ conventional defensive forces as a screening force
- Maintain organizational morale

Note: For the sake of brevity, this example will only examine the Operational COG.

5. Critical Requirements

- Radar Sites
- Communication nodes
- Iraqi Air Force
- Class IX for IAD systems
- Resupply for Class V
- Morale of fixed site crews

6. Check CV's

- Radars vulnerable to jamming
- Air force
- LOCs & Class IX

Decisive Points

(Note—Selected Samples, not an exhaustive listing)

- APODS & SPODS in Saudi Arabia
- Strait of Hormuz
- APODS in Turkey
- Kuwait SPOD

7. Key-Step — 7: Conduct Initial Force Structure Analysis (Apportioned and Assigned Forces)

a. Conduct an Analysis of Available Forces and Assets.

- Review forces that have been provided for planning and their locations (apportioned and assigned).
- Determine the status of reserve forces and the time they will be available.
- Referring back to your specified and implied tasks, now determine what broad force structure and capabilities are necessary to accomplish these tasks (e.g., is a Carrier Battle Group or a forcible entry capability required?).
- Identify shortfalls between the two.
  - Example:

    Task: Seize APOD.

    Observation: No forced entry capability (MEU, Airborne).

b. Force Planning. The purpose of force planning is to identify all forces needed to accomplish the supported commander's CONOPS and plan the phase of the forces into the theater. It consists of determining the force requirements by operation phase, mission, mission priority, mission sequence, and OA. It includes major force phasing; integration planning; force list structure development; followed by force list development.

(1) Force planning is the responsibility of the CCDR supported by the Service component commanders, the Services, JS and JS JFC/JFP and/or the JFM (see Chapter 7, GFM).

(2) The primary objectives of force planning are to apply the right force to the mission while ensuring force visibility, force mobility, and adaptability.

(3) Force planning begins during CONOPS development. The supported commander determines force requirements, develops a TPFDD and letter of instruction (LOI), and designs force modules to align and time-phase the forces in accordance with (IAW) the CONOPS.

(4) Major combat forces are selected for planning purposes from those apportioned for planning and included in the supported commander's CONOPS by operation phase, mission, and mission priority. The supported CCMD Service components then collaboratively make tentative assessments of the combat support (CS) and combat service support (CSS) required IAW the CONOPS.

(5) During the planning stage, these forces are planning assumptions called "preferred forces" (see Chapter 7, GFM).

(6) The commander identifies and resolves or reports shortfalls.

(7) For some high priority plans, the JS JFC/JFP may be tasked to contingency source the plan. Contingency sourcing is based on planning assumptions and contingency sourcing guidance. To the degree that the contingency sourcing planning assumptions and sourcing guidance mirrors the conditions at execution, contingency sourcing may validate the supported CCMD preferred forces and delineate associated risk. If the plan is directed to be executed, the force requirements are forwarded to the JS for validation and tasked to the JS JFC/JFPs to provide sourcing recommendations to the SECDEF. When the SECDEF accepts the sourcing recommendation, the FP is ordered to deploy the force to meet the requested force requirement. After the forces are sourced, the CCDR refines the force plan to ensure it supports the CONOPS, provides force visibility, and enables flexibility. However, the planner must realize that if the resources for the proposed plan are not available, the plan must change (see Chapter 21, Assessment).

(a) The CONOP for a plan generally covers how the combined forces will work together to accomplish the mission. When forwarding force requirements, the Force Providing

Community, JS JFC/JFPs, Services and FPs, need to know the CONOP for each unit and what tasks and missions the specific requested force will be expected to accomplish. This information is necessary so the identified force can be trained, organized and equipped to maximize the chances of mission success. The overall operation CONOP and the CONOP for each unit are usually not the same. The GFM Chapter provides additional details.

c. Availability of Forces for Joint Operations Planning.

(1) Availability of forces for joint operation planning utilizes three terms as defined in the GEF, JSCP and GFMIG — *assigned, allocated, and apportioned* — to define the availability of forces and resources for planning and conducting joint operations. For deliberate planning, contingency plans assigned through the JSCP process utilize force assignment and force apportionment (see GFM Chapter 7 for an in-depth discussion of assigned and allocated forces).

(a) Apportionment. Pursuant to the JSCP, "apportioned forces are types of combat and related support forces provided to CCDRs as a *starting point for planning purposes only.*" Forces apportioned for planning purposes may not be those allocated for execution.

1 *Force Apportionment Guidance*. Apportioned forces are combat and related support capabilities provided to CCDR's for planning purposes only. Apportionment supports the overlapping requirements of the NDS and the NMS. "Available forces" are apportioned without consideration to readiness status; however, apportioned forces are what a CCDR can reasonably expect to be available, but not necessarily allocated for use when a contingency plan transitions to execution.

2 *Force Apportionment*. Forces are grouped into three categories:

- Forces committed (named operations). The SECDEF can dip into this category for high priority planning efforts.
- Forces available for planning.
- Homeland Defense.

3 *Preferred Force Identification*. Preferred force identification spans the entire planning process from Strategic Guidance through Plan Assessment. Preferred forces are forces that are identified by the supported CCDR in order to continue employment, sustainment and transportation planning and assess risk. These forces are *planning assumptions* only, are not considered "sourced" units, and does not indicate that these forces will be contingency or execution sourced. CCDR Service and functional components are encouraged to work with the JS JFC/JFPs and their Service components to make the best possible assumptions with respect to generating preferred forces. The preferred forces identified for a plan by the CCDR should not be greater than the number of forces apportioned for planning unless the CJCS or his designee either grants permission or has provided amplifying planning guidance. To the degree the CCDR is able to make good planning assumptions when selecting preferred forces determines the feasibility of a plan, and may assist the JS JFC/JFP in identifying forces should the plan be designated for contingency sourcing or transitions to execution. Preferred forces are based on planning assumptions; these units are still *not* considered "sourced."

d. Force Sourcing and the GFM Process. Force sourcing covers a wide range of sourcing methodologies providing CCDRs with requested capabilities. The intent is to provide the CCDR with the most capable forces based on stated capability requirements, balanced against risks (operational, future challenges, force management, institutional) and global priorities. Force sourcing is broken into three broad categories: *execution and contingency sourced forces, and preferred force identification*, which was previously discussed under force apportionment guidance.

8. Key-Step — 8: Conduct Initial Risk Assessment

Strategic Risk: The potential impact upon the United States – to include our population, territory, and interests – of current and contingency events given their estimated consequences and probabilities.

Global Force Management Implementation Guidance

a. The concept of assessing and balancing risks has a long pedigree in defense planning. By its very nature, from the most basic tactical level to the realm of global grand strategy, military planning involves assessing risks. To the extent that resources are not sufficient to provide certainty of success, military plans and operations involve accepting varying degrees of risk and deciding whether one COA is, in view of what is at stake, worth the likely cost in blood and treasure, or is too risky to undertake, or whether another COA would entail more or less risk.[45]

b. Throughout the history of armed conflict, government and military leaders have tried to reckon with the effect of casualties on policy, strategy, and mission accomplishment. Government and military leaders consider battle losses from different perspectives. However, both must balance the following against the value of national objectives:

- Effects of casualties.
- Impact on civilians.
- Damage to the environment.
- Loss of equipment.
- Level of public reaction.
- Weakening of a force/capability in one CCMD to strengthen another.

What SecDef Decisions Consider Risk?

DoD Management Functions	SecDef Decisions
Strategic Assessments	• Approve the NDS • Approve QDR decisions about strategy, resources, capability & capacity.
Employ the Force	• Approve GEF Force Allocation priorities
Manage the Force	• Approve GFM Allocation Plan (GFMAP) • Approve DEPORDs
Develop the Force	• Approve GFM Rotational policies • Approve GDF
DoD Corporate Support	• Approve DoD Presidents budget request • Approve DoD Strategic Management Plan

How can the CJCS provide "best military advice" from a global-strategic Perspective to support SecDef decisions?

[45] *CRS, Quadrennial Defense Review 2010: Overview and Implications for National Security Planning.*

c. Joint Strategic Planning System (JSPS). The JSPS is the primary means by which the CJCS carries out statutory responsibilities assigned in Titles 6, 10, 22 and 50 of the USC. The primary roles are to 1) conduct independent assessments; 2) provide independent advice to the President, SECDEF, NSC, and Homeland Security Council (HSC); and 3) assist the President and SECDEF in providing unified strategic direction to the Armed Forces. The JSPS is a system that enables the Chairman to effectively assess, advise, direct, and execute in fulfillment of these statutory responsibilities. One of the CJCS's major statutory responsibilities is to provide the CJCS's assessment to Congress on the nature and magnitude of the strategic and military risks associated with executing the missions called for under the current NMS (10 USC 153(b)). Another major statutory responsibility is to provide assessment on each QDR, to include an assessment of risk, an assessment of the review, and an assessment of the assignment of roles and missions to the Armed Forces (10 USC 118(e)).

(1) Title 10 USC, Section 153, requires the CJCS to submit to Congress through the SECDEF a report providing an assessment of the nature and magnitude of the strategic and military risks associated with executing the missions called for under the current NMS. In accordance with Title 10, USC, Section 163, the CJCS is required to confer with and obtain information from the CCDRs, and evaluate and integrate that information into his advice to the President and the SECDEF.

(2) The Chairman's Assessments. The CJCS is required to submit to the SECDEF an annual assessment of "the nature and magnitude of the strategic and military risks associated with executing the missions called for under the current NMS." The SECDEF is then required to forward the assessment, with or without comments, to the congressional defense committees. Conclusions of the report, known as the "Chairman's Risk Assessment," are often reflected in subsequent congressional testimony by the CJCS and other senior officers, and the report is often reflected in comments by Members of Congress. The Chairman's Risk Assessment generally focuses on short-term military readiness.

(a) The CJCS conducts both deliberate and continuous assessments. These assessments include the Comprehensive Joint Assessment (CJA), the Joint Combat Capability Assessment (JCCA), the Joint Strategy Review (JSR) process, the Chairman's Risk Assessment (CRA), the Capabilities Gap Assessment (CGA) process that produces the Chairman's Program Recommendation (CPR), the Chairman's Program Assessment (CPA), plans assessments, and the GFM process to create a common annual review of the strategic environment and friendly/threat capabilities over time. The CRA, CGA, and CPR are components of the JSR process. These processes comprise the Chairman's assessment component of the JSPS.

(3) Joint Strategic Planning System. The JSPS provides formal structure to the Chairman's statutory responsibilities and considers the strategic environment and the alignment of ends, ways, means, risk, and risk mitigation over time to provide the best possible assessments, advice, and direction of the Armed Forces in support of senior leaders and processes at the national and OSD level. The JSPS aligns with established and emerging OSD and JS processes and documents (see CJCSI 3100.01B, JSPS).

d. War is inherently complex, dynamic, and fluid. It is characterized by uncertainty, ambiguity, and friction. Uncertainty results from unknowns or lack of information. Ambiguity is the blurring or fog that makes it difficult to distinguish fact from impression about a situation and the enemy. Friction results from change, operational hazards, fatigue, and fears brought on by danger. These characteristics cloud the operating environment; they create risks that affect a commander's ability to fight and win. In uncertainty, ambiguity, and friction, both danger and opportunity exist. Hence, a leader's ability to adapt and take risks are key traits.

e. Principles. The basic principles that provide a framework for implementing the risk management process are:

(1) Integrating Risk Management into Mission Planning, Preparation, and Execution. The commander and staff continuously: identify hazards and assess risks; develop and coordinate control measures; determine the level of residual risk for accident hazards in order to evaluate COAs; and integrate control measures into staff estimates, OPLANs, OPORDs, and missions. Commanders assess the areas in which they might take risks and approve control measures that will reduce risks. Leaders ensure that all subordinates understand and properly execute risk controls. They continuously assess variable hazards and implement risk controls.

(2) Making Risk Decisions at the Appropriate Level in the Chain of Command. The commander should address risk guidance in their commander's guidance. He bases his risk guidance on established Joint and other appropriate policies and on his higher commander's direction, and then gives guidance on how much risk he is willing to accept and delegate. Subordinates seek the higher commander's approval to accept risks that might imperil the next higher commander's intent.

(3) Accepting no Unnecessary Risk. Commanders compare and balance risks against mission expectations and accept risks only if the benefits outweigh the potential costs or losses. Commanders alone decide whether to accept the level of residual risk to accomplish the mission.

f. Accepting Risk. Accepting risk is a function of both risk assessment and risk management. The approach to accepting risk entails an identification and assessment of threats, addressing risk, defining indicators, and observing and evaluating.

g. Identify Threats to the Force. Consider all aspects of mission, enemy, terrain, and weather, time, troops available and civilian (METT-TC) for current and future situations. Sources of information about threats include reconnaissance, intelligence, experience/expertise of the commander and staff, etc.

h. Assess Threats. Assess each threat to determine the risk of potential loss based on probability and severity of the threat. Probability may be ranked as frequent: occurs often, continuously experienced; likely: occurs several times; occasional: occurs sporadically; seldom: unlikely, but could occur at some time; or unlikely: can assume it will not occur. Severity may be catastrophic: mission is made impossible; critical: severe mission impact; marginal: mission possible using alternate options; or negligible: minor disruptions to mission. Determining the risk is more an art than a science. Use historical data, intuitive analysis, and judgment to estimate the risk of each threat. Probability and severity levels are estimated based on the user's knowledge of probability of occurrence and the severity of consequences once the occurrence happens. The level of risk is assessed by a combination of the threat, its probability of occurring, and degree of severity. The levels of risk are:

- Extremely High: loss of ability to accomplish mission.
- High: significantly degrades mission capabilities in terms of required mission standard.
- Moderate: degrades mission capabilities in terms of required mission standards.
- Low: little or no impact on accomplishment of the mission.

i. Address Risk, Determine Residual Risk, and Make Risk Decision. For each threat, develop one or more options that will eliminate or reduce the risk of the threat. Specify who, what, where, when, and how. Determine any residual risk and revise the evaluation of the level of risk remaining. The commander alone then decides whether or not to accept the level of residual risk. If the commander determines that the risk is too great to continue the mission or a COA, then the commander directs the development of additional measures to account for the risk, or the COA is modified or rejected.

j. Define Indicators. Think through the threat—what information will provide indication that the risk is no longer acceptable? Ensure that subordinates and staff are informed of the importance of communicating the status of those indicators.

k. Observe and Evaluate. In execution, monitor the status of the indicators and enact further options as warranted. After the operation, evaluate the effectiveness of each option in reducing or eliminating risk. For options that were not effective, determine why and what to do the next time the threat is identified.

l. Applying Risk Management. Risk management requires a clear understanding of what constitutes unnecessary risk, when the benefits actually do outweigh costs, and guidance as to the command level to make those decisions. When a commander decides to accept risk, the decision must be coordinated with the affected units; where and how the commander is willing to accept risk are detailed in each COA.

- Risk management assists commanders in conserving lives and resources and avoiding or mitigating unnecessary risk, making an informed decision to execute a mission, identifying feasible and effective control measures where specific standards do not exist, and providing reasonable alternatives for mission accomplishment.

Risk management is the process of identifying, assessing, and controlling risks arising from strategic and operational factors and making decisions that balance risk costs with mission benefits.

- While risk cannot be totally eliminated, it can be managed by a systematic approach that weighs the costs—time, personnel, and resources—against the benefits of mission accomplishment. Commanders have always risk-managed their actions: intuitively, by their past experiences, or otherwise. Risk-management won't prevent losses but, properly applied, it allows the commander to take necessary and prudent risks without arbitrary restrictions and while maximizing combat capabilities.
- Risk management does not inhibit commander's flexibility and initiative, remove risk altogether (or support a zero defects mindset), require a GO/NO-GO decision, sanction or justify violating the law, or remove the necessity for development of standing operating procedures (SOPs). Risk management should be applied to all levels of war, across the range of military operations, and all phases of an operation to include any branches and sequels of an operation. To alleviate or reduce risk, commanders may change the CONOPS or concept of fire support, execute a branch plan, or take other measures to reduce or bypass enemy capabilities.
- *Safety* is crucial to successful training and operations and the preservation of military power. High-tempo operations may increase the risk of injury and death due to mishaps. Command interest, discipline, risk mitigation measures, and training lessen those risks. The JFC reduces the chance of mishap by conducting risk assessments, assigning a safety officer and staff, implementing a safety program, and seeking advice from local personnel. Safety planning factors could include the geospatial and weather data, local road conditions and driving habits, uncharted or uncleared mine fields, and special equipment hazards.
- To assist in risk management, commanders and their staffs may develop or institute a risk management process tailored to their particular mission or operational area. Figure XIV-11 is a generic model that contains the likely elements of a risk management process.[46]

[46] *JP 3-0, Joint Operations.*

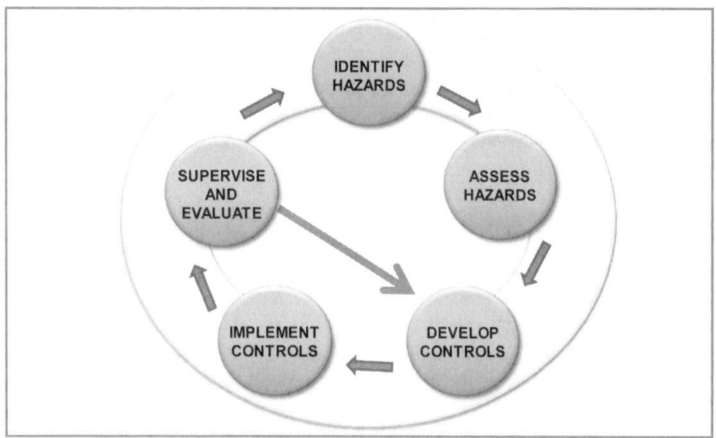

Figure XIV -11. Risk Management Process (JP 5-0).

m. JP 5-00.2, JTF *Planning Guidance and Procedures* and the U.S. Army Field Manual No. 100-14, *Risk Management* utilizes a Risk Management Process that have broken risk management into a five-step process of identifying and controlling hazards to conserve combat power and resources. Also, the *United Kingdom Joint Warfare Publication* 5-00, *Joint Operations Planning* has an excellent Annex on Risk Management. The five steps of risk management are:

- Step 1. Identify hazards.
- Step 2. Assess hazards to determine risks.
- Step 3. Develop controls and make risk decisions.
- Step 4. Implement controls.
- Step 5. Supervise and evaluate.

(1) Steps 1 and 2 together comprise the *risk assessment*. In Step 1, commanders identify the hazards that may be encountered in executing a mission. In Step 2, commanders determine the direct impact of each hazard on the operation. The risk assessment provides commanders with enhanced situational awareness. This awareness builds confidence and allows units to take timely, efficient, and effective measures.

(2) Steps 3 through 5 are the essential follow-through actions to effectively *manage* risk. In these steps, leaders balance risk against costs—political, economic, environmental, and to combat power— and take appropriate actions to eliminate unnecessary risk. During execution, as well as during planning and preparation, leaders continuously assess the risk to the overall mission and to those involved in the task.

Finally, leaders and individuals evaluate the effectiveness of controls and provide lessons learned so that others may benefit from the experience.

> *Sizing up opponents to determine victory, assessing dangers and distances is the proper course of action for military leaders.*
>
> *Sun Tzu, The Art of War, "Terrain"*

(3) Risk decisions should be based upon awareness rather than mechanical habit. Leaders should act on a keen appreciation for the essential factors that make each situation unique instead of from conditioned response. Throughout the entire operational continuum, the commander must consider U.S. Government civilians and contract support personnel in his risk management process. Hazards can exist, regardless of enemy or

adversary actions, in areas with no direct enemy contact and in areas outside the enemy's or adversary's influence.

n. When you conduct a preliminary risk assessment, you must determine what *obstacles* or *actions* may preclude mission accomplishment. The first two steps of the three-step risk management process are the *identification of risk* and the *assessment of its hazard.*

(1) *Identify risk*: Assess the probability and severity of loss linked to known or assumed hazards.

(2) *Assessment of the risk(s) hazard*: The condition with the potential to cause injury, illness or death of personnel; damage to, or loss of, equipment or property; or mission degradation.

(3) *Risk Management*. The process by which decision makers reduce or offset risk.[47]

o. Operational risk is the commander's conceptual balance between danger and opportunity; it considers the resources available, the component's mission and the operational environment. The rewards of meeting the desired objectives or effects must outweigh the potential costs associated with mission accomplishment. While the commander must ultimately make the decisions what risks the strategy and forces will assume, the staff's role is to identify critical decision and risk points, provide supporting information and ensure the commander's risk decisions are considered throughout operational planning and execution.

> *Military Risk: The ability of U.S. Armed Forces to adequately resource and execute military operations in support of the strategic objectives of the National Military Strategy.*
>
> *Global Force Management Implementation Guidance*

p. The commander expresses guidance regarding risk in several ways. The commander's risk estimate is based on the mission, his or her experience, higher headquarters' guidance and staff estimates. With these considerations, the commander formulates initial staff guidance, followed by an intent statement during the mission analysis step. The commander expresses an estimate of risk every time he or she provides guidance. Some risk factors permit quantitative analysis, while others will be wholly qualitative. Probability and statistics support risk analysis, but the commander will have to address operational risk subjectively when supporting information is unavailable. See the following chart for example operational risk matrix.

q. Risk vs. Gamble. The difference between a risk and a gamble is that you can recover from a risk, but you can't from a gamble. Ensure that when conducting your Risk Analysis that if your risk mitigation fails, you still will not.

> *"If you can recover from the loss, it's a risk. If not, it's a gamble"*
>
> *Field Marshal Erwin Rommel*

[47] *JP 5-00.2, JTF Planning Guidance and Procedures.*

Risk Identification	Mission Risk Assessment			Controls (mitigation)	Residual Mission Risk*
	Probability	Severity	Risk		
Outside Interference with theater LOCs	Likely	Critical	High (RED)	-Patrol -ISR Escorts	Moderate (YELLOW)
Blue infrastructure inadequate to support deployed forces	Likely	Marginal	Moderate (YELLOW)	-Site surveys -LOG ADVON -LOG afloat	Low
Blue forces conduct active asymmetric ops	Likely	Negligible	Moderate (YELLOW)	-Force protection -IO/MISO -CMO	Low
Yellow/Orange conduct coordinated attacks with Red	Frequent	Critical	High (RED)	-Deter/ Dissuade (DIME) -Force Planning	Low
Air Strike and/or missile strike against Blue	Likely	Marginal	Moderate (YELLOW)	-ISR -Air Defense -Deter / Dissuade	Low

Example: Operational Risk Matrix.

9. Key-Step — 9: Determine Commander's CCIR: CDR's Critical Information Requirements

CCIRs are elements of information required by the commander that directly affect decision-making. CCIRs are a key information management tool for the commander and help the commander assess the operational environment and identify decision points throughout the conduct of operations. *CCIRs are established by the commander* and should be developed and recommended by staffs as part of the planning process.

a. Characteristics of CCIRs result from the analysis of information requirements (Figure XIV-12, (JP 5-0)) in the context of a mission, commander's intent, and the concept of operation. Commanders designate CCIRs to let their staffs and subordinates know what information they deem necessary for decision-making. In all cases, the fewer the CCIRs, the better the staff can focus its efforts and allocate scarce resources. Staffs may recommend CCIRs; however, they keep the number of recommended CCIRs to a minimum. CCIRs *are not static*. Commanders add, delete, adjust, and update them throughout an operation based on the information they need for decision-making. To assist in managing CCIRs, commanders should adopt a process to guide the staff. This process should include specific responsibilities for development, validation, dissemination, monitoring, reporting, and maintenance (i.e., modifying/deleting).[48]

[48] *JP 3-0, Joint Operations.*

Information Requirements Categories

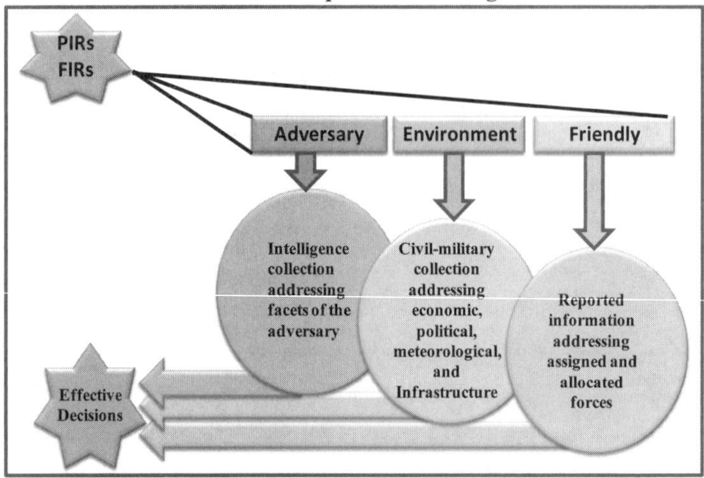

Figure XIV-12. Information Requirements Catergories.

b. Commanders Critical Information Requirement (CCIR).

(1) CCIRs comprise information requirements identified by the commander as being critical to timely information management and the decision-making process that affect successful mission accomplishment. CCIRs result from an analysis of information requirements in the context of the mission and the commander's intent. Described in the following sections are the two key subcomponents: Friendly Force Information Requirements (FFIR) and Priority Intelligence Requirement (PIRs).

(a) *PIR*. Those intelligence requirements for which a commander has an anticipated and stated priority in the task of planning and decision making.[49]

(b) *FFIR*. Information that the commander and staff need about the forces available for the operation.[50]

"Commanders Critical Information Requirements"

***Commanders Critical Information Requirement*.. An information Requirement identified by the commander as being critical to facilitating Timely decision-making. The two key elements are friendly force information Requirements and priority intelligence requirements.**

JP 3-0, *Joint Operations*

•**Two Categories**:

- **Priority Intelligence Requirement (PIR)** – An intelligence requirement, stated as a priority for intelligence support, that the commander and staff need to understand the adversary or the environment.
- **Friendly Force Information Requirements (FFIR)** – Information the commander and staff need to understand the status of friendly force and supporting capabilities.

[49] *Joint Pub 1-02, DOD Dictionary of Military and Associated Terms.*

[50] *Ibid.*

(c) The information needed to verify or refute a planning assumption is an example of a CCIR. As mentioned earlier, CCIRs are not static. They are situation-dependent, focused on predictable events or activities, time-sensitive, beg a decision by the commander when answered, and always established by an order or plan.

(d) CCIR *previously* had three categories; (1) PIR, (2) EEFI, and (3) Friendly Force Information Requirement (FFIR). As discussed previously, PIR remains a category of CCIR, as does FFIR. The third category of CCIR, EEFI, has been eliminated as a category of CCIR. EEFI was defined as "*key questions likely to be asked by adversary officials and intelligence systems about specific friendly intentions, capabilities, and activities, so they can obtain answers critical to their operational effectiveness.*" In reality, one can still expect to hear the term used in the operating forces because many commanders believe it is still necessary to provide guidance to his staff concerning operational security, therefore it is mentioned here.

(e) There is a proverbial saying that "a CCIR is so important that when answered, one must wake up the commander." There are general criteria that apply to CCIR that planners must consider when proposing them to the commander for approval: (1) answering a CCIR begs a decision by the commander, and not necessarily by the staff, and (2) the information or intelligence necessary to answer or satisfy a CCIR must be critical to the success of the mission. Keep in mind that a CCIR cannot be a CCIR unless approved by the commander. This criterion also pertains to PIRs, which is the subject of the next section.

c. Priority Intelligence Requirement.

(1) PIRs are normally identified during JIPOE and refined during mission analysis and COA development. JIPOE provides basic intelligence oriented on understanding the battlespace and adversary. The intelligence enables the J-2 to identify PIRs based on what intelligence critical to mission accomplishment is lacking.

Priority Intelligence Requirements

- *PIRs are the commander's statements of the force's critical intelligence needs.*
  - *The J-2 is responsible for assisting the commander in determining PIRs.*
- *Keep PIRs to a minimum.*

(2) The J-2 will normally recommend PIRs to the commander for approval. IO planners can have input into PIRs. Input can be submitted through several personnel, including J-2 personnel supporting the IO cell, the J-2 or Deputy J-2, the J-2 Joint Intelligence Operations Center (JIOC) commander, J-2 Watch Officer in the Joint Operations Center (JOC), or Joint Intelligence Support Element (JISE) Watch Officer or Senior Analysis.

(3) PIRs are meant to *focus intelligence operations.* The J-2 will rarely, if ever, have all the intelligence collection assets and personnel necessary to satisfy all requirements. The preponderance of the level of effort for the J-2's organization and assets will direct their focus of effort on answering PIR. Therefore, it is important that the number of PIRs remain low. Otherwise everything becomes a priority and PIRs no longer serve their intended purpose since there is no longer a focus for intelligence operations.

d. A Conceptual look at Intelligence Requirements. More often than not, intelligence requirements are worded in a manner that does not enable a collection asset or resource to answer it directly.

Intelligence requirement- "any subject, general or specific, upon which there is a need for the collection of information, or the production of intelligence," and "a requirement for intelligence to fill a gap in the command's knowledge or understanding of the battlespace or threat forces."

Joint Pub 1-02

e. If an intelligence requirement cannot be collected on directly, then it must be dissected into *information requirements.*

> *Information requirement- "those items of information regarding the adversary and the environment that need to be collected and processed in order to meet the intelligence requirements of a commander."*
> *Joint Pub 1-02*

f. Information requirements are worded in a manner that enables a collection manager to task an asset to collect on it directly. It is basically an intelligence requirement broken down into "*bite-size chunks*," and when an asset collects on and answers these "bite-size chunks," an answer to the intelligence requirement can be determined. If an intelligence requirement is worded in a manner that enables the collection manager to task a collection asset to collect directly on it, then it is not required to be broken down into information requirements. Otherwise, one or more information requirements will normally be drafted to support answering an intelligence requirement.

(1) There are various methods that can be used to break down an intelligence requirement into information requirements. Either the submitter of a Request for Information (RFI) can attempt to break it down, or the J-2 analysts can do so upon receipt of a validated intelligence requirement. Either way can be effective, but if the first method is attempted, one can expect J-2 personnel to modify the information requirements to facilitate collection or processing of the intelligence necessary to answer the intelligence requirement.

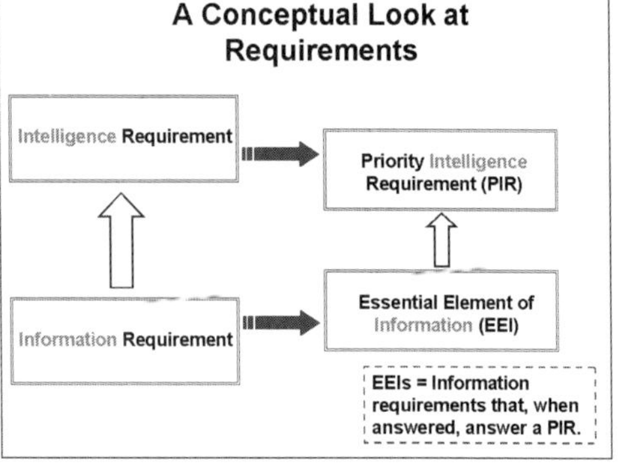
A Conceptual Look at Requirements
Intelligence Requirement
Priority Intelligence Requirement (PIR)
Information Requirement
Essential Element of Information (EEI)
EEIs = Information requirements that, when answered, answer a PIR.

(2) Since a PIR is an intelligence requirement, it too may have to be dissected in the same manner as a standard intelligence requirement. The information requirements that must be answered to answer a PIR are called *Essential Elements of Information*, or EEI. They are simply information requirements that receive a special title due to their importance, i.e., because they are tied to a PIR. According to the Joint Pub 2-0, *Joint and National Intelligence Support to Military Operations*, "those information requirements that are most critical or that would answer a PIR are known as essential elements of information (EEIs)."

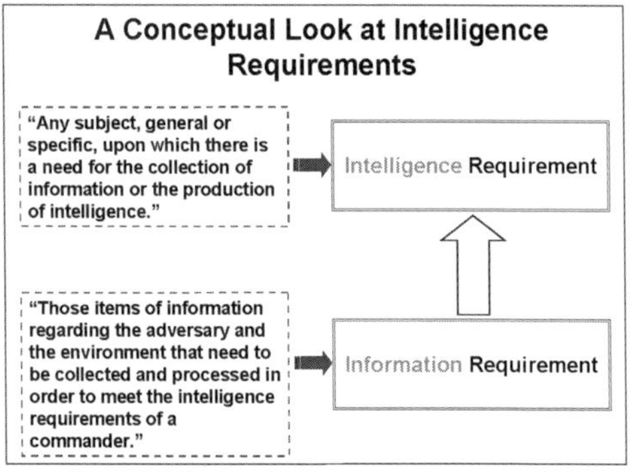
A Conceptual Look at Intelligence Requirements
"Any subject, general or specific, upon which there is a need for the collection of information or the production of intelligence."
Intelligence Requirement
"Those items of information regarding the adversary and the environment that need to be collected and processed in order to meet the intelligence requirements of a commander."
Information Requirement

(3) Useful unclassified reference documents for drafting intelligence and information requirements are the *Intelligence Requirements Handbooks* published by the Marine Corps Intelligence Activity (MCIA). They are also available electronically on the MCIA portal or by contacting MCIA at DSN 278-6146 or COMM 703-784-6146. MCIA has produced a *Generic Intelligence Requirements Handbook* (GIRH) specifically for IO.

(4) Examples pulled from doctrine and from the Information Operation GIRH are as follows. These examples contain a select sample of EEIs; however, there could be any number of EEIs necessary to support answering a PIR.

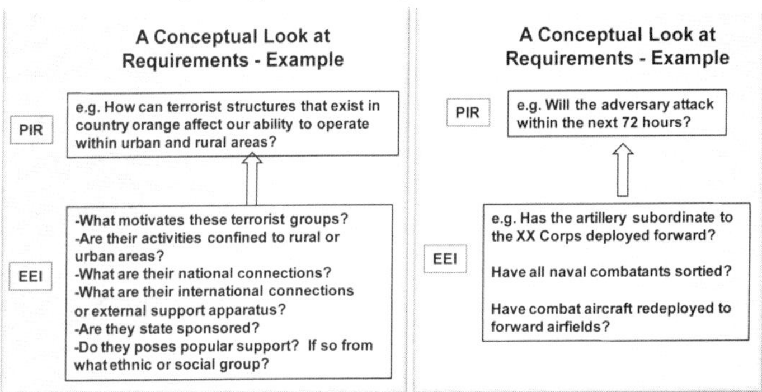

A Conceptual Look at Requirements - Example
PIR
e.g. How can terrorist structures that exist in country orange affect our ability to operate within urban and rural areas?
EEI
-What motivates these terrorist groups?
-Are their activities confined to rural or urban areas?
-What are their national connections?
-What are their international connections or external support apparatus?
-Are they state sponsored?
-Do they poses popular support? If so from what ethnic or social group?
A Conceptual Look at Requirements - Example
PIR
e.g. Will the adversary attack within the next 72 hours?
EEI
e.g. Has the artillery subordinate to the XX Corps deployed forward?
Have all naval combatants sortied?
Have combat aircraft redeployed to forward airfields?

g. Requests for Information (RFI). Many RFI's are between supporting and supported headquarters. It is important to remember that RFI's are broader than just intelligence based, however, we will discuss intelligence RFI's since they seem prevalent.

(1) If we cannot answer intelligence or information requirements with existing resources, then we must generate an RFI. However, keep in mind that RFIs must be satisfied at the LOWEST possible level – that is you! If you cannot find the information that satisfies your intelligence or information requirement, then begin generating an RFI.

(a) If you determine that you must submit an RFI, there are certain responsibilities associated with doing so. *First*, you must conduct the initial research. While this sounds trivial, it is often overlooked. We've already mentioned that you must answer the question at the lowest level. That level is yours! The answer to your question may lie in a readily available document or on the Internet - whether it is classified or unclassified. While it may be easier to simply "pass the buck" and let someone else do the work for you, it will always delay the answer and slow down the entire process.

> ***Request for Information***
>
> *"Any specific time-sensitive ad hoc requirement for intelligence information or products to support an ongoing crisis or operation not necessarily related to standing requirements or scheduled intelligence production. A request for information can be initiated to respond to operational requirements and will be validated in accordance with the theater command's procedures."*

> *Two excellent sources that can immediately be visited to begin initial research are the INTELINK homepage on both SIPRNET and JWICS. Each page contains a Google search capability just as the NIPRNET does if you choose to use open sources. Depending on your search string, you can significantly increase the timeframe by which you can find an answer to your RFI.*

(b) *Second*, if you are unsuccessful in your initial search, then you must clearly articulate your requirement. Be as specific as possible, but avoid using language associated with a specific occupation that may appear foreign to the person who must review and validate your RFI. Additionally, *avoid asking for a particular intelligence collection asset to satisfy your requirement.* It is the Collection Manager's job to determine the most effective manner to satisfy your requirement.

(c) *Third*, you must justify your request. Clearly articulate why your request is so important compared to the hundreds of others? Stating, "because the boss wants it" is not good enough. You must clearly and accurately justify your request for it to be accepted and prioritized. If you can tie your requirement to a PIR, it may be prioritized higher than other requirements which may not be.

(d) *Fourth*, you must truthfully determine and document the "Latest Time Information is of Value," or LTIOV. We have already mentioned this is the time beyond which the information will no longer be useful to you. When identifying this time, ensure it is accurate. Obviously, you will want your answer immediately, but if you really don't need it for a week, then state that. This helps the collection manager to prioritize requests and free collection assets to perform other priority missions until assets or resources are available to support your requirement.

RFI Responsibilities

- *Conduct Initial Research*
- *Clearly State the Requirement*
- *Justify the Request*
- *Provide an Accurate LTIOV*

(e) Prior to submitting an RFI via the means established by the J-2, you should ask four questions to validate your efforts. The questions represent validation criteria that should be met prior to submitting the RFI. When this criterion is met, then your RFI can be submitted.

1 Requestor's Questions:

- Is the answer to the question relevant to executing or planning my mission?
- Have I thoroughly searched existing, accessible repositories for this information?
- Have I provided concise, yet sufficient detail to enable the RFI Manager to understand the intelligence requirement that I need answered?
- Is my stated LTIOV realistic?

2 The RFI Manager will receive your RFI and ask at least five questions to determine if it will be validated or not.

3 RFI Manager's Questions:

- Does the information justify the dedication of scarce intelligence resources?
- Does it duplicate an existing requirement?
- Has the requirement been previously satisfied?
- Does the RFI pertain to only one intelligence requirement?
- Can I retrieve the information required by the LTIOV?

g. If the RFI Manager decides it cannot be validated, you should expect a phone call or email explaining why. If your RFI is validated, then it will be prioritized against other validated "intelligence" or "information" requirements. The RFI will then evolve into either a production requirement or a collection requirement. A *production requirement* is submitted when new, finished intelligence derived from original research is required to satisfy all or a portion of the RFI. A *collection requirement* is submitted when insufficient information exists to answer an RFI. A production requirement is submitted to the "Analysis & Production Cell," and a collection requirement is submitted to the "Collection Management Cell" for action. The RFI Manager will normally confer with each cell (or section) to receive their estimate of supportability prior to submitting it either to one cell or the other as a whole or a portion thereof to each of them. If it takes the efforts of both cells to satisfy the requirement, the RFI Manager will work with personnel from each cell to determine which portion of the requirement each will satisfy. Each cell will in turn take their portion of the RFI for action.

h. The RFI Manager could be co-located with the Collection Management Cell, or they could occupy a separate and distinct RFI Management Cell. To learn more about the RFI process and the organization that supports it, planners should direct their questions to either the J-2, Deputy J-2, or directly to the RFI Manager or Collection Manager. Each can offer insight into how the process is managed and how the organization was established to support it.

i. Requesting National Intelligence Support. If organic or attached intelligence capabilities within a JTF cannot satisfy an RFI, it will likely be forwarded to the higher headquarters, i.e., to the CCMD JIOC, requesting that they satisfy the RFI with their organic or attached intelligence capabilities. RFI's are normally submitted to higher via *Community On-line Intelligence System for End Users and Managers* (COLISEUM). Likewise, if the CCDR JIOC cannot satisfy the RFI, it is then submitted via COLISEUM to the Defense JIOC (DJIOC), requesting that it be satisfied with national intelligence capabilities.

COLISEUM is a production management tool used throughout the Defense Intelligence Production Community. The tool allows all users to register and track requests for information and production requirements, perform research for existing requirements or answers, and manage/account for production resources. COLISEUM is a web-based application available throughout the Intelligence Community.

j. The CCMD JIOC is the primary focal point for providing intelligence support to the CCMD. The CCMDs JIOCs are the primary intelligence organizations providing support to joint forces at the operational and tactical levels. The JIOC fuses the in-theater capabilities of all Director of National Intelligence (DNI), Service, combatant support agency, and combat command ISR assets into a central location for intelligence planning, collection management, tasking, analysis and support. The JIOC concept seamlessly combines all intelligence functions, disciplines, and operations in a single organization, ensures the availability of all sources of information from both CCMD ISR assets and national intelligence resources, and fully synchronizes and integrates intelligence with operation planning and execution. Although a particular JIOC cannot be expected to completely satisfy every RFI, it can coordinate support from other intelligence organizations, both lower, higher, and laterally.

k. As the lead DOD intelligence organization for coordinating intelligence support to meet CCMD requirements, the DJIOC coordinates and prioritizes military intelligence requirements across the CCMDs, combat support agencies (CSA's), reserve component, and Service intelligence centers. The DJIOC formulates recommended solutions to de-conflict requirements for national intelligence with USSTRATCOM's Joint Functional Component Command for Intelligence, Surveillance and Reconnaissance (JFCC-ISR) and DNI representatives to ensure an integrated response to CCMD needs. It is the channel through which joint force's crisis-related and time-sensitive intelligence requirements are tasked to the appropriate national agency or command, when they cannot be satisfied using assigned or attached assets.

l. National Augmentation Support. One of the most effective means that a CCDR or JTF J-2 can facilitate having crisis-related RFI's satisfied in a relatively rapid manner is to request and exploit the capabilities of a National Intelligence Support Team, or NIST. At the request of a CCDR, the DJIOC may deploy a NIST to support a commander, JTF (CJTF), during a crisis or deliberate operation. The NIST is a nationally sourced team composed of intelligence and communications experts from DIA, CIA, National Geospatial-Intelligence Agency (NGA), NSA, or other IC agencies as required. The NIST mission is to provide a tailored, national level, all-source intelligence team to deployed commanders during crisis or deliberate operations. NIST supports intelligence operations at the JTF HQ and is traditionally collocated with the J-2. In direct support of the JTF, the NIST will perform functions as designated by the J-2. The NIST is designated to provide a full range of intelligence support to a CJTF, from a single agency element with limited ultra-high frequency voice connectivity to a fully equipped team with the Joint Deployable Intelligence Support System (JDISS) and Joint Worldwide Intelligence Communications System (JWICS) video-teleconferencing capabilities. NIST provides coordination with national intelligence agencies, analytical expertise, Indications and Warning (I&W), special assessments, targeting support, and access to national data bases, and facilitates RFI management.

m. Decision Support.[51] CCIR support the commander's future decision requirements and are often related to Measures of Effectiveness and Measures of Performance. PIR are often expressed in terms of the elements of PMESII while FFIR are often expressed in terms of DIME. All are developed to support specific decisions the commander must make.

10. Key-Step — 10: Develop Mission Statement

a. Mission Statement. One product of the mission analysis process is the mission statement. Your initial mission analysis as a staff will result in a *tentative mission statement.* This tentative mission statement is a *recommendation* for the commander based on mission analysis. It will serve to identify the broad options open to the commander and to orient the staff. This recommendation is presented to the commander for approval normally during the mission analysis brief. It must be a clear, concise statement of the essential tasks to be accomplished by the command and the purpose of those tasks. Although several tasks may have been identified during the mission analysis, the proposed mission includes only those that are essential to the overall success of the mission. The tasks that are routine or inherent responsibilities of a commander are not included in the proposed mission. The proposed mission becomes the focus of the commander's staff's estimates. It should be continually reviewed during the planning process to ensure planning is not straying from this critical focus (or that the mission requires adjustment). It is contained in paragraph 1 of the commander's estimate and paragraph 2 of the basic OPLAN or OPORD.

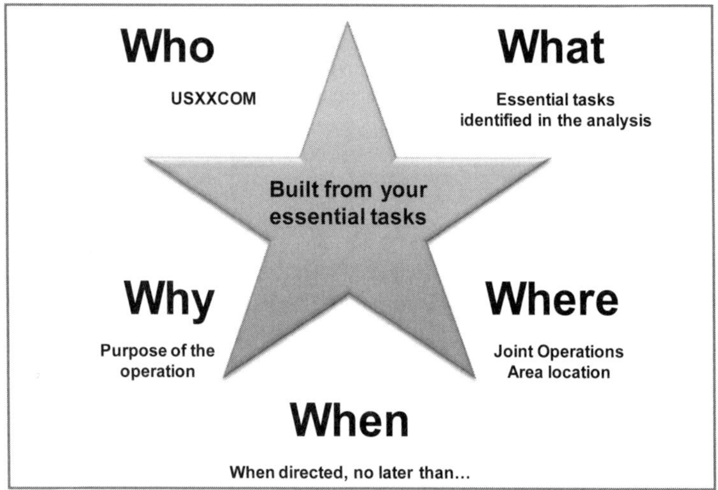

b. The mission statement should be a short sentence or paragraph that describes the organization's essential task (or tasks) and purpose — a clear (*brevity and clarity*) statement of the action to be taken and the reason for doing so. The mission statement contains the elements of who, what, when, where, and why, but seldom specifies how. These five elements of the mission statement answer the questions:

[51] *JP 2-01, Joint and National Support to Military Operations.*

- Who will execute the tasks (unit/organization)?
- What is the essential task(s) (mission task)?
- When will the operation begin (time/event i.e., O/O, when directed)?
- Where will the operation occur (AO, objective)?
- Why will we conduct the operation (for what purpose)?

c. Clarity of the joint force mission statement and its understanding by subordinates, before and during the joint operation, is vital to success. The mission statement along with the commanders' intent, provide the primary focus for subordinates during planning, preparations, execution and assessment.[52]

d. No mission statement should be written and not revisited thereafter; it's important to revisit it during the entire plan development process to ensure that it meets the needs of the commander and the national leadership. A sample CCDR's mission statement could look like this:

> *"When directed, CDRUSXXCOM deters regional aggressors; if deterrence fails, CDRUSXXCOM defends the country of X and defeats external aggressors and conducts stability and support operations in order to protect U.S. interests and the Government of X."*

e. The who, where, when of the mission statement is straightforward. The *what* and *why*, however, are more challenging to write clearly and can be confusing to subordinates.

The *what* is a task and is expressed in terms of action verbs (for example, deter, defeat, deny, conduct, provide, contain, isolate, etc.). These tasks are measurable and can be grouped by actions by friendly forces and effects on adversary forces/ capabilities. The what in the mission statement is the essential task(s) to be accomplished. It may be expressed in terms of either actions by a friendly force or effects on an adversary force. Commanders should utilize doctrinal approved tasks. These tasks have specific meaning, are measurable, and often describe results or effects of the tasks relationship to the adversary and friendly forces.

The *why* puts the task(s) into context by describing the reason for conducting the task(s). It provides the mission purpose to the mission statement-why are we doing this task(s)? The purpose normally describes using a descriptive phrase and is often more important than the task because it provides clarity to the task(s) and assists with subordinate initiatives.

Example: Task Lists

Terrain:	Enemy:	Friendly:
• Seize	• Disrupt	• Screen
• Secure	• Defeat	• Guard
• Clear	• Destroy	• Cover
• Occupy	• Block	• Withdraw
• Retain	• Contain	• Attack by Fire
• Recon	• Fix	• Support by Fire
	• Canalize	• Follow & Assume
	• Delay	• Follow & Support
	• Interdict	• Breach
	• Isolate	• Disengage
	• Penetrate	• Exfiltrate
	• Suppress	• Infiltrate
	• Neutralize	
	• Feint	
	• Demonstration	
	• Ambush	
	• Bypass	

Purpose (in order to...)

• Allow	• Influence
• Cause	• Open
• Create	• Prevent
• Deceive	• Protect
• Deny	• Restore
• Divert	• Support
• Enable	• Surprise

[52] *JP 3-0, Joint Operations.*

11. Key-Step — 11: Develop and Conduct Mission Analysis Brief

Upon conclusion of the Mission Analysis and JIPOE, the staff will present a Mission Analysis Brief to the commander. The purpose of the Mission Analysis Brief is to provide the commander with the results of the preliminary staff analysis, offer a forum to surface issues that have been identified, and an opportunity for the commander to give his guidance to the staff and to approve or disapprove of the staff's analysis. However, modifications to this brief may be necessary based on the commander's availability of relevant information. There is no set format for the Mission Analysis Brief, however Figure XIV-13 contains two *tested examples* of Mission Analysis Briefings that work.

a. The mission analysis briefing should not be a unit readiness briefing. Staff officers must know the status of subordinate and supporting units and brief relevant information as it applies to the situation.

b. The mission analysis briefing is given to both the commander and the staff. This is often the only time the entire staff is present, and the only opportunity to ensure that all staff members are starting from a common reference point. Mission analysis is critical to ensure thorough understanding of the task and subsequent planning.

c. The briefing focuses on relevant conclusions reached as a result of the mission analysis. This helps the commander and staffs develop a shared vision of the requirements for the OPLAN and execution.

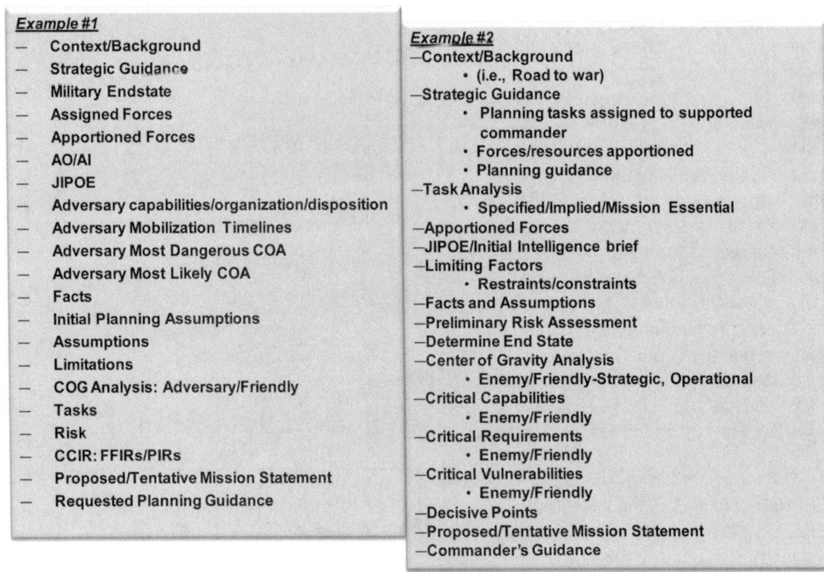

Example #1
– Context/Background
– Strategic Guidance
– Military Endstate
– Assigned Forces
– Apportioned Forces
– AO/AI
– JIPOE
– Adversary capabilities/organization/disposition
– Adversary Mobilization Timelines
– Adversary Most Dangerous COA
– Adversary Most Likely COA
– Facts
– Initial Planning Assumptions
– Assumptions
– Limitations
– COG Analysis: Adversary/Friendly
– Tasks
– Risk
– CCIR: FFIRs/PIRs
– Proposed/Tentative Mission Statement
– Requested Planning Guidance
Example #2
–Context/Background
• (i.e., Road to war)
–Strategic Guidance
• Planning tasks assigned to supported commander
• Forces/resources apportioned
• Planning guidance
–Task Analysis
• Specified/Implied/Mission Essential
–Apportioned Forces
–JIPOE/Initial Intelligence brief
–Limiting Factors
• Restraints/constraints
–Facts and Assumptions
–Preliminary Risk Assessment
–Determine End State
–Center of Gravity Analysis
• Enemy/Friendly-Strategic, Operational
–Critical Capabilities
• Enemy/Friendly
–Critical Requirements
• Enemy/Friendly
–Critical Vulnerabilities
• Enemy/Friendly
–Decisive Points
–Proposed/Tentative Mission Statement
–Commander's Guidance

Figure XVI-13. Sample: Mission Analysis Briefs.

12. Key-Step — 12: Prepare Initial Staff Estimates

As discussed earlier, the development of an effective commander's estimate must be supported by mission analysis, planning guidance, and *staff estimates*.

a. Battle rhythm is a deliberate daily cycle of command, staff, and unit activities intended to synchronize current and future operations.[53] The battle rhythm facilitates integration and collaboration. The COS normally manages the headquarters battle rhythm. This battle rhythm serves several important functions, to include: establishing a routine for staff interaction coordination, facilitating interaction between CDR and staff, synchronizing activities of the staff in time and purpose and facilitating planning by the staff and decision-making by the CDR.

b. Early staff estimates are frequently given as oral briefings to the rest of the staff. They are continually ongoing and updated based on changes in the situation. In the beginning, they tend to emphasize information collection more than analysis. CJCSM 3122.01 contains sample formats for staff estimates. (Also, see the JOPP Chapter of this publication.)

c. The role of the staff is to support the commander in achieving situational understanding, making decisions, disseminating directives, and following directives through execution. The staff's effort during planning focuses on developing effective plans and orders and helping the commander make related decisions. The staff does this by integrating situation-specific information with sound doctrine and technical competence. The staff's planning activities initially focus on mission analysis, which develops information to help the commander, staff, and subordinate commanders understand the situation and mission. Later, during COA development and comparison, the staff provides recommendations to support the commander's selection of a COA. Once the commander approves a COA, the staff coordinates all necessary details and prepares the plan or order.

d. Throughout planning, staff officers prepare recommendations within their functional areas, such as system, weapons, and munitions capabilities, limitations, and employment; risk identification and mitigation; resource allocation and synchronization of supporting assets; and multinational and interagency considerations. Staff sections prepare and continuously update staff estimates that address these and other areas continuously throughout the JOPP. The staff maintains these estimates throughout the operation, not just during pre-execution planning.

e. Not every situation will require or permit a lengthy and formal staff estimate process. During CAP, the commander may review the assigned mission, receive oral staff briefings, develop and select a COA informally, and direct that plan development commence. However, deliberate planning will demand a more formal and thorough process. Staff estimates should be shared collaboratively with subordinate and supporting commanders to help them prepare their supporting estimates, plans, and orders. This will improve parallel planning and collaboration efforts of subordinate and supporting elements and help reduce the planning times for the entire process.[54]

13. Key-Step — 13: Approval of Mission Statement, Develop Commander's Intent and Publish Initial Planning Guidance

a. Restated Mission Statement. Immediately after the mission analysis briefing, the commander approves a restated mission. This can be the staff's recommended mission statement, a modified version of the staff's recommendation, or one that the commander has developed personally. Once approved, the restated mission becomes the unit mission.

[53] *JP 3-33, Joint Task force Headquarters.*

[54] *JP 5-0, Joint Operations Planning.*

> *"It is the one overriding expression of will by which everything in the order and every action by every commander and soldier in the army must be dominated, it should, therefore, be worded by the commander himself."*
>
> *Field Marshal Viscount William Joseph Slim, Defeat into Victory, Batting Japan in Burma and India, 1942-1945, Cooper Square Press, p. 211.*

b. Commander's Intent. The intent statement is the commander's personal vision of how the campaign will unfold. Generally, the commander will write his own intent statement. Frequently the staff will provide substantial input(s). The commander's intent is a clear and concise expression of the purpose of the operation and the military endstate. It provides focus to the staff and helps subordinate and supporting commanders take actions to achieve the military endstate without further orders, even when operations do not unfold as planned. It also includes where the commander will accept risk during the operation. At the theater strategic level, commander's intent must necessarily be much broader – it must provide an overall vision for the campaign that helps staff and subordinate commanders understand the intent for integrating all elements of national power and achieving unified action. The CCDR must *envision* and *articulate* how military power and joint operations will dominate the adversary and support or reinforce the interagency and our allies in accomplishing strategic success. Through his intent, the commander identifies the major unifying efforts during the campaign, the points and events where operations must dominate the enemy and control conditions in the OE, and where other elements of national power will play a central role. The intent must allow for decentralized execution.

(1) It provides the link between the mission and the concept of operations by stating the method that, along with the mission, are the basis for subordinates to exercise initiative when unanticipated opportunities arise or when the original concept of operations no longer applies. If the commander wishes to explain a broader purpose beyond that of the mission statement, he may do so. The mission and the commander's intent must be understood two echelons down. The intent statement at any level must support the intent of the next higher commander.

(2) The initial intent statement normally contains the purpose and military endstate as the initial impetus for the planning process. It could be stated verbally when time is short. The commander refines the intent statement as planning progresses. The commander's approved intent is written in the "*Execution*" paragraph as part of the operation plan or order.

(3) A well-devised intent statement enables subordinates to decide how to act when facing unforeseen opportunities and threats, and in situations where the concept of operations no longer applies. This statement deals primarily with the military conditions that lead to mission accomplishment, so the commander may highlight selected objectives and effects. The statement also can discuss other instruments of national power as they relate to the mission and the potential impact of military operations on these instruments. The commander's intent may include the commander's assessment of the adversary commander's intent and an assessment of where and how much risk is acceptable during the operation.

(4) Remember, the commander's intent is not a summary of the CONOPs. It should not tell specifically how the operation is being conducted, but should be crafted to allow subordinate commanders sufficient flexibility and freedom to act in accomplishing their assigned mission(s) even in the "*fog of war.*"

While there is no specified joint format for commander's intent, a generally accepted construct includes the purpose, method, and endstate:

- Purpose: The purpose is the *reason* for the military action with respect to the mission of the next higher echelon. It explains why the military action is being conducted. This helps the force pursue the mission without further orders, even when actions do not unfold as planned. Thus, if an unanticipated situation arises, participating commanders understand the purpose of the forthcoming action well enough to act decisively and within the bounds of the higher commander's intent.
- Method: The "*how*," in doctrinally concise terminology, explains the offensive form of maneuver, the alternative defense, or other action to be used by the force as a whole. Details as to specific subordinate missions are not discussed.
- Endstate: The endstate describes what the commander wants to see in military terms after the *completion of the mission* by the friendly forces.

Commander's Intent (Purpose)

Maintain Green's sovereignty and territorial integrity.

Commander's Intent (Method)

USXXCOM forces will secure LOCs to ensure a rapid build-up of forces in the JOA. The utilization of HN support will be maximized. Forces will deploy into theater under the auspices of participation in C/J (Coalition/Joint) exercises demonstrating C/J force capabilities. IO will be optimized to communicate capability and coalition resolve against Red aggression. During these exercises, C/J forces will be positioned throughout the JOA with the capability to rapidly project full spectrum combat power against Red forces violating Green sovereignty.

Commander's Intent (Endstate)

Red forces withdrawn from forward staging bases and postured at their peacetime locations. These forces will be incapable of conducting rapid force build-up (>7 days) threatening Green.

c. Initial Planning Guidance. After approving the mission statement and issuing their intent, commanders provide the staff (and subordinates in a collaborative environment) with enough additional guidance (including preliminary decisions) to focus the staff and subordinate planning activities during COA development. As a minimum, the initial planning guidance should include the mission statement; assumptions; operational limitations; a discussion of the national strategic endstate; termination criteria; military endstate military objectives; and the commander's initial thoughts on desired and undesired effects. The planning guidance should also address the role of agencies and multinational partners in the pending operation and any related special considerations as required.[55]

(1) The commander approves the derived mission and gives the staff (and normally subordinate commanders) initial *planning guidance.* This guidance is essential for timely and effective COA development and analysis. The guidance should precede the staff's preparation for conducting their respective staff estimates. The commander's responsibility is to *implant a desired vision* of the forthcoming combat action into the minds of the staff. Enough guidance (preliminary decisions) must be provided to allow the subordinates to

[55] *JP 5-0, Joint Operations Planning.*

plan the action necessary to accomplish the mission consistent with his and the SECDEF's intent. The commander's guidance must focus on the essential tasks and associated objectives that support the accomplishment of the assigned national objectives. It emphasizes in broad terms when, where, and how the commander intends to employ combat power to accomplish the mission within the higher commander's intent.

(2) The commander may provide the planning guidance to the entire staff and/or subordinate commanders or meet each staff officer or subordinate unit commander individually as the situation and information dictates. The guidance can be given in a written form or orally. No format for the planning guidance is prescribed. However, the guidance should be sufficiently detailed to provide a clear direction and to avoid unnecessary efforts by the staff or subordinate commanders.

(3) The content of planning guidance varies from commander to commander and is dependent on the situation and time available. Planning guidance may include:

- Situation.
- The derived mission, including essential task(s) and associated objectives.
- Purpose of the forthcoming military action.
- Information available (or unavailable) at the time
- Forces available for planning.
- Limiting factors (constraints and restraints) – including time constraints for planning.
- Pertinent assumptions.
- Tentative Courses of Action (COAs) under consideration: friendly strengths to be emphasized or enemy weaknesses the COAs should attack, or specific planning tasks.
- Preliminary guidance for use (or non-use) of nuclear weapons.
- Coordinating instructions.
- Acceptable level of risk to own and friendly forces.
- Information Operations guidance.

(4) Planning guidance can be very explicit and detailed, or it can be very broad, allowing the staff and/or subordinate commander's wide latitude in developing subsequent COAs. However, no matter its scope, the content of planning guidance must be arranged in a logical sequence to reduce the chances of misunderstanding and to enhance clarity. Moreover, one must recognize that all the elements of planning guidance are *tentative only*. The commander may issue successive planning guidance during the decision-making process. Yet, the focus of his staff should remain upon the framework provided in the initial planning guidance. The commander should provide subsequent planning guidance during the rest of the plan development process.

(5) Initial planning guidance includes *Termination Criteria* and *Mission Success Criteria*. These criteria become the basis for assessment and include measures of performance (MOP) and measures of effectiveness (MOE).

(a) *Termination*. As discussed earlier, but worth a quick review, when and under what circumstances to suspend or terminate a military operation is a political decision. Effective planning cannot occur without a clear picture of the military endstate and termination criteria. Knowing when to terminate military operations and how to preserve achieved advantages is essential to achieving the national strategic endstate. Even so, it is essential that the CJCS and the supported JFC advise the President and SECDEF during the decision-making process. The supported JFC should ensure that political leaders understand the implications, both immediate and long term, of a suspension of hostilities at any point in the conflict. Once established, the national strategic objectives enable the supported commander to develop the military endstate, recommended termination criteria, and supporting military objectives. Termination criteria typically apply to the *end of a joint operation and disengagement by joint forces*. This often signals the end of the use of the military instrument of national power.

Mission Success Criteria

As measured, utilizing MOE/MOPs discussed earlier in this chapter describe the standards for determining mission accomplishment. The JFC includes these criteria in the initial planning guidance so that the joint force staff and components better understand what constitutes mission success. Mission success criteria can apply to any joint operation, phase, and joint force component operation. These criteria help the JFC determine if and when to move to the next major operation or phase.

Measure of Performance (MOP)

Used to assess friendly actions tied to measuring task accomplishment.

Measure of Effectiveness (MOE)

A criterion used to assess changes in system behavior, capability, or operational environment that is tied to measuring the attainment of an endstate, achievement of an objective, or creation of an effect. The CCDR should develop MOE to serve as tools in determining the degree to which mission objectives are met. MOE's should be developed for quantitative or qualitative standards as a means to evaluate operations and guide decision-making. Accurate and effective MOE contribute to mission effectiveness in many ways. MOE's will vary with mission; however, planners should ensure that MOE's possess the following characteristics:

- *Appropriate*. Correlate to the audience objectives.
- *Mission-related.* Must reflect the commander's desired endstate and the specific military objectives to reach the endstate.
- *Measurable*. Quantitative MOEs reflect reality more accurately than non-quantitative MOEs, and hence, are generally the measure of choice when the situation permits their use.
- *Numerically Reasonable*. MOEs should be limited to the minimum required to effectively portray the measure of the attainment.
- *Universally Understood and Accepted.* MOEs should be clear and concise based among the various government agencies, HN, and others to ensure that all concerned focus on efforts desired as well as the criteria for transition and termination of the military role.
- *Useful.* MOEs should detect situation changes quickly enough for the commander to immediately and effectively respond.

If the mission is unambiguous and limited in time and scope, mission success criteria could be readily identifiable and linked directly to the mission statement. For example, if the JFC's mission is to *evacuate all U.S. personnel from the U.S. embassy in Grayland*, then mission analysis could identify two primary success criteria: (1) all U.S. personnel are evacuated and (2) established ROE are not violated.

Possible MOEs in FHA operations could include:

• Drops in mortality rates in the affected population, below a specified level per day.

• Increase in water available to each disaster victim per day

to various levels established for human consumption, to support sanitation measures, and for livestock consumption.

• Decreases in population of displaced persons in camps to a level sustainable by the effected country or HN military organizations.

• Decrease in incidence of disease to an acceptable or manageable level. An increase in the presence and capabilities of NGOs, PVOs, and IOs.

4 However, more complex operations will require MOEs and MOPs for each task, effect, and phase of the operation. For example, if the JFC's specified tasks are to *ensure friendly transit through the Straits of Gray, eject Redland forces from Grayland, and restore stability along the Grayland-Redland border,* then mission analysis should indicate many potential success criteria — measured by MOEs and MOPs — some for each desired effect and task.

5 Measuring the status of tasks, effects, and objectives becomes the basis for reports to senior commanders and civilian leaders on the progress of the operation. The CCDR can then advise the President and SECDEF accordingly and adjust operations as required. Whether in a supported or supporting role, JFCs at all levels must develop their mission success criteria with a clear understanding of termination criteria established by the CJCS and SECDEF.[56]

II. In-Progress Discussion (IPR)

IPRs occur throughout the APEX enterprise: The APEX plan review process consists of four core planning phases: Strategic Guidance, Concept Development, Plan Development, and Plan Assessment (Figure XVI-14). Each of these planning phases will include as many IPRs as necessary to complete the plan. During each phase, CCDRs ensure planning is synchronized with the SECDEF's intent and consistent with current national objectives and assumptions. This is accomplished by maintaining an on-going dialog with the JS and OSD, as the guidance and assessments may change during the planning process and over the life of the plan (Chapter VIII, IPRs).

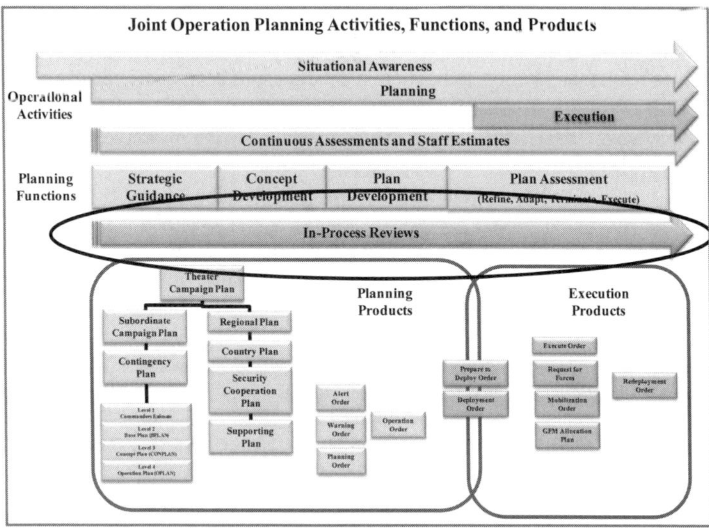

Figure XIV-14. IPR Process (CJCSG 3130).

[56] *JP 5-0, Joint Operations Planning.*

Concept Development: Function II 15

1. Function II — Concept Development

a. During the concept development step, CCDRs develop, analyze, and compare viable COAs and develop staff estimates that are coordinated with the Military Departments when applicable. Analysis includes wargaming, operational modeling, and initial feasibility assessments.

b. As you work through the Concept Development Function you will be visualizing and thinking through the entire operation or campaign from end to start, start to end. It's important to emphasize here, as discussed in Chapter II and III, operations and campaigns are broken into phases which are a way to view and conduct a complex joint operation in manageable parts. You will determine requirements in terms of forces, resources, time, space and purpose. Doctrine now standardizes phasing in OPLANs within all CCMDs. The main purpose of phasing is to integrate and synchronize related activities, thereby enhancing flexibility and unity of effort during execution. Reaching the endstate often requires arranging a major operation or campaign in several phases. Phasing assists CCDRs and staffs by helping them to visualize and think through the entire operation or campaign and to define requirements in terms of forces, resources, time, space, and purpose. Phases are designed to be conducted sequentially, but activities from a phase may continue into subsequent phases.

c. The staff writes (or graphically portrays) the CONOPS in sufficient detail so that subordinate and supporting CDRs understand their mission, tasks, and other requirements and can develop their supporting plans accordingly. During CONOPS development, the CDR determines the best arrangement of simultaneous and sequential actions and activities to accomplish the assigned mission consistent with the approved COA. This arrangement of actions dictates the sequencing of forces into the OA, *providing the link between the CONOPS and force planning*. The link between the CONOPS and force planning is preserved and perpetuated through the TPFDD structure. This structure must ensure unit integrity, force mobility, and force visibility as well as the ability to rapidly transition to branches or sequels as operational conditions dictate. Planners ensure that the CONOPS, force plan, deployment plans, and supporting plans provide the flexibility to adapt to changing conditions, and are consistent with the CCDR's intent.

d. If the scope, complexity, and duration of the military action you contemplate warrants a campaign, then the staff outlines the series of military operations and associated objectives and develops the CONOPS for the preliminary part of the campaign in sufficient detail to impart a clear understanding of the CDR's concept of how the assigned mission will be accomplished.

e. During CONOPS development, the CCDR must assimilate many variables under conditions of uncertainty to determine the essential military conditions, sequence of actions, and application of capabilities and associated forces to create effects and achieve objectives. CCDRs and their staffs must be continually aware of the higher-level objectives and associated effects that influence planning at every juncture. If operational objectives are not linked to strategic objectives, the inherent linkage or "nesting" is broken and eventually tactical considerations can begin to drive the overall strategy at cross-purposes.

2. COA Development Preparation and Considerations

- Time Available.
- Political Considerations.
- Flexible Deterrent Options.
- Lines of Operation.

a. Time Available. The CDR and the nature of the mission will dictate the number of COAs to be considered. Staff sections continually inform COA development by an ongoing staff estimate process to ensure suitability, feasibility, acceptability, and compliance with Joint Doctrine (deviations from Joint Doctrine should be conscious decisions and not the result of a lack of knowledge of doctrinal procedures). Additionally, staffs ensure completeness (answers Who, What, When, Where, Why, How).

b. Political Considerations. Planning for the use of military forces includes a discussion of the political implications of their transportation, staging, and employment. The CCDR's political advisor is a valuable asset in advising the CCDR and staff on issues crucial to the planning process, such as overflight and transit rights for deploying forces, basing, and support agreements. Multinational and coalition force concerns and sensitivities must also be considered.

(1) Political objectives drive the operation at every level from strategic to tactical. There are many degrees to which political objectives influence operations: ROE restrictions and basing access and overflight rights are examples. Two important factors about political primacy stand out. *First*, all military personnel should understand the political objectives and the potential impact of inappropriate actions. Having an understanding of the political objective helps avoid actions which may have adverse political effects. It is not uncommon today in the GWOT for junior leaders to make decisions which have significant political implications. *Secondly*, CDRs should remain aware of changes not only in the operational situation, but also to changes in political objectives that may warrant a change in military operations. These changes may not always be obvious.

(2) The integration of U.S. political and military objectives and the subsequent translation of these objectives into action have always been essential to success at all levels of operation. The global environment that is characterized by regional instability, failed states, increased weapons proliferation, global terrorism, and unconventional threats to U.S. citizens, interests, and territories, requires even greater cooperation.

(3) Today's adversary is a dynamic, adaptive foe who operates within a complex, interconnected operational environment. Attaining our national objectives requires the efficient and effective use of the diplomatic, informational, military, and economic (DIME) instruments of national power and systems taxonomy of the multi-dimensional Political, Military, Economic, Social, Information and Infrastructure (PMESII). This situational understanding supported by and coordinated with that of our allies and various intergovernmental, nongovernmental, and regional security organizations is critical to success.

(4) Military operations must be strategically integrated and operational and tactically coordinated with the activities of other agencies of the USG, IGOs, NGOs, regional organizations, the operations of foreign forces, and activities of various HN agencies. Sometimes the CDR draws on the capabilities of other organizations; sometimes the CDR provides capabilities to other organizations; and sometimes the CDR merely deconflicts his activities with those of others. These same organizations may be involved in pre-hostilities operations, activities during combat, and in the transition to post-hostilities activities. Roles and relationships among agencies and organizations, CCMDs, U.S. state and local governments, and overseas with the U.S. Chief of Mission (COM), and country team in a U.S. embassy, must be clearly understood. Interagency coordination forges the vital link between the military and the diplomatic, informational, and economic instruments of national power. Successful interagency, IGO, and NGO coordination helps enable the USG to build

international support, conserve resources, and conduct coherent operations that efficiently achieve shared goals.

c. Flexible Deterrent Options (FDOs). Flexible deterrent options are preplanned, deterrence-oriented actions carefully tailored to send the right signal and influence an adversary's actions. They can be established to dissuade actions before a crisis arises or to deter further aggression during a crisis. FDOs are developed for each instrument of national power — diplomatic, informational, military, economic, and others (financial, intelligence and law enforcement-DIMEFIL) — but they are most effective when used in combination with other instruments of national power.

(1) FDOs facilitate early strategic decision-making, rapid de-escalation and crisis resolution by laying out a wide range of interrelated response paths. Examples of FDOs for each instrument of national power are listed in Figures XVII-1 through XVII-4. *Key goals of FDOs are:*

- Deter aggression through communicating the strength of U.S. commitments to treaty obligations and peaceful development.
- Confront the adversary with unacceptable costs for its possible aggression.
- Isolate the adversary from regional neighbors and attempt to split the adversary coalition.
- Rapidly improve the military balance of power in the Operational Area (OA).

Example of Requested Economic Flexible Deterrent Options

- ➢ Freeze or seize real property in the US where possible.
- ➢ Freeze monetary assets in the US where possible.
- ➢ Freeze international financial institutions to restrict or terminate financial transactions.
- ➢ Encourage US and international financial institutions to restrict or terminate financial transactions.
- ➢ Encourage US and international corporations to restrict transactions.
- ➢ Embargo goods and services.
- ➢ Enact trade sanctions.
- ➢ Enact restrictions on technology transfer.
- ➢ Cancel or restrict US-funded programs.
- ➢ Reduce security assistance programs.

Figure XV-1. Examples of Requested Economic Flexible Deterrent Options.

(2) FDOs Implementation. The use of FDOs must be consistent with U.S. national security strategy (i.e., the instruments of national power are normally used in combination with one another), therefore, continuous coordination with interagency partners is imperative. All operation plans have FDOs, and CCDRs are tasked by the JSCP to plan requests for appropriate options using all instruments of national power.[1]

(3) Military FDOs. Military FDOs underscore the importance of early response to a crisis. Deployment timelines, combined with the requirement for a rapid, early response, generally requires military FDO force packages to be light; however, military FDOs are not intended to place U.S. forces in jeopardy if deterrence fails (risk analysis should be an inherent step in determining which FDOs to use, and how and when to use them). Military

[1] *JP 3-0, Joint Operations.*

FDOs are carefully tailored to avoid the classic "too much, too soon" or "too little, too late" responses. They rapidly improve the military balance of power in the OA, especially in terms of early warning, intelligence gathering, logistic infrastructure, air and maritime forces, information operations, and force protection assets, without precipitating armed response from the adversary. Military FDOs are most effective when used in concert with the other instruments of power. They can be initiated before or after, and with or without unambiguous warning (Figure XV-2).[2]

Example of Requested Military Flexible Deterrent Options

- Increase readiness posture of in-place forces.
- Upgrade alert status.
- Increase intelligence, surveillance, and reconnaissance.
- Initiate or increase show-of-force actions.
- Increase training and exercise activities.
- Maintain and open dialogue with the news media.
- Take steps to increase US public support.
- Increase defense support to public diplomacy.
- Increase information operations.
- Deploy forces into or near the potential operational area.
- Increase active and passive protection measures.
- Ensure consistency of strategic communications messages.

Figure XV-2. Examples of Requested Military Flexible Deterrent Options.

Example of Requested Informational Flexible Deterrent Options

- Promote US policy objectives through public statements.
- Ensure consistency of strategic communications themes and messages.
- Encourage Congressional support.
- Gain US and international public confidence and popular support.
- Maintain open dialogue with the news media.
- Keep selected issues as lead stories.
- Increase protection of friendly critical information structure.
- Impose sanctions on communications systems technology transfer.
- Implement psychological operations.

Figure XV-3. Examples of Requested Informational Flexible Deterrent Options.

[2] *JP 5-0, Joint Operations Planning.*

Example of Requested Diplomatic Flexible Deterrent Options

- Alert and introduce special teams (e.g., public diplomacy).
- Reduce international diplomatic ties.
- Increase cultural group pressure.
- Promote democratic elections.
- Initiate noncombatant evacuation procedures.
- Identify the steps to peaceful resolution.
- Restrict activities of diplomatic missions.
- Prepare to withdraw or withdraw US embassy personnel.
- Take actions to gain support of allies and friends.
- Restrict travel of US citizens.
- Gain support through the United Nations.
- Demonstrate international resolve.

Figure XV-4. Example of Requested Diplomatic Flexible Deterrent Options.

So far within Mission Analysis we've discussed the process of operational design in the following steps:

- *Endstate* (in terms of desired strategic political-military outcomes).
- *Objectives* that describe the conditions necessary to meet the endstate.
- Desired *effects* that support the defined objectives.
- Friendly and enemy *center(s) of gravity (COG)* using a systems approach.
- *Decisive points* that allow the joint force to affect the enemy's COG and look for decisive points necessary to protect friendly COGs.

Now let's look at identifying *lines of operation* that describe how decisive points are to be achieved and linked together in such a way as to overwhelm or disrupt the enemy's COG.

3. Lines of Operations (LOO)

A *line of operations* is a line that defines the directional orientation of a force in time and space in relation to the enemy and links the force with its base of operations and objectives. LOO can be thought of as the analytical bridge between the outcomes of the mission analysis process and the development of COA's. It is important to conduct LOO analysis prior to COA development to ensure COAs achieve military objectives. As CDRs visualize the design of the operation, they may use several LOO to help visualize the intended progress of the joint force toward achieving operational and strategic objectives. LOOs connect a series of decisive points that lead to control of a geographic or force-oriented objective (see Figure XV-5). Operations designed using lines of operations generally consist of a series of actions executed according to a well-defined sequence. Major combat operations are typically designed using lines of operations. These lines tie offensive and defensive tasks to the geographic and positional references in the operational area. Commanders synchronize activities along complementary lines of operations to achieve the endstate. LOO may be either interior or exterior.

a. In operational design, LOO describe how decisive points are linked to operational objectives. Joint doctrine defines *LOO* as "lines that define the orientation of the force in time, space, and purpose in relation to an adversary or objective."[3] They connect the force with its base of operations and its objectives.

- CDRs establish the military conditions and endstate for each operation, developing LOO that focus efforts to create the conditions that produce the endstate.
- Subordinate CDRs adjust the level of effort and missions along each line of operation. LOO are formulated during COA development and refined through continual assessment.[4]

b. LOO must be Derived from Decisive Points. The kinds of decisive points related to a LOO define the description of the line of operation. This is why decisive points must be determined first before defining LOO. LOO are the least understood portion of operational design and therefore tend to be misapplied. The importance of well-defined and understood LOO is basic to linking decisive points, COG, objectives, and endstate. Properly defined, LOO provide clarity and distinction and provide the rationale for everything that the joint force does. Therefore, poorly defined lines of operation weaken the plan and lead to confusion. LOO should be broadly defined to encompass a more flexible way of thinking.

> *"Having determined in order, endstate, objectives and effects, center(s) of gravity, decisive points, and lines of operation, planners then can link lines of operation to decisive points and examine the how and where certain decisive points support multiple lines of operation."*
>
> *Dr. Keith D. Dickson, Operational Design: A Methodology for Planners, Professor of Military Studies, Joint Forces Staff College*

c. Normally, joint operations require CDRs to synchronize activities along multiple and complementary LOO working through a series of military strategic and operational objectives to attain the military endstate. There are many possible ways to graphically depict LOO, which can assist planners to visualize/conceptualize the joint operation from beginning to end and prepare the OPLAN or OPORD accordingly.

d. From the perspective of unified action, there are many diplomatic, economic, and informational activities that can affect the sequencing and conduct of military operations along both physical and logical LOO. Planners should consider depicting relevant actions or events of the other instruments of national power on their LOO diagrams.

(1) A LOO connects a series of decisive points over time that lead to control of a geographic objective or defeat of an enemy force (Figures XV-5 and XV-6). *CDRs use LOO to connect the force with its base of operations and objectives when positional reference to the enemy is a factor.*

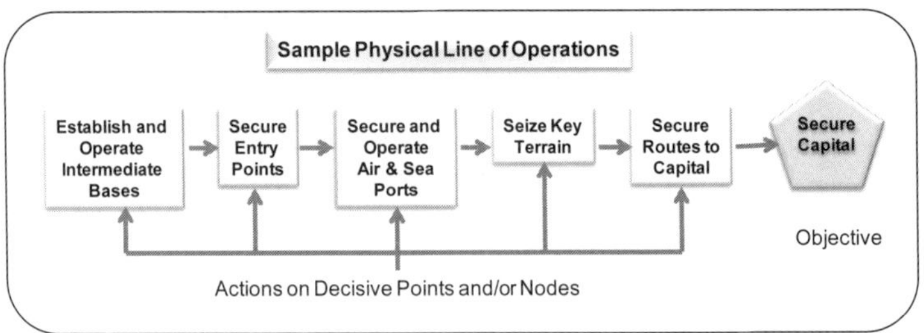

Figure XV-5. Sample Physical Line of Operation (JP 5-0).

[3] *JP 1-02, DOD Dictionary of Military and Associated Terms.*

[4] *FM 3-0, Operations.*

Operations designed using LOO generally consist of a series of cyclic, short-term events executed according to a well-defined, finite timeline. Major combat operations are typically designed using LOOs. These tie offensive and defensive operations to the geographic and positional references of the AO. CDRs synchronize activities along complementary lines of operation to attain the endstate. LOO may be either *interior* or *exterior*.[5]

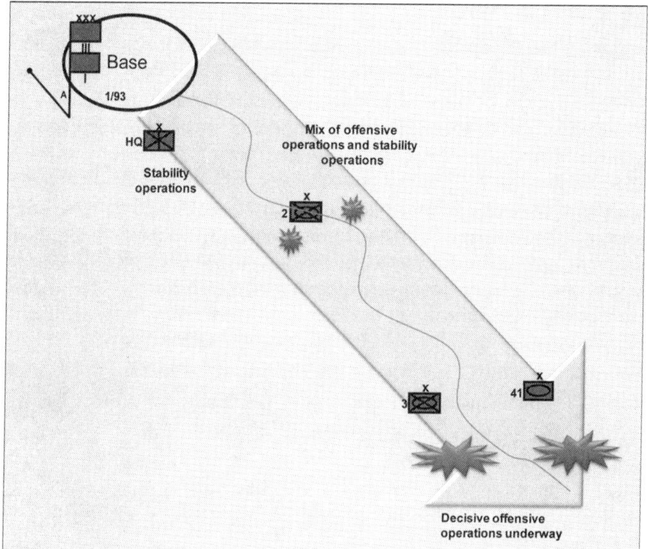

Figure XV-6. Lines of Operation.

(a) Interior and Exterior Lines. The concept of interior and exterior lines applies to both maneuver and logistics. If a force is interposed between two or more adversary forces, it is said to be operating on interior lines. Thus the force is able to move against any of the opposing forces, or switch its resources over a shorter distance than its adversary. Such a concept depends on the terrain and the state of mobility of both sides. In Figure XV-7, the defending force (Force B) has a shorter distance to move in order to reinforce its force elements in contact. The attacking force (Force A) has a greater distance to travel to switch resources across its three operations (A1, A2, A3).[6]

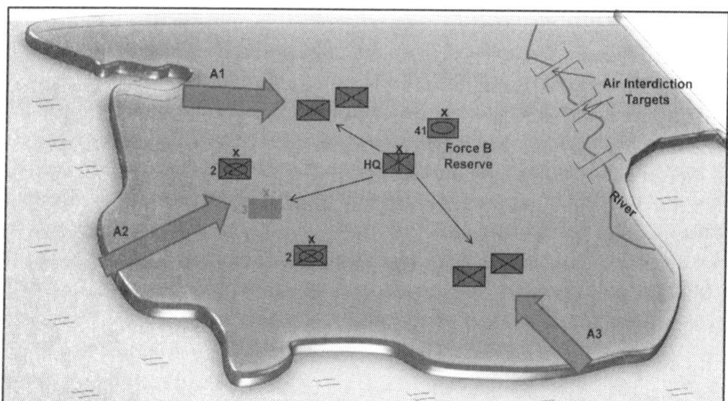

Figure XV-7. Interior and Exterior Lines.[7]

[5] FM 3-0, Operations.

[6] Joint Doctrine Publication 01 (JDP 01), United Kingdom.

[7] FM 3-0, Operations.

(b) A force operates on *interior lines* when its operations diverge from a central point and when it is therefore closer to separate adversary forces than the latter are to one another. Interior lines benefit a weaker force by allowing it to shift the main effort laterally more rapidly than the adversary, and provide increased security to logistical support operations. Interior lines usually represent central position, where a friendly force can reinforce or concentrate its elements faster than the enemy force can reposition. With interior lines, friendly forces are closer to separate enemy forces than the enemy forces are to one another. Interior lines allow an isolated force to mass combat power against a specific portion of an enemy force by shifting capabilities more rapidly than the enemy can react.

(c) A force operates on *exterior lines* when its operations converge on the adversary. Successful operations on exterior lines require a stronger or more mobile force, but offer the opportunity to encircle and annihilate a weaker or less mobile opponent. Assuring strategic mobility enhances exterior LOO by providing the JFC greater freedom of maneuver.[8]

1 The relevance of interior and exterior physical lines depends on the relationship of time and distance between the opposing forces. Although an adversary force may have interior lines with respect to the friendly force, this advantage disappears if the friendly force is more agile and operates at a higher operational tempo. Conversely, if a smaller force maneuvers to a position between larger but less agile adversary forces, the friendly force may be able to defeat them in detail before they can react effectively.

2 A joint operation may have *single* or *multiple* physical LOO.

a A *single LOO* has the advantage of concentrating forces and simplifying planning.

b *Multiple LOO*, on the other hand, increase flexibility and create opportunities for success. Multiple LOO also make it difficult for an adversary to determine the objectives of the campaign or major operation, forcing the adversary to disperse resources to defend against multiple threats. The decision to operate on multiple lines will depend to a great extent on the availability of resources.

(2) LOO reflect the more traditional linkage of decisive points, objectives, and endstate. However, using LOO alone does not project the operational design beyond the defeat of the enemy force. Combining LOO with Lines of Effort (LOE) allows CDRs to project operational design beyond the current phase of the operation to set the conditions for an enduring peace. It allows them to consider the less tangible aspects of the operational environment in which the other instruments of national power are predominant. CDRs can visualize post-hostility operations from a far more conceptual perspective. The resulting operational design reflects the thorough integration of full spectrum operations across the spectrum of conflict.[9]

4. Lines of Effort (LOE)

A LOE links multiple tasks and missions using the logic of purpose—cause and effect—to focus efforts toward establishing operational and strategic conditions. LOE are essential to operational design when positional references to an enemy or adversary have little relevance. In operations involving many nonmilitary factors, lines of effort may be the only way to link tasks, effects, conditions, and the desired endstate (see Figure XV-8 and XV-9). LOE are often essential to helping commanders visualize how military capabilities can support the other instruments of national power. They are a particularly valuable tool when used to achieve unity of effort in operations involving multinational forces and civilian organizations, where unity of command is elusive, if not impractical.

a. CDRs use LOE to link the logic of purpose – conditions and effects – with a series of conceptual decisive points or objectives to the conditions that define the endstate. Operations designed using LOE are typically focused on conditions rather than physical objectives. LOE combine the complementary, long-term effects of stability tasks with the cyclic, short-term events characteristic of combat operations. LOE also help CDRs visualize how military means can support nonmilitary instruments of national power.

[8] *JP 5-0, Joint Operations Planning.*

[9] *FM 3-0, Operations.*

b. Commanders use LOE to describe how they envision their operations creating the more intangible endstate conditions. These LOE show how individual actions relate to each other and to achieving the endstate. Ideally, LOE combine the complementary, long-term effects of stability or civil support tasks with the cyclic, short-term events typical of offensive or defensive tasks. Using LOE, CDRs develop tasks and missions, allocate resources, and assess the effectiveness of the operation. The CDR may specify which LOE represent the decisive operation and which are shaping operations. CDRs synchronize activities along multiple LOE to achieve the conditions that compose the desired endstate. [10]

c. Commanders at all levels may use LOE to develop missions and tasks and to allocate resources. Commanders may designate one LOE as the decisive operation and others as shaping operations. Commanders synchronize and sequence related actions along multiple LOE. Seeing these relationships helps commanders assess progress toward achieving the endstate as forces perform tasks and accomplish missions.

d . Commanders typically visualize stability and civil support operations along LOE (Figure XV-8 and XV-9). For stability operations, commanders may consider linking primary stability tasks to their corresponding Department of State post-conflict technical sectors. These stability tasks link military actions with the broader interagency effort across the levels of war. A full array of LOE might include offensive and defensive lines, as well as a line for information operations. Information operations typically produce effects across multiple LOE.

e. As operations progress, commanders may modify the LOE after assessing conditions and collaborating with multinational military and civilian partners. LOE typically focus on integrating the effects of military operations with those of other instruments of national power to support the broader effort. Each operation, however, is different. Commanders develop and modify LOE to keep operations focused on achieving the endstate, even as the situation changes.

f. Success in one LOE reinforces successes in the others. There is no list of LOE that applies in all cases. LOE are directly related to one another. They connect objectives that, when accomplished, support achieving the endstate. Operations designed using LOE typically employ an extended, event-driven timeline with short-, mid-, and long-term goals.

(e) CDRs determine which LOE apply to their AO and how the LOE connect with and support one another.

(f) CDRs at all levels should select the LOE that relate best to achieving the desired endstate in accordance with the CDR's intent. The following list of possible LOE is not all inclusive. However, it gives us a place to start:

- Conduct information operations.
- Conduct combat operations/civil security operations.
- Train and employ HN security forces.
- Establish or restore essential services.
- Support development of better governance.
- Support economic development.

[10] *FM 3-0, Operations.*

[11] *FM 3-24/MCWP 3-33.5, Counterinsurgency.*

Example Lines of Effort

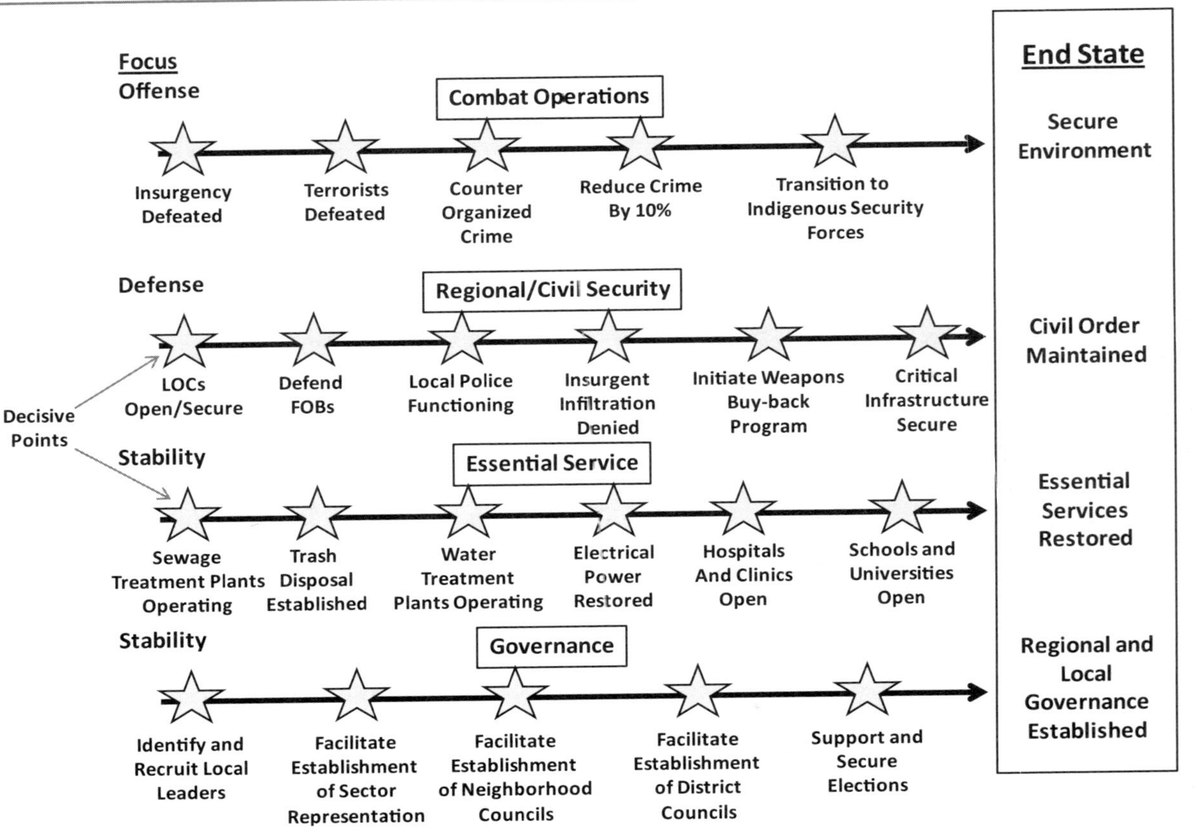

Figure XV-8. Example Lines of Effort (Stability) (FM 3-0).

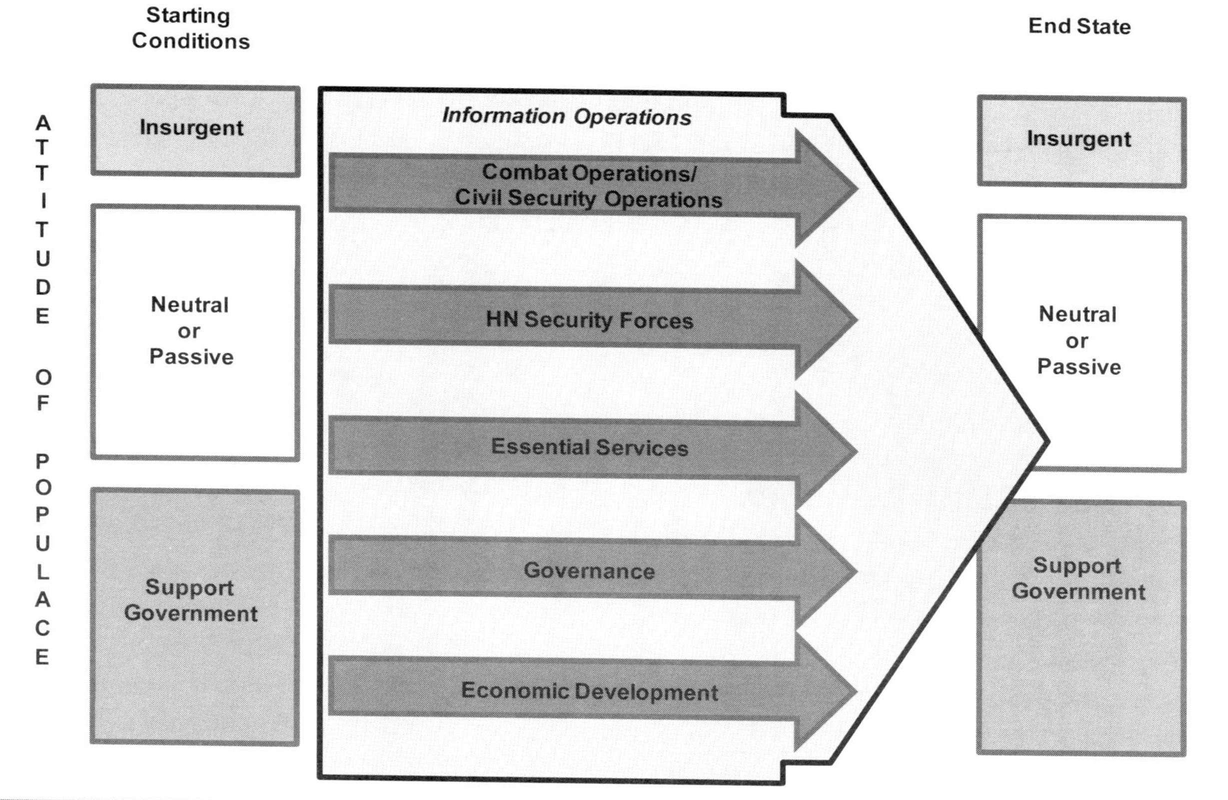

Figure XV-9. Example Lines of Effort (Counterinsurgency) (FM 3-24).

(g) These lines can be customized, renamed, changed altogether, or simply not used. CDRs may combine two or more LOE or split one LOE into several. For example, IO is integrated into all LOE; however, CDRs may designate a separate LOE for IO if necessary to better describe their intent (XV-10). Likewise, some CDRs may designate separate LOE for combat operations and civil security operations.[11]

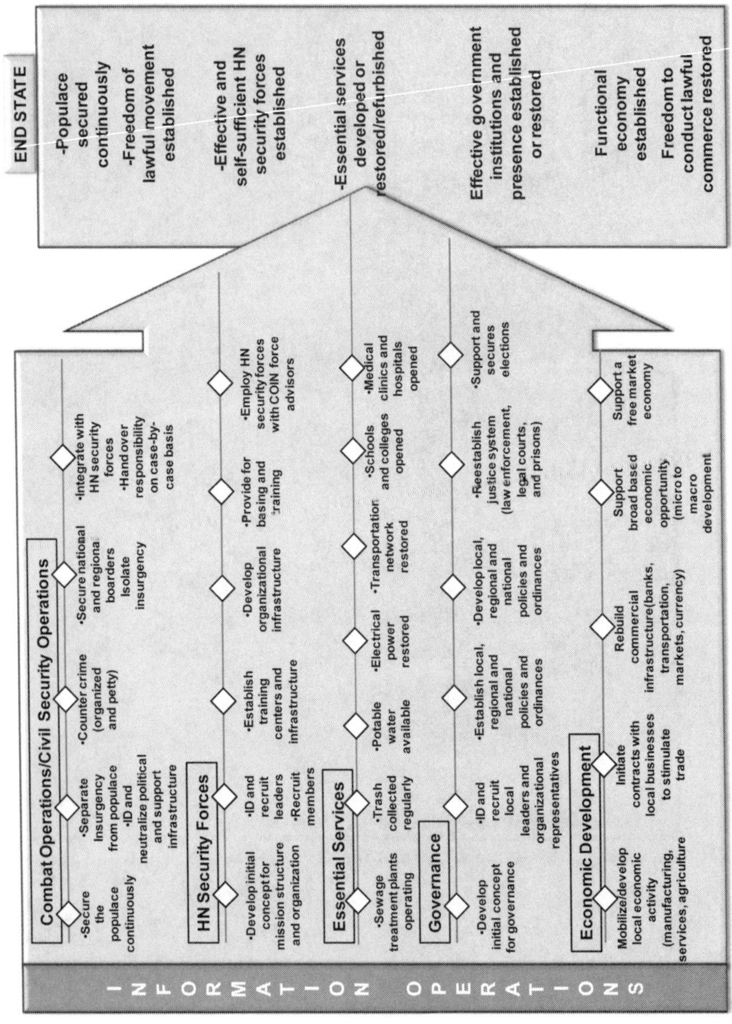

Figure XV-10. Example of Goals and Objectives along Lines of Effort (FM 3-24).

5. Combining LOO and LOE

Commanders may use both LOO and LOE to connect objectives to a central, unifying purpose. They link objectives to the endstate. Continuous assessment gives commanders the information required to revise and adjust lines of operations and effort. LOO portray the more traditional links between objectives, decisive points, and centers of gravity. However, LOO do not project the operational design beyond defeating enemy forces and seizing terrain. Combining LOO and LOE allows commanders to include nonmilitary activities in their operational design. This combination helps commanders incorporate stability tasks that set the endstate conditions into the operation. It allows commanders to consider the less tangible aspects of the OE where the other instruments of national power dominate. Commanders can then visualize concurrent and post-conflict stability activities. Making these connections relates the tasks and purposes of the elements of full spectrum operations with joint effects identified in the campaign plan. The resulting operational design effectively combines full-spectrum operations throughout the campaign or major operation.

For example, CDRs may conduct offensive and defensive operations along a LOO to form a shield behind which LOE for simultaneous stability operations can maintain a secure environment for the populace. Accomplishing the objectives of combat operations/civil security operations sets the conditions needed to achieve essential services and economic development objectives. When the populace perceives that the environment is safe enough to leave families at home, workers will seek employment or conduct public economic activity. Popular participation in civil and economic life facilitates further provision of essential services and development of greater economic activity. Over time such activities establish an environment that attracts outside capital for further development. Neglecting objectives along one LOO or LOE risks creating vulnerable conditions along another that the enemy can exploit. Achieving the desired endstate requires linked successes along all LLO.

a. Commanders may describe an operation along LOO, LOE, or a combination of both. Irregular warfare, for example, typically requires a deliberate approach using LOO complemented with LOE; the combination of them may change based on the conditions within the operational area. An operational approach using both LOO and LOE reflects the characteristics and advantages of each. With this approach, commanders synchronize and sequence actions, deliberately creating complementary and reinforcing effects. The lines then converge on the well-defined, commonly understood endstate outlined in the commander's intent.[12]

b. An operational design composed of both LOO and LOE reflects the characteristics and advantages of each. With this approach, it is vital for the CDR to synchronize actions across the LOO, creating complementary and reinforcing effects. This ensures that the LOO converge on a well-defined, commonly understood endstate that is composed of the set of conditions initially outlined in the CDR's intent (XV-11).[13]

[12] *FM 3-0, Operations.*

[13] *FM 3-0, Operations.*

Operational Design Schematic

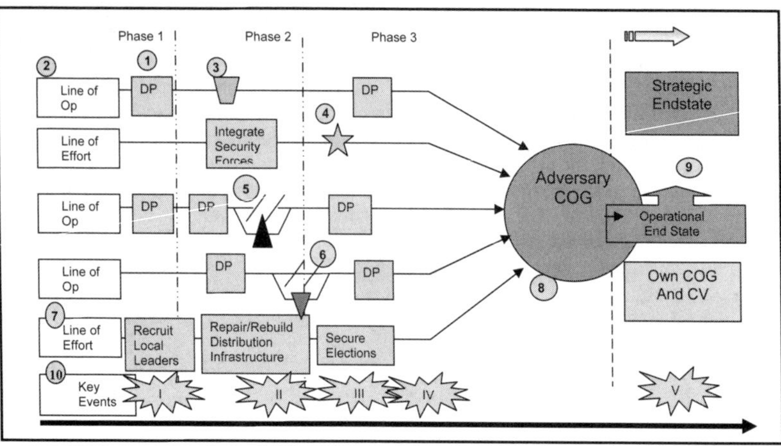

Figure XVII-11. Operational Design Schematic.

1. DPs are sequenced in time and space along LOO. This sequencing can be assisted by Phases. DPs are the key to unlocking the COG, and without proper identification, the COG cannot be defeated nor neutralized.

2. Lines of Operation can be environmental or functional or a mixture of both. They should not be decided until the Decisive Points have been derived and the critical path identified.

3. Operational Pauses may be introduced where necessary. Momentum must be maintained elsewhere.

4. Culmination Point is reached when an operation or battle can just be maintained, but not developed to any great advantage.

5. Branches are deliberate plans which can be introduced to Lines of Operation whenever necessary, and are continuously refined as the campaign develops.

6. Sequels are contingency plans introduced when phases are not completed as planned.

7. Lines of Effort typically visualize stability and civil support operations, although CDRs select the lines of effort that relate best to achieving the desired endstate.

8. The adversary COG at the operational level is that which most resists the end-state. Without the neutralization or destruction of the adversary's COG, the end-state cannot be reached. Activity, necessary to finally achieve the end-state conditions, may take place after its destruction or neutralization, but this will not be decisive or critical. It may be useful to show own COG as it is the thing that needs protecting most, and is therefore that which the adversary is likely to direct his efforts against.

9. The end-state provides the focus for campaign planning and all activities should be judged against their relevance to its achievement. The operational end-state will usually be given by the Military Strategic Authority and may be a list of objectives or a statement. It needs analysis in order to identify measurable conditions which together indicate that the end-state has been achieved.

10. It may be useful to include a line showing key events. These might be the deadline for compliance with a UN resolution, the date an adversary 2nd Echelon force might be ready for combat, the estimated time for the completion of mobilization, or the holding of the first free and fair elections.

Course of Action (COA) Development 16

1. Step 3 to JOPP — COA Development

a. COA Development. A COA is any force employment option open to a CDR that, if adopted, would result in the accomplishment of the mission of the campaign. For each COA, the CDR must envision the employment of own/friendly forces and assets as a whole, taking into account externally imposed limitations, the area of operations, and the conclusions previously drawn during the mission analysis and the CDR's guidance. However, there are a few more considerations prior to actually developing your COA's. You must first consider your time available, political considerations, any flexible deterrent options and off course, lines of operations (see Chapter 15).

b. Defining the COA. Each COA is a broad statement of a possible way to accomplish the mission. A COA consists of the following information:

- WHO (type of forces) will execute the tasks/take the action?
- WHAT type of action or tasks are contemplated?
- WHEN will the tasks begin?
- WHERE will the tasks occur?
- WHY (for what purpose) the action is required (relate to endstate)?
- HOW will the available forces be employed?

The staff converts the approved COA into a concept of operations. COA determination consists of four primary activities: *COA development, analysis and wargaming, comparison, and approval.*

2. Tentative Courses of Action

The output of COA development are *tentative* COAs in which the CDR describes for each COA, in broad but clear terms, what is to be done, the size of forces deemed necessary, and time in which force needs to be brought to bear. Tentative COAs allow for initial conceptualization and broad descriptions of potential approaches to the conduct of operations that will accomplish the desired endstate. The CCDR gives the staff their preliminary thoughts on possible and acceptable military actions early in the planning process to provide focus to their efforts, allowing them to concentrate on developing COAs that are the most appropriate (Figure XVI-1).

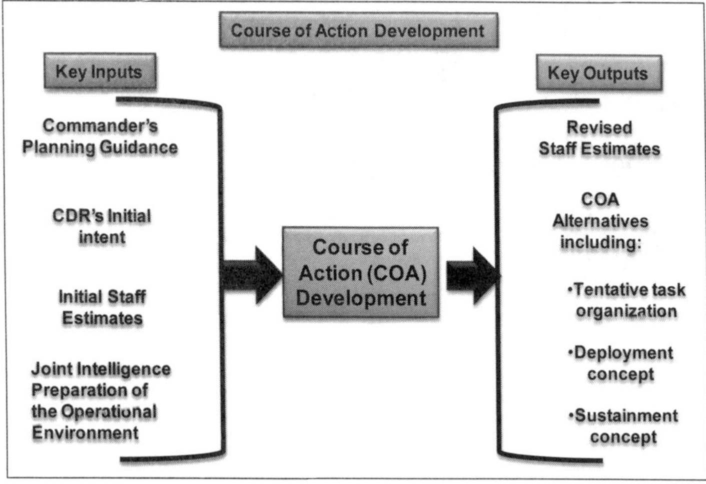

Figure XVI-1. Course of Action Development.

a. A *tentative* COA should be simple, brief, yet complete. The COA sketch contains the general arrangement of forces, the anticipated movement or maneuver of those forces, a brief description of the concept of operation, and major tasks to components (Figure XVI-2). The COA's should have descriptive titles. Distinguishing factors of the COA may suggest titles that are descriptive in nature. A COA should answer the following questions:

- How much force is required to accomplish the mission?
- Generally, in what order should coalition forces be deployed?
- Where and how should coalition air, naval, ground and special operation forces be employed in theater?
- What major tasks must be performed and in what sequence?
- How is the coalition to be sustained for the duration of the campaign?
- What are the command relationships?

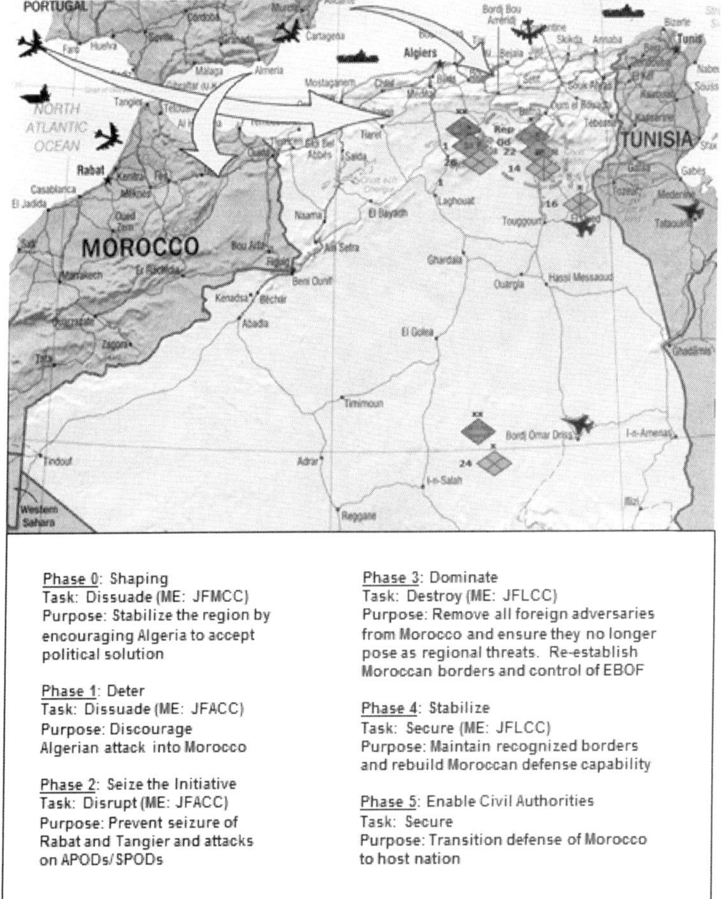

PORTUGAL
NORTH ATLANTIC OCEAN
MOROCCO
TUNISIA
Algiers
Rabat
Western Sahara
Phase 0: Shaping
Task: Dissuade (ME: JFMCC)
Purpose: Stabilize the region by encouraging Algeria to accept political solution
Phase 1: Deter
Task: Dissuade (ME: JFACC)
Purpose: Discourage Algerian attack into Morocco
Phase 2: Seize the Initiative
Task: Disrupt (ME: JFACC)
Purpose: Prevent seizure of Rabat and Tangier and attacks on APODs/SPODs
Phase 3: Dominate
Task: Destroy (ME: JFLCC)
Purpose: Remove all foreign adversaries from Morocco and ensure they no longer pose as regional threats. Re-establish Moroccan borders and control of EBOF
Phase 4: Stabilize
Task: Secure (ME: JFLCC)
Purpose: Maintain recognized borders and rebuild Moroccan defense capability
Phase 5: Enable Civil Authorities
Task: Secure
Purpose: Transition defense of Morocco to host nation

Figure XVI-2. Course of Action Example.

b. To develop *tentative* COAs, the staff must focus on key information necessary to make decisions and assimilate the data in mission analysis. Usually, the staff develops no more than three COAs to focus their efforts and concentrate valuable resources on the most likely scenarios. All COAs selected for analysis must be valid. A valid COA is one that is *adequate, feasible, acceptable, distinguishable, and complete.*

(1) Adequate - Can accomplish the mission within the CDR's guidance.

(2) Feasible - Can accomplish the mission within the established time, space, and resource limitations.

(3) Acceptable - Must balance cost and risk with the advantage gained.

(4) Distinguishable - Must be sufficiently different from the other courses of action.

(5) Complete - Must incorporate:

- Objectives (including desired effects) and tasks to be performed.
- Major forces required.
- Concepts for deployment, employment, and sustainment.
- Time estimates for achieving objectives.
- Military endstate and mission success criteria

c. The staff should reject potential *tentative* COAs that do not meet all five criteria. A good COA accomplishes the mission within the CDR's guidance and positions the joint force for future operations and provides flexibility to meet unforeseen events during execution. It also gives components the maximum latitude for initiative. Embedded within COA development is the application of operational art. Planners can develop different COAs for using joint force capabilities (operational fires and maneuver, joint force organization, etc.) by varying the elements of operational design (such as phasing, line of operations, and so forth).

d. Risk. During COA development, the CDR and staff continue risk assessment, focusing on identifying and assessing hazards to mission accomplishment. The staff also continues to revise intelligence products. Generally, at the theater level, each COA will constitute a theater strategic or operational concept and should outline the following per *JP 5-0, Joint Operations Planning.*

(1) Major strategic and operational tasks to be accomplished in the order in which they are to be accomplished.

(2) Capabilities required.

(3) Task organization and related communications systems support concept.

(4) Sustainment concept.

(5) Deployment concept.

(6) Estimate of time required to reach mission success criteria or termination criteria.

(7) Concept for maintaining a theater reserve.

e. Tentative Courses of Action TTP's. Following is listed a logical flow of TTP's that will help focus the staff while conceptualizing Tentative COA's:

(1) Review. Review information contained in the mission analysis and CDRs' guidance. The staff should review once again the mission statement it developed and the CDR approved during mission analysis. All staff members should understand the mission and the tasks that must be accomplished to achieve mission success. Following this review or upon the receipt of new information or tasking(s) from higher headquarters, if the mission statement appears inadequate or outdated, then the staff should recommend appropriate changes. The CDR has also given the staff his planning guidance. This guidance is directly linked to the CDR's operational design, how he visualizes the operation unfolding.

(2) Determine the COA Development Technique. A critical first decision in COA development is whether to conduct simultaneous or sequential development of the COAs. Each approach has distinct advantages and disadvantages. The advantage of simultaneous development of COAs is potential time savings. Separate groups are simultaneously working on different COAs. The disadvantage of this approach is that the synergy of the JPG may be disrupted by breaking up the team. The approach is manpower intensive and requires component and directorate representation in each COA group, and there is an increased likelihood that the COAs will not be distinctive. While there is potential time to be saved, experience has demonstrated that it is not an automatic result. The simultaneous COA development approach can work, but its inherent disadvantages must be addressed and some risk accepted up front. The recommended approach if time and resources allows is the sequential method.

(3) Planning cells with land, maritime, air, space and special operations planners as well as Joint Interagency Coordination Group (JIACG) reps (and others as necessary) should initially develop ways to accomplish the *essential tasks.*

(a) Regardless of the eventual COA, the staff should plan to accomplish the higher CDR's intent by understanding its essential task(s) and purpose and the intended contribution to the higher CDR's mission success. The staff must ensure that all the COAs developed will fulfill the command mission and the purpose of the operation by conducting a review of all essential tasks developed during mission analysis.

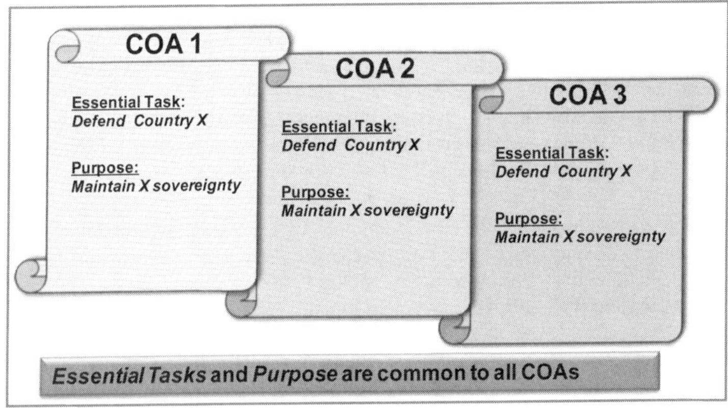

(b) They should then consider ways to accomplish the other tasks. A technique is for these planners to think two levels down (e.g., how could the MARFOR's component commands, MEF, or appropriate subordinate, accomplish the assigned tasks?).

(4) Once the staff has begun to visualize a *tentative* COA, it should see how it can best synchronize (arrange in terms of time, space, and purpose) the actions of all the elements of the force. The staff should estimate the anticipated duration of the operation. One method of synchronizing actions is the use of phasing as discussed earlier. Phasing assists the CDR and staff to visualize and think through the entire operation or campaign and to define requirements in terms of forces, resources, time, space, and purpose. Planners should then *integrate and synchronize* these ideas (which will essentially be Service perspectives) by using the joint architecture of maneuver, firepower, protection, support, and command and control (see the taxonomy used in the Universal Joint Task List). See the questions below:

(a) Land Operations. What are ways land forces can integrate/synchronize maneuver, firepower, protection, support, and command and control with other forces to accomplish their assigned tasks? Compare friendly against enemy forces to see if there are sufficient land forces to accomplish the tasks.

(b) Air Operations. What are ways air forces can integrate/synchronize maneuver, firepower, protection, support, and command and control with other forces to accomplish their assigned tasks? Compare friendly against enemy forces to see if there are sufficient air forces to accomplish the tasks.

(c) Maritime. What are ways maritime forces can integrate/synchronize maneuver, firepower, protection, support, and command and control with other forces to accomplish their assigned tasks? Compare friendly against enemy forces to see if there are sufficient maritime forces to accomplish the tasks.

(d) Special Operations. What are ways special operations forces can integrate/synchronize maneuver, firepower, protection, support, and command and control with other forces to accomplish their assigned tasks? Compare friendly against enemy forces to see if there are sufficient special operations forces to accomplish the tasks.

(e) Space Operations. What are the major ways that space operations can support maneuver, firepower, protection, support and establishment of command and control?

(f) Information Operations (IO). What are the ways joint forces can integrate the core capabilities of electronic warfare, computer network operations, psychological operations, military deception, and operations security, in concert with specified supporting and related capabilities, to influence, disrupt, corrupt or usurp adversarial human and automated decision making while protecting our own.

(5) The *tentative* COAs should focus on where COGs and DPs (or vulnerabilities, e.g., "keys to achieving desired effect on centers of gravity") may occur. The CDR and the staff review and refine their COG analysis begun during mission analysis based on updated intelligence, JIPOE products and initial staff estimates. The refined enemy and friendly COGs and critical vulnerabilities are used in the development of the initial COAs. The COG analysis helps the CDR orient on the enemy and compare his strengths and weakness to those of the enemy. The staff takes the CDR's operational design, reviews it, and focuses on the friendly and enemy COGs and critical vulnerabilities.

(6) By looking at friendly COG's and vulnerabilities, the staff understands the capabilities of their own force and those critical vulnerabilities that will require protection. Protection resource limitations will probably mean that the staff cannot plan to protect each asset individually, but rather look at developing overlapping protection techniques. The strength of one asset or capability may provide protection from the weakness of another.

(7) Identify the Sequencing (simultaneous/sequential/or combination) of the operation for each COA.

(8) Identify Main and Supporting Efforts, by phase, the purposes of these efforts, and key supporting/supported relationships within phases.

(9) Identify Component Level Mission/Tasks (who, what and where) that will accomplish the stated purposes of main and supporting efforts. Think of component tasks from the perspective of movement and maneuver, firepower, protection, support and C2. Display them with graphic control measures as much as possible. The lines of operation/effort that you completed earlier will help identify these tasks.

(10) Recognize MILDEC (Military Deception) Planning. Results of deception operations may influence any COA. The CCDR has a resident MILDEC planner on staff charged with developing the CDR's Deception Plan. Access to this plan will be as required and as directed by CDR only. Recognize that by design MILDEC planning is just behind operational planning (see JP 3-13.4 Military Deception).

(11) Task-Organization. The staff should develop a detailed task organization (two levels down) to execute the COA. The CDR and staff determine appropriate command relationships to include operational mission assignments and support relationships.

(12) Logistics. No COA is complete without a plan to sustain it properly. The logistic concept is more than just gathering information on various logistic functions. It entails the organization of capabilities and resources into an overall theater campaign or operation sustainment concept. It concentrates forces and material resources strategically so that the right force is available at the designated times and places with the essential equipment to conduct decisive operations. It assists thinking through a cohesive sustainment for joint, single Service and supporting forces relationships, in conjunction with multinational, interagency, non-governmental, or international organizations.

(13) Develop Initial COA Sketches and Statements. Answer the questions:

- WHO (type of forces) will execute the tasks?
- WHAT is the task?
- WHERE will the tasks occur? (Start adding graphic control measures, e.g., areas of operation, amphibious objective areas).
- WHEN will the tasks begin?
- HOW (but do not usurp the components' prerogatives) should the CCDR provide "operational direction," so the components can accomplish "tactical actions?"
- WHY (for what purpose) will each force conduct its part of the operation?

(14) Test the Validity of each *Tentative* COA.

See facing page for further discussion.

(14) Test the Validity of each Tentative COA

(a) Tests for adequacy/suitability

- Does it accomplish the mission?
- Does it meet the CCDRs intent?
- Does it accomplish all the essential tasks?
- Does it meet the conditions for the endstate?
- Does it take into consideration the enemy and friendly centers of gravity?

(b) Preliminary test for feasibility

- Does the CCDR have the force structure and lift assets (means) to carry it out? The COA is feasible if it can be carried out with the forces, support, and technology available, within the constraints of the physical environment and against expected enemy opposition.
- Although this process occurs during COA analysis and the test at this time is preliminary, it may be possible to declare a COA infeasible (for example, resources are obviously insufficient). However, it may be possible to fill shortfalls by requesting support from the CCDR or other means.

(c) Preliminary test for acceptability

- Does it contain unacceptable risks? (Is it worth the possible cost?) A COA is considered acceptable if the estimated results justify the risks. The basis of this test consists of an estimation of friendly losses in forces, time, position, and opportunity.
- Does it take into account the limitations placed on the CCDR (must do, cannot do, other physical limitations)?
- Acceptability is considered from the perspective of the CCDR by reviewing the strategic objectives.
- COAs are reconciled with external constraints, particularly ROE.
- Requires visualization of execution of the COA against each enemy capability. Although this process occurs during COA analysis and the test at this time is preliminary, it may be possible to declare a COA unacceptable if it violates the CCDRs definition of acceptable risk.

(d) Test for variety

Is it fundamentally different from other COAs? They can be different when considering:

- The focus or direction of main effort.
- The scheme of maneuver (land, air, maritime, and special operation).
- Sequential vs. simultaneous maneuvers.
- The primary mechanism for mission accomplishment.
- Task organization.
- The use of reserves.

(e) Test for completeness

Does it answer all of the questions WHO, WHAT, WHERE, WHEN, HOW and WHY?

(15) Determine Command Relationships and Organizational Options. (See Appendix E).

(a) Joint Force Organization and Command Relationships. Organizations and relationships are based on the campaign design, complexity of the campaign, and degree of control required. Establishing *command relationships* includes determining the types of subordinate commands and the degree of authority to be delegated to each. Clear definition of command relationships further clarifies the intent of the CCDR and contributes to decentralized execution and unity of effort. The CCDR has the authority to determine the types of subordinate commands from several doctrinal options, including Service components, functional components, and subordinate joint commands. The options for delegating authority emanate from COCOM and range from command to support relationships. Regardless of the Command or Support relationships selected, it is the CCDR (or JFC) responsibility to ensure that these relationships are understood and clear to all subordinate, adjacent and supporting HQs. The following are considerations for establishing Joint Force Organizations:

1 CCDRs (JFCs) will normally designate JFACCs and organize special operations forces into a functional component.

2 Joint Forces will normally be organized with a combination of Service and functional components with operational responsibilities.

3 Functional component staffs should be joint with Service representation in approximate proportion to the mix of subordinate forces. These staffs will be required to be organized and trained prior to employment in order to be efficient and effective, which will require advanced planning.

4 CCDRs may establish supporting/supported relationships between components to facilitate operations.

5 CCDRs define the authority and responsibilities of functional component CDRs based on the strategic concept of operations and may alter their authority and responsibility during the course of an operation.

6 CCDRs must balance the need for centralized direction with decentralized execution.

7 Major changes in the joint force organization are normally conducted at phase changes.

(b) Operational Objectives and Subordinate Tasks.

1 The theater and supporting *operational objectives* assigned to subordinates are critical elements of the theater-strategic design of the campaign. They establish the conditions necessary to reach the desired endstate and achieve the national strategic objectives. The CCDR carefully defines the objectives to ensure clarity of theater and operational intent, and identify specific tasks required to achieve those objectives. Tasks are shaped by the concept of operations—intended sequencing and integration of air, land, sea, special operations, and space forces. Tasks are prioritized in order of criticality while considering the enemy's objectives and the need to gain advantage.

2 One of the fundamental purposes of a campaign plan is to synchronize employment of all available military (land, sea, air, and special operations, as well as space, information and protection) forces and capabilities. This overwhelming application of military capabilities can be achieved by assigning the appropriate tasks to components for each phase, though supporting CDRs will also contribute with their own capabilities. These tasks can be derived from an understanding of how component and supporting forces interrelate, not only among themselves, but also with respect to the enemy.

(16) Refining the theater design/operational area and initial battlespace architecture (e.g., control measures). The Theater Design is normally a legally/politically binding document which will initiate planning and negotiations throughout the CCMD, interagency and internationally. It will provide flexibility/options and/or limitations to the CCDR. The theater design must be *precise*. Specifics are required to negotiate basing and overflight. DOS will be the lead agency here. Theater design is also *resource sensitive*. Limited infrastructure resources must be optimized, i.e., APOE/DS/SPOE/DS, and when utilizing a host nation's resources, negotiations for sharing those resources is common.

(a) Operational area. *An operational area* is an overarching term encompassing more descriptive terms for geographic areas in which military operations are conducted. Operational areas include, but are not limited to, such descriptors as AOR, theater of war, theater of operations, JOA, amphibious objective area (AOA), joint special operations area (JSOA), and area of operations (AO). Except for AOR, which is normally assigned in the UCP, the GCC and other JFCs designate smaller operational areas on a temporary basis. Operational areas have physical dimensions comprised of some combination of air, land, and maritime domains. JFCs define these areas with geographical boundaries, which facilitate the coordination, integration, and deconfliction of joint operations among joint force components and supporting commands. The size of these operational areas and the types of forces employed within them depend on the scope and nature of the crisis and the projected duration of operations.

(b) CCMD-Level Areas. GCCs conduct operations in their assigned AORs across the range of military operations. When warranted, the President, SECDEF, or GCCs may designate a theater of war and/or theater of operations for each operation. GCCs can elect to control operations directly in these operational areas, or may establish subordinate joint forces for that purpose, allowing themselves to remain focused on the broader AOR.[1]

1 *Area of Responsibility*. An AOR is an area established by the President and SECDEF on an enduring basis that defines geographic responsibilities for a GCC. A GCC has authority to plan for operations within the AOR and conduct those operations approved by the President or SECDEF.

2 *Theater of War*. A theater of war is a geographical area comprised of some combination of air, land, and maritime domains established for the conduct of major operations and campaigns involving combat. A theater of war is established primarily when there is a formal declaration of war or it is necessary to encompass more than one theater of operations (or a JOA and a separate theater of operations) within a single boundary for the purposes of C2, logistics, protection, or mutual support. A theater of war does not normally encompass a GCC's entire AOR, but may cross the boundaries of two or more AORs.

3 *Theater of Operations*. A theater of operations is a geographical area comprised of some combination of air, land, and maritime domains established for the conduct of joint operations. A theater of operations is established primarily when the scope of the operation in time, space, purpose, and/or employed forces exceeds what can normally be accommodated by a JOA. One or more theaters of operations may be designated. Different theaters of operations will normally be geographically separate and focused on different missions. A theater of operations typically is smaller than a theater of war, but is large enough to allow for operations in depth and over extended periods of time. Theaters of operations are normally associated with major operations and campaigns.

4 *Combat Zones and Communications Zones (COMMZ)*. Geographic CCDRs also may establish combat zones and COMMZs. The combat zone is an area required by forces to conduct combat operations. It normally extends forward from the land force rear boundary. The COMMZ contains those theater organizations, LOCs, and other agencies required to support and sustain combat forces. The COMMZ usually includes the rear portions of the theaters of operations and theater of war (if designated) and reaches back to the CONUS base or perhaps to a supporting CCDR's AOR. The COMMZ includes airports and seaports that support the flow of forces and logistics into the operational area. It usually is contiguous to the combat zone but may be separate — connected only by thin LOCs — in very fluid, dynamic situations.[2]

(17) Prepare the COA Concept of Operations Statement (or Tasks), Sketch, and Task Organization.

(a) COA concept of operations statements (or tasks) answer WHO, WHAT, WHERE, WHEN, HOW, and WHY.

[1] *JP 3-0, Joint Operation Planning.*

[2] *JP 3-0, Doctrine for Joint Operation.*

(b) Finalize COA sketches.

(c) Finalize the task organization.

(18) Conduct COA Development Brief to CCDR. Figure XVII-3 is a suggested sequence (see facing page).

(19) CCDR Provides Guidance on COAs.

(a) Review and approve COAs for further analysis.

(b) Direct revisions to COAs, combinations of COAs, or development of additional COA(s).

(c) Directs priority for which enemy COA will be used during wargaming of friendly COA(s).

(20) Continue the Staff Estimate Process. The staff must continue to conduct their staff estimates of supportability for each COA.

(21) Conduct Vertical and Horizontal Parallel Planning.

(a) Discuss the planning status of staff counterparts with both CCDR's and JFC components' staffs.

(b) Coordinate planning with staff counterparts from other functional areas.

(c) Permit adjustments in planning as additional details are learned from higher and adjacent echelons, and permit lower echelons to begin planning efforts and generate questions (e.g., Requests for Information/Intelligence).[3]

f. There are several planning techniques available to you during COA development. See following page of a step-by-step approach utilizing the *backwards planning technique* (reverse planning).

3. Planning Directive Published

The Planning Directive identifies planning responsibilities for developing CCMD plans. It provides guidance and requirements to the HQ staff and subordinate commands concerning coordinated planning actions for plan development. The CCDR normally communicates initial planning guidance to the staff, subordinate CDRs, and supporting CDRs by publishing a planning directive to ensure that everyone understands the CDR's intent and to achieve unity of effort. Generally, the J-5 coordinates staff action for deliberate planning. The J-5 staff receives the CCDR's initial guidance and combines it with the information gained from the initial staff assessments. The CCDR, through the J-5, may convene a preliminary planning conference for members of the JPEC who will be involved with the plan. This is the opportunity for representatives to meet face-to-face. At the conference, the CCDR and selected members of the staff brief the attendees on important aspects of the plan and may solicit their initial reactions. Many potential conflicts can be avoided by this early exchange of information.[4]

4. Staff Estimates

Continuous staff estimates are the foundation for the CCDR's selection of a COA. Up until this point our COAs have been developed from initial impression based on limited knowledge. Staff estimates determine whether the mission can be accomplished and also determine which COA can best be supported. The staff divisions analyze and refine each COA to determine its supportability. These estimates can form the cornerstone for staff annexes to orders and plans.

a. Staff estimates must be comprehensive and continuous and must visualize the future, but at the same time they must optimize the limited time available and not become

[3] *CJCSM 3500.05A, JTFHQ Master Training Guide.*

[4] *CJCSM 3122.01A, Joint Operation Planning and Execution System Vol I: Planning, Policies, and Procedures, Enclosure T, Appendix A, contains sample formats for the Planning Directive.*

COA Development Brief to CCDR (Recommended Briefing Sequence)

J3

- Context/Background- i.e., road to war
- Initiation- review guidance for initiation in general
- Strategic guidance-planning tasks assigned to supported CDR, forces/resources apportioned, planning guidance, updates, defense agreements, USG Security Cooperation Plan, Theater Security Plan, JSCP
- Forces Apportioned/Assigned

J2

- Joint Intelligence Preparation of the Operational Environment (JIPOE)
- ECOA- Most dangerous, most likely; strengths and weaknesses

J3

- Update Facts and Assumptions
- Mission Statement
- Commanders Intent (purpose, method, endstate)
- Endstate: political/military
  - Termination criteria
- Center of Gravity Analysis results: Critical Factors Strategic/Operational
- JOA/Theater of Ops/COMMZ sketch
- Phase 0 Shaping Activities recommended (for current theater security plan)
- FDO's with desired effect (DIME)
- COA Sketch and Statement by phase (3 COA's minimum)
  - Task organization
  - Component tasking
  - Timeline (I+5=W Day/MEU (SOC) in Gulf, etc.)
  - Recommended C2 by phase
  - Operational Design (LOO)
  - COA Risks
- COA summarized distinctions (COA 1, 2, and 3)
- COA Priority for Wargaming

Commanders Guidance

Figure XVI-3. Recommended Briefing Sequence.

Backwards Planning Technique (Reverse Planning)

There are several planning techniques available to you during COA development. The step-by-step approach below utilizes the backwards planning technique (reverse planning):

Step 1

Determine how much force will be needed in the theater at the end of the campaign, what those forces will be doing, and how those forces will be postured geographically. Use troop to task analysis. Draw a sketch to help you visualize the forces and their location.

Step 2

Looking at your sketch and working backwards, determine the best way to get the forces you just postured in Step 1 from their ultimate locations at the end of the campaign to a base in friendly territory. This will help you formulate your desired basing plan.

Step 3

Using your mission statement as a guide, determine the tasks the force must accomplish enroute to their ultimate positions at the end of the campaign. Draw a sketch of the maneuver plan. Make sure your force does everything the SECDEF has directed the CCDR to do (refer to specified tasks from the mission analysis steps).

Step 4

Determine the basing required to posture the force in friendly territory, and the tasks the force must accomplish to get to these bases. Sketch this as part of a deployment plan.

Step 5

Determine if the force you just considered is enough to accomplish all the tasks the SECDEF has given you. Adjust the force strength to fit the tasks. You should now be able to answer the first question.

Step 6

Given the tasks to be performed, determine in what order you want the force to be deployed into theater. Consider force categories such as combat, C4ISR, protection, sustainment, theater enablers, and theater opening. You can now answer the second question.

Step 7

You now have all the information necessary to answer the rest of the questions regarding force employment, major tasks and their sequencing, sustainment and command relationships.[5]

[5]*Army War College, Campaign Planning Primer.*

overly time-consuming. *Comprehensive* estimates consider both the quantifiable and the intangible aspects of military operations. They translate friendly and enemy strengths, weapons systems, training, morale, and leadership into combat capabilities. The estimate process requires a clear understanding of weather and terrain effects and, more important, the ability to visualize the battle or crisis situations requiring military forces. Estimates must provide a timely, accurate evaluation of the unit, the enemy, and the unit's area of operations at a given time.

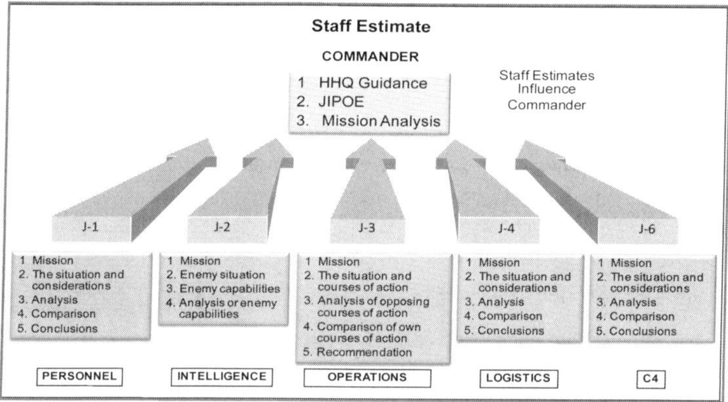

Figure XVI-4. Staff Estimates.

b. Estimates must be as thorough as time and circumstances permit. The CDR and staff must constantly collect, process, and evaluate information. They update their estimates:

- When the CDR and staff recognize new facts.
- When they replace assumptions with facts or find their assumptions invalid.
- When they receive changes to the mission or when changes are indicated.

c. Estimates for the current operation can often provide a basis for estimates for future missions as well as changes to current operations. Technological advances and near-real-time information estimates ensure that estimates can be continuously updated. Estimates must *visualize the future* and support the CDR's battlefield visualization. They are the link between current operations and future plans. The CDR's vision directs the endstate. Each subordinate unit CDR must also possess the ability to envision the organization's endstate. Estimates contribute to this vision. Failure to make staff estimates can lead to errors and omissions when developing, analyzing, and comparing COA's.

d. Not every situation will allow or require an extensive and lengthy planning effort. It is conceivable that a CDR could review the assigned task, receive oral briefings, make a quick decision, and direct writing of the plan to commence. This would complete the process and might be suitable if the task were simple and straightforward.

e. Most CDRs, however, are more likely to demand a thorough, well-coordinated plan that requires a complex staff estimate process. Written staff estimates are carefully prepared, coordinated, and fully documented.

f. Again, the purpose of the staff estimates is to determine whether the mission can be accomplished and to determine which COA can best be supported. This, together with the supporting discussion, gives the CDR the best possible information to select a COA. Each staff division:

- Reviews the mission and situation from its own staff functional perspective.
- Examines the factors and assumptions for which it is the responsible staff.
- Analyzes each COA from its staff functional perspective.

- Concludes whether the mission can be supported and which COA can be best supported from its particular staff functional perspective.

g. Because of the unique talents of each joint staff division, involvement of all is vital. Each staff estimate takes on a different focus that identifies certain assumptions, detailed aspects of the COAs, and potential deficiencies that are simply not known at any other level, but nevertheless must be considered. Such a detailed study of the COAs involves the corresponding staffs of subordinate and supporting commands.

h. The form and the number of COAs under consideration may change during this step. These changes result in refined COAs.

i. The product of this step is the sum total of the individual efforts of the staff divisions. Complete, fully documented staff estimates are extremely useful to the J-5 staff, which extracts information from them for the CDR's estimate. The estimates are also valuable to planners in subordinate and supporting commands as they prepare supporting plans. Although documenting the staff estimates can be delayed until after the preparation of the CDR's estimate, they should be sent to subordinate and supporting CDRs in time to help them prepare annexes for their supporting plans.

j. The principal elements of the staff estimates normally include *mission, situation and considerations, analysis of opposing COAs, comparison of friendly COAs, and conclusions.* The coordinating staff and each staff principle develop facts, assessments, and information that relate to their functional field. Types of estimates generally include, but are not limited to, operations, personnel, intelligence, logistics, civil-military operations, special staff, etc. The details in each basic category vary with the staff performing the analysis. The principal staff divisions have a similar perspective — they focus on friendly COAs and their supportability. However, the Intelligence Directorate (J-2) estimates on intelligence (provided at the beginning of the process) concentrate on the adversary: adversary situation, including strengths and weaknesses, adversary capabilities and an analysis of those capabilities, and conclusions drawn from that analysis. The analysis of adversary capabilities includes an analysis of the various COAs available to the adversary according to its capabilities, which include attacking, withdrawing, defending, delaying, etc. The J-2's conclusion will indicate the adversary's most likely COA and identify adversary COGs.[6]

k. In many cases the steps in the concept development phase are not separate and distinct, as the evolution of the refined COA illustrates. During planning guidance and early in the staff estimates, the initial COAs may have been developed from initial impressions and based on limited staff support. But as concept development progresses, COAs are refined and evolve to include many of the following considerations:

- What military operations are considered?
- Where will they be performed?
- Who will conduct the operation?
- When is the operation planned to occur?
- How will the operation be conducted?

l. *An iterative process of modifying, adding to, and deleting from the original tentative list is used to develop these refined COAs.* The staff continually evaluates the situation as the planning process continues (XVI-4). Early staff estimates are frequently given as oral briefings to the rest of the staff. In the beginning, they tend to emphasize information collection more than analysis. It is only in the later stages of the process that the staff estimates are expected to indicate which COAs can be best supported.

[6]*CJCSM 3122.01, Joint Operation Planning and Execution System Vol. I: Planning, Policies, and Procedures, Enclosure S contains sample formats for staff estimates.*

Course of Action (COA) Analysis & Wargaming

17

1. Step 4 to JOPP — COA Analysis and Wargaming

COA analysis and wargaming allows the CDR, his staff and subordinate CDRs and their staffs to gain a common understanding of friendly and threat COAs. This common understanding allows them to determine the advantages and disadvantages of each COA and forms the basis for the CDR's comparison (Step 5) and approval (Step 6). COA Wargaming involves a detailed assessment of each COA as it pertains to the enemy and the operational environment. Each friendly COA is wargamed against selected threat COAs. The CDR will select the COAs he wants wargamed and provide wargaming guidance along with *governing factors (evaluation criteria).*

Wargaming Process

1. Assemble the necessary tools and information.

2. Establish specific analysis rules to follow.

3. Wargame the COAs.

4. Record vital wargaming activities.

5. Identify advantages/disadvantages.

6. Refine each COA based on wargaming results.

(1) While time consuming, this procedure reveals strengths and weaknesses of each friendly COA, anticipates battlefield events, synchronizes warfighting functions, determines task organization for combat, identifies decision points, informs potential branches and sequels, and identifies cross-Service or component support requirements. Wargaming should also answer these questions:

- Does the COA achieve the purpose of the mission?
- Is the COA supportable?
- What if?

2. Analysis of Opposing COAs

The heart of the commander's estimate process is the *analysis of opposing COAs.* Analysis is nothing more than wargaming—either manual or computer assisted. In the previous steps of the estimate, ECOAs and COAs were examined relative to their basic concepts. ECOAs were developed based on enemy capabilities, objectives, and our estimate of the enemy's intent, and COAs developed based on friendly mission and capabilities. In this step we conduct an analysis of the probable effect *each ECOA has on the chances of success of each friendly COA.* The aim is to develop a sound basis for determining the *feasibility and acceptability* of the COAs. Analysis also provides the planning staff with a greatly improved understanding of their COAs and the relationship between them. The COA analysis identifies which COA best accomplishes the mission while best positioning the force for future operations. It helps the commander and staff to:

- Determine how to maximize combat power against the enemy while protecting the friendly forces and minimizing collateral damage.
- Have as near an identical visualization of the operation as possible.
- Anticipate events in the operational environment and potential reaction options.
- Determine conditions and resources required for success.
- Determine when and where to apply the force's capabilities.
- Focus intelligence collection requirements.
- Determine the most flexible COA.

3. The Conduct of Analysis and Wargaming

The wargame is a disciplined process, with rules and steps that attempt to visualize the flow of the operation. The process considers friendly dispositions, strengths, and weaknesses; enemy assets and probable COAs; and characteristics of the physical environment. It relies heavily on joint doctrinal foundation, tactical judgment, and operational experience. It focuses the staff's attention on each phase of the operation in a logical sequence. It is an iterative process of action, reaction, and counteraction. Wargaming stimulates ideas and provides insights that might not otherwise be discovered. It highlights critical tasks and provides familiarity with operational possibilities otherwise difficult to achieve. Wargaming is a critical portion of the planning process and should be allocated more time than any other step. Each retained COA should, at a minimum, be war gamed against both the most likely and *most dangerous* ECOAs.

4. During the wargame, the staff takes a COA statement and begins to add more detail to the concept, while determining the strengths or weaknesses of each COA. Wargaming tests a COA and can provide insights that can be used to improve upon a developed COA. The commander and his staff (and subordinate commanders and staffs if the war game is conducted collaboratively) may change an existing COA or develop a new COA after identifying unforeseen critical events, tasks, requirements, or problems.

4. Examining and Testing the COAs

For the wargame to be effective, the CDR should indicate what aspects of the COA he desires to be examined and tested. Wargaming guidance may include a list of friendly COAs to be wargamed against specific threat COAs (e.g., COA 1 against the enemy's most likely, most dangerous), the timeline for the phase or stage of the operations, a list of critical events and level of detail (i.e., two levels down).

a. Analysis of the proposed COAs should reveal a number of Factors including:

- Potential decision points.
- Task organization adjustment.
- Identification of plan branches and sequels.
- Identification of high-value targets.
- Risk assessment.
- COA advantages and disadvantages.
- Recommended CCIR.

b. COA Analysis Considerations. Before we move to making key decisions for wargaming, we need to collate and review a few important items. A few are listed below:

- Governing factors.
- Friendly and enemy forces.
- RFI's.
- Assumptions.
- Known critical events.

c. In the context of COA Analysis, we will look at governing factors and critical events.

(1) Governing Factors (Figure XVII-1). Determining the governing factors or better said, the evaluation criteria, to be used is a critical requirement that begins with wargaming. The CDR and staff choose the evaluation criteria during wargaming that will be used to select the COA that will become the CONOPS.CDRs establish evaluation criteria based on judgment, personal experience, METT-T and those criteria the staff uses to measure the effectiveness and efficiency of one COA relative to other COAs following the war game. These evaluation criteria help focus the wargaming effort and provide the framework for data collection by the staff. They are those aspects of the situation (or externally imposed factors) that the commander deems *critical* to the accomplishment of his mission. Potential influencing factors include elements of the Commander's Guidance and/or Commander's Intent, selected principles of war, external constraints, and even anticipated future operations for involved forces or against the same objective. Governing Factors change from mission to mission. Though these factors will be applied in the next step when the COAs are compared, it will be helpful during this wargaming step for all participants to be familiar with the factors so that any insights into a given COA which influence a factor are recorded for later comparison. The criteria may include anything the commander desires. If not received directly from the commander, they are often derived from his intent statement.

(a) The factors should look at both what will create success and what will cause failure. They may be used to determine the criteria of success for comparing the COAs in Step 5.

(2) Developing Governing Factors. Governing factors do not stand alone. Each must have a clearly-defined definition. Defining the criteria in precise terms reduces subjectivity and ensures the interpretation of each remains constant. The following list provides a good starting point for developing a COA comparison criteria list. See Appendix G for more examples of evaluation criteria.

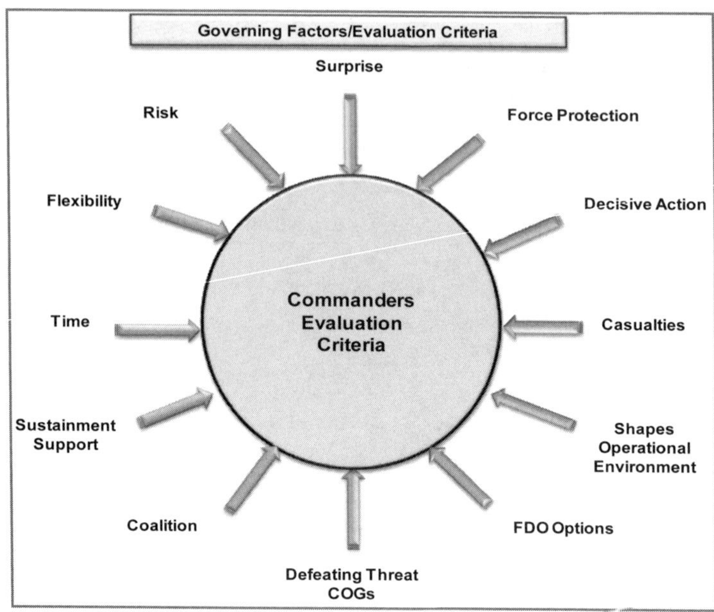

Figure XVII-1. Possible Commander's Governing Criteria.

(a) Some possible sources for determining criteria are:

- The Commander's Guidance and Commander's Intent.
- Mission accomplishment at an acceptable cost.
- The principles of war-Joint operations/SSTR (MOOSEMUSS).
- Doctrinal fundamentals for the type of operation(s) being conducted.
- The level of residual risk in the COA.
- Implicit significant factors relating to the operation (e.g., need for speed, security).
- Each staff member may identify factors relating to that staff function.
- Elements of Operational Art.
- Other factors to consider: political constraints, risk, financial costs, flexibility, simplicity, surprise, speed, mass, sustainability, C2, infrastructure survivability, etc.

Operational Design Elements		
Termination	Simultaneity and Depth	Endstate and Objectives
Timing and Tempo	Effects	Forces and Functions
Center of Gravity	Leverage	Decisive points
Balance	Direct versus Indirect	Anticipation
Lines of Operation	Synergy	Operational Reach
Arranging Operations	Culmination	

> ***Possible Evaluation Criteria for Humanitarian Assistance Operation***
> *-Protects the force -Delivers resources -Complements HN*
> *-Facilitates enterprise transition -Targets COG's -Enduring Partnerships*
> *-Meet critical partner nation stabilization/development needs that only US/DOD can/will provide*

(3) List Known Critical Events. These are essential tasks, or a series of critical tasks, conducted over a period of time that require detailed analysis (e.g., the series of component tasks to be performed on D-Day). This may be expanded to review component tasks over a phase(s) of an operation (e.g., lodgment phase) or over a period of time (C-Day through D-Day). The planning staff may wish at this point to also identify Decision Points (those decisions in time and space that the commander must make to ensure timely execution and synchronization of resources). These decision points are most likely linked to a critical event (e.g., commitment of the JTF Reserve force).

5. Wargaming Decisions

There are two key decisions to make before COA analysis begins. The *first* decision is to decide *what type of wargame will be used*. This decision should be based on CDR's guidance, time and resources available, staff expertise, and availability of simulation models. Note that at this point in the planning process, there may be no phases developed for the COA; Pre-hostilities, Hostilities, and Post-hostilities may be your only considerations at this point. Phasing comes later when the planner begins to flesh out the selected COA into a strategic concept. Some considerations are:

- Information Review: Mission Analysis, CDR's intent, planning guidance, CCDR's orders.
- Gather tools, materials, personnel and data:
  - Friendly courses of action to be analyzed.
  - Enemy courses of action against which you will evaluate the friendly COAs.
  - Representations of the operational area such as maps, overlays, etc.
  - Representations of friendly and enemy force dispositions and capabilities.
  - Subject matter experts (INTEL, SJA, POLAD, Log, IW, C4, PAO, etc.).
  - Red cell.
  - Scribe/recorder.
- Keep discussions elevated to the theater level.
- Balance between stifling creativity and making progress.

The *second* decision is to prioritize the enemy COAs the wargame is to be analyzed against. In time-constrained situations, it may not be possible to wargame against all courses of action.

6. Wargaming

Wargaming provides a means for the CDR and participants to analyze a tentative COA and obtain insights that otherwise might not have occurred. An objective, comprehensive analysis of tentative COAs is difficult even without time constraints. Based upon time available, the CDR should wargame each tentative COA against the most probable and the most dangerous adversary COAs (or most difficult objectives in non-combat operations) identified through the JIPOE process.

Some considerations are:

- Refine wargaming methodology.
- Pre-conditions or start points and endstate for each phase.
- Advantages/disadvantages of the COA.
- Unresolved issues.
- COA modifications or refinements.
- Estimated duration of critical events/phases.
- Major tasks for components.
- Identify critical events and decision points.
- Identify branches and sequels.
- Identify risks.
- Recommended EEIs and supporting collections plan priorities.
- Highlight ROE requirements.

a. Wargaming is a conscious attempt to visualize the flow of the operation, given joint force strengths and dispositions, adversary capabilities and possible COAs, the operational area (OA), and other aspects of the operational environment. Each critical event within a proposed COA should be wargamed based upon time available using the action, reaction, counteraction method of friendly and/or opposition force interaction. Here, the friendly force will make two moves because this activity is intended to validate and refine the friendly forces COA, not the adversaries. The basic wargaming method (modified to fit the specific mission and environment) can apply to noncombat as well as combat operations.

b. Wargaming stimulates thought about the operation so the staff will obtain ideas and insights that otherwise might not have occurred. This process highlights tasks that appear to be particularly important to the operation and provides a degree of familiarity with operational-level possibilities that might otherwise be difficult to achieve.

c. The wargaming process can be as simple as a detailed narrative effort that describes the action, probable reaction, counteraction, assets, and time used. A more comprehensive version is the "sketch-note" technique, which adds operational sketches and notes to the narrative process in order to gain a clearer picture. The most sophisticated form of wargaming is modern, computer-aided modeling and simulation. Figure XVII-2 on the facing page provides a list of key inputs and outputs for wargaming.

d. <u>Where to Begin</u>. See "Wargaming Steps" on facing page.

(a) Manual Wargaming:

<u>1</u> Deliberate Timeline Analysis- Consider actions day by day or in other discrete blocks of time. This is the most thorough method when time permits for detailed analysis.

<u>2</u> Operational Phasing- Used as a framework for COA analysis. Identify significant actions and requirements by functional area and/or JTF component.

<u>3</u> Critical Events/Sequence of Essential Tasks- The sequence of essential tasks, also known as the critical events method, highlights the initial shaping actions necessary to establish a sustainment capability and to engage enemy units in the deep battle area. At the same time, it enables the planners to adapt if the enemy executes a reaction that necessitates the reordering of the essential tasks. This technique also allows wargamers to concurrently analyze the essential tasks required to execute the CONOPS.

- Focus on specific critical events that encompass the essence of the COA. If time is particularly limited, focus only on the principal defeat mechanism. It is important to identify a measure of effectiveness that attempts to quantify the achievement of that defeat mechanism. This measure of effectiveness should enable a consistent comparison of

Wargaming Steps

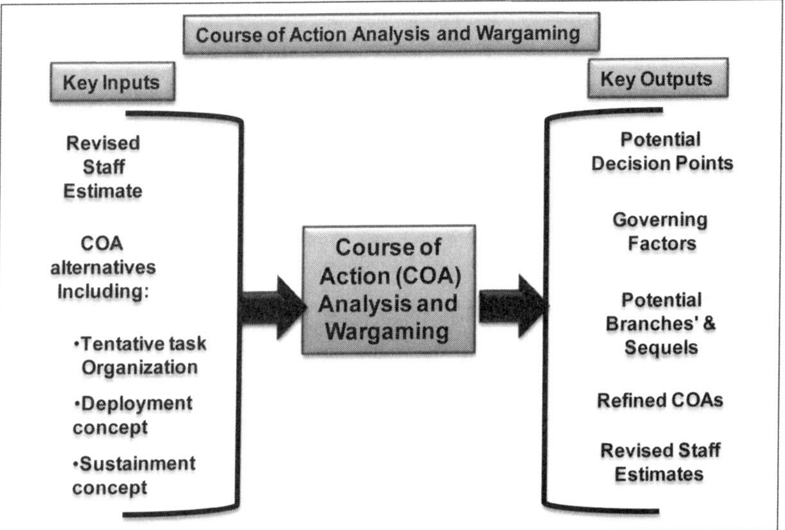

Figure XVII-2. Course of Action Analysis and Wargaming.

Step 1-Prepare for the Wargame

- Gather tools.
- List and review friendly forces.
- List and review enemy forces.
- List known critical events.
- Determine participants.
- Determine enemy COA to oppose.
- Select wargame method. (manual, computer).
- Select method to record and display wargaming results (narrative, sketch and notes, wargame worksheets, synchronization matrix).

Step 2 - Conduct the Wargame and Assess the Results

- Purpose of the Wargame (identify gaps, visualization, etc.).
- Basic methodology (action, reactio, counteraction).
- Type of method to be used: Based on CDR's guidance.

Step 3-Output of Wargame

- Results of the Wargame Brief: potential decision points, governing factors (those factors that the CDR deems critical to mission accomplishment), potential branches and sequels.
- Revised staff estimates.
- Refined COAs.
- Feedback through the COA Decision Brief.

each COA against each enemy COA for each specific critical event. If necessary, different measures of effectiveness should be developed for assessing different types of critical events (e.g., destruction, blockade, air control, neutralization, ensure defense). As with the focus on operational phasing, the critical events discussion identifies significant actions and requirements by functional area and/or by JTF component.

4 Computer assisted. There are many forms of computer-assisted wargames. Most wargames require a significant spin-up time to load scenarios and then to train users. However, the potential to utilize the computer model for multiple scenarios or blended scenarios makes it valuable.

e. Preparing for the Wargame.

(1) Room Organization. Contemplate how to organize the players in a logical manner. Below (Figure XVII-3) is an *example* of a manual wargame room organization. Note that along one wall are blown-up slides for CCIRs, Facts and Assumptions, Friendly Forces, Enemy COA. On the screens in the front of the wargame table are the working Synchronization Matrix and Decision Support Template. There is no limit to the information you can display around the room for quick reference. However, it's important to limit those items displayed to essential items that facilitate the Wargame, not distract from it.

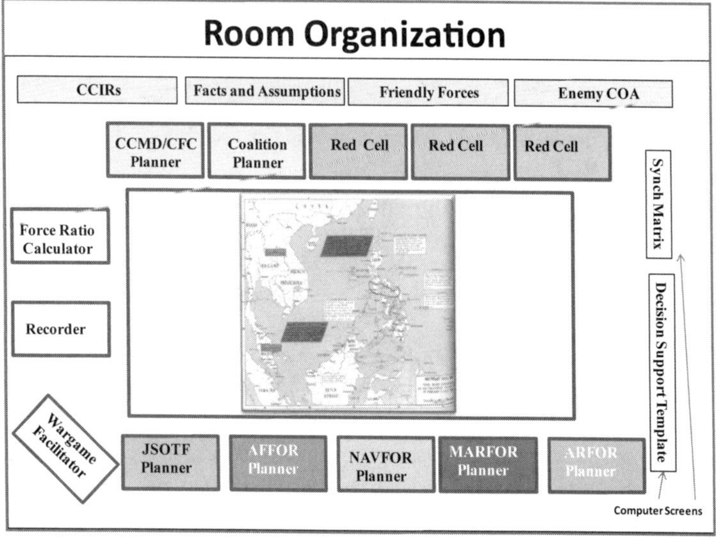

Figure XVII-3. Example Wargame Room Organization.

(2) Briefing Sequence. The following briefing sequence is an example of how the Blue Team/facilitator choreographs their Wargame:

- Blue Team/Facilitator: reviews conditions and timing for particular COA (includes current *Friendly* Intel picture)
- Red Cell Orientation.
- CCMD / CFC Planner.
- CJSOTF Planner.
- AFFOR Planner.
- NAVFOR Planner.
- MARFOR Planner.
- ARFOR Planner.

- Coalition Planner.
- Force Ratio Assessment.
- Issues from Recorder.
- Facilitator: highlights issues and concludes game turn.

(3) Rules of the Road.

- Facilitator has final say!!
- Facilitator is the only one who brings problems to arbitrators.
- Don't argue with Red Cell.
- Come prepared…try not to ad lib.
- Keep pace and pay attention.
- Take notes.
  - Submit appropriate information to Recorder.

Name: *Major Pane*
Component: *ARFOR*
COA#: *1*
Critical Event: *Commence CATK*
Topic Item: *ARFOR commence CATK from lodgment in Tunis with 85% strength in personnel Due to planned force flow...need 40 sorties of strike aircraft...*

Decision Point: *CFC-T will launch ARFOR CATK simultaneously with MARFOR once ARFOR Has at least 85% strength in combat forces and sustainment capability for 30 DOS in Classes I, III & IV.*

Assumption: *Once CTAK has been launched, there will be no interruption of planned force flow From N+30 - N+45.*

Figure XVII-4. Submission Cards to Recorder.

f. Wargame Process.

- Begins with time or event.
  - *Could be an enemy attack.*
- Next is friendly action, then threat reaction, then friendly counteraction.
  - If necessary, may continue beyond three moves.
- Continually assess feasibility.

(1) The facilitator and the red cell commander get together to agree on the rules of the war game. The wargame begins with an event designated by the facilitator. It could be an enemy offensive/defensive action or it could be a friendly offensive/defensive action. They decide where (in the AO) and when (H-Hour or L-Hour) it will begin. They review the initial array of forces. Of note, they must come to an agreement on the effectiveness of *intelligence collection* and *shaping actions* by both sides prior to the war game. The facilitator must ensure that all members of the wargame know what critical events will be wargamed and what techniques will be used. This coordination within the friendly cell and between the friendly and the red cell must be done well in advance.

(a) Within each wargaming method the war game normally has three total moves. If necessary, that portion of the wargame may be extended beyond the three moves. The facilitator decides how many moves are made in the wargame.

(b) During the war game the players must continually assess the COA's feasibility. Can it be supported? Can we do this? Do we need more combat power, more intelligence assets, more battlespace, more time, do we have the necessary logistics and communications? Has the threat successfully countered a certain phase or stage of a friendly COA? If we find that many of the aforementioned questions were answered "yes," then we may have to make major revisions to our friendly COA. We don't make major revisions to a COA in the midst of a wargame. Instead, we stop the wargame, make the revisions, and start over at the beginning.

(2) The war game is for comparing and contrasting friendly COAs with the threat COAs. We compare and contrast friendly COAs with each other in the fifth step of the JOPP, comparison/decision. Avoid becoming emotionally attached to a friendly COA; this will lead to the overlooking of the COA shortcomings and weaknesses.

- *Avoid comparing* one friendly COA with another friendly COA during the war game.
- Must remain unbiased.
- Do not violate timelines.

(a) A wargame for one COA at the JTF level may take six to eight hours. The facilitator must allocate enough time to ensure the war game will thoroughly test a COA.

(b) The wargame considers friendly dispositions, strengths, and weaknesses; threat assets and probable COAs; and characteristics of the area of operations. Through a logical sequence it focuses the players on essential tasks to be accomplished.

(c) When the wargame is complete and the worksheet and sync matrix are filled out, there should be enough detail to "flesh out" the bones of the COA and begin orders development (once the COA has been selected by the commander in comparison/decision).

(d) Additionally, the wargame will produce a refined event template and the initial decision support template which we call "decision support tools." These are similar to a football coach's game plan. The tools can help predict what the threat will do. The tools also provide the commander options for employing his forces to counter a threat action. The tools will prepare the commander (coach) and the staff (team) for a wide spectrum of possibilities and a choice of immediate solutions.

(e) The wargame relies heavily on doctrinal foundation, tactical judgment, and experience. It stimulates new ideas and provides insights that might have been overlooked. The dynamics of the war game require the red cell commander and the red cell members to be aggressive, but realistic, in the execution of threat activities.

- Records advantages and disadvantages of each COA as they become evident.
- Creates *decision support tools* (a game plan) (Figures XVII 12-15).
- Focuses OPT on the threat *and* commander's evaluation criteria.

g. <u>Wargaming Products</u>. We seek certain things to come out of the wargame in addition to wargamed COAs. We enter the wargame with a "rough" event template and must complete the wargame with a "refined," more accurate event template. Remember, the event template with its NAIs and TPLs will help the J-2 focus the intelligence collection effort. An event matrix can be used as a "script" for the intelligence rep during the wargame. It can also tell us if we are relying too much on one or two collection platforms and if we have overextended these assets.

(1) A first draft of a decision support template (DST) and decision support matrix (DSM) should also come out of the COA wargame. The first time we see decision points (DP) and target areas of interest (TAI) should be on the first draft DST as developed in the wargame. As more information about friendly forces and threat forces becomes available, the DST and DSM may change.

(2) Below the products in RED (italics) are produced by the intelligence section. Products in BLUE (underlined) are produced by the friendly or the operations section.

- Refines existing *named areas of interest (NAI)* and *time phase lines (TPL)* and identifies new *NAIs*. Identifies likely *target areas of interest (TAI)* and *decision points (DP)*.
- Refines the J-2's *situation and event templates* and creates first draft *decision support template and matrix*.
- *High value targets (HVT)* from J-2 source the OPT's *high-payoff targets (HPT)*.

h. Wargaming Outputs. There are inputs to the COA war game. Similarly, there are outputs from the COA war game. We must have the modified COAs, both sketch and narrative, ready to present to the Commander at the COA war game back brief.

(1) The critical events are associated with the essential tasks which we identified in mission analysis. The decision points are tied to points in time and space when and where the commander must make a critical decision. Decision points will be tied to the commander's critical information requirements (CCIR). Remember, CCIRs generate three types of information requirements: PIRs, FFIRs, and EEFIs. The commander approves CCIR's. From a threat perspective, PIRs tied to a decision point will require an intelligence collection asset to gather information about the threat. In IPB we tie PIRs to NAIs which are linked to threat COAs. The sync matrix is a tool which will help us determine if we have adequate resources.

- Wargamed COAs with graphic and narrative. Branches and sequels identified.
- Information on Commander's Evaluation Criteria.
- Initial task organization.
- Critical events and decision points.
- *New* resource shortfalls.
- *Refined/new* CCIRs and event template/matrix.
- *Initial* decision support template/matrix.
- "Fleshed out" synchronization matrix.
- *Refined* staff estimates.

(2) The outputs of the COA war game will be used in the comparison/decision step, orders development, and transition. The outputs on the slide are "products." The results of the wargame are the strengths and weaknesses of each friendly COA, the core of the back brief to the CDR.

i. Red Team. Once the enemy COA has been selected, it's time to build your Red Team.

(1) One of the most critical elements of your wargaming will be the synchronization of the Red Team. Without an aggressive and forward-leaning Red Team, the plan as a whole will be lesser for it. The most important element of wargaming is not the tool used, but the people who participate. Staff members who participate in wargaming should be the individuals who were deeply involved in the development of COAs. A robust cell that can aggressively pursue the adversary's point of view when considering adversary counter-action is essential. This *"red cell"* role-plays the adversary CDR and staff. The red cell develops critical decision points relative to the friendly COAs, projects adversary reactions to friendly actions, and estimates adversary losses for each friendly COA. By trying to accurately portray the capabilities of the enemy, the red cell helps the staff fully address friendly responses for each adversary COA.

(a) The primary purpose of the Red Team is to provide additional operational analysis on the adversary. The Red Team process normally is a peer review of the selected COA against adversary most dangerous (MD) or most likely (ML) COA. The Red Team will fight the MD and/or ML COA, but the goal is not to see if the Red Team can come up with a plan to defeat Blue's COA. The most important piece here is for the Red team to have enough time to develop a Red CONOPS prior to the wargame itself. The Red Team leader and team should be extremely knowledgeable of the Blue plan and should have been in on the development of that plan. The Red Team will, among other duties:

- Project adversary reactions to friendly actions.
- Attempt to accurately portray the capabilities of the adversary.
- Develop critical decision points relative to the friendly COAs.
- Analyze friendly critical vulnerabilities.
- ID adversary HVTs' key terrain, avenues of approach, doctrine and tactics, unconventional threats, reconnaissance and surveillance plans, etc.
- Depict adversary NAIs, TAIs, and DPs and in refining situation templates (threat COAs).

(b) If subordinate functional and Service components establish similar cells that mirror their adversary counterparts, this Red Cell network can collaborate to effectively wargame the adversary's full range of capabilities against the joint force. In addition to supporting the wargaming effort during planning, the Red Cell can continue to view friendly joint operations from the adversary's perspective during execution. The Red Cell process can be applied to noncombat operations to help determine unforeseen or most likely obstacles to, as well as the potential results of planned operations.

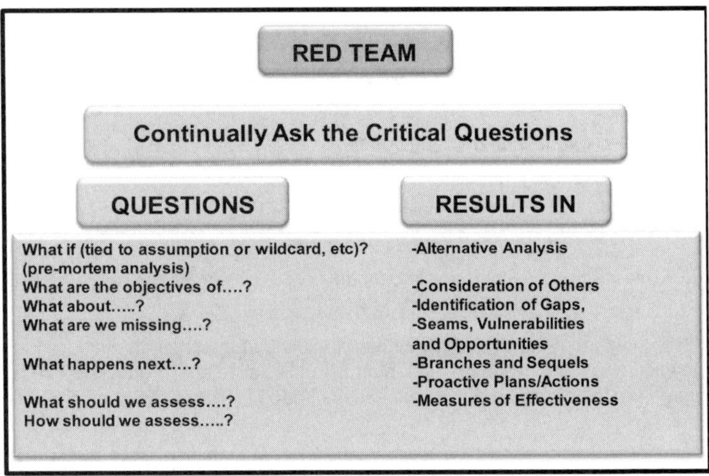

RED TEAM
Continually Ask the Critical Questions
QUESTIONS
RESULTS IN
What if (tied to assumption or wildcard, etc)?
(pre-mortem analysis)
What are the objectives of....?
What about.....?
What are we missing....?
What happens next....?
What should we assess....?
How should we assess.....?
-Alternative Analysis
-Consideration of Others
-Identification of Gaps,
-Seams, Vulnerabilities
and Opportunities
-Branches and Sequels
-Proactive Plans/Actions
-Measures of Effectiveness

Figure XVII-5. Red Team Critical Questions.

j. White Team. A small team of arbitrators normally composed of senior individuals familiar with the plan is a smart investment to ensure that the wargame does not get bogged down in unnecessary disagreement or arguing. The White Team will provide overall oversight to the wargame and any adjudication required between participants. The White Team may also include the facilitator and/or senior mentors as required.

> *"To be practical, any plan must take account of the enemy's power to frustrate it; the best chance of overcoming such obstruction is to have a plan that can be easily varied to fit the circumstances met; to keep such adaptability, while still keeping the initiative, the best way to operate is along a line which offers alternative objectives."*
>
> *B.H. Liddell Hart*

k. What a Wargame Provides.

- COA Wargame helps the commander determine how to *best apply friendly strengths against threat critical vulnerabilities (CV)* while *protecting* friendly CVs.
- COA Wargame provides *decision support tools* which facilitate the *transition to current operations* and subsequent execution.
- The red cell and the war game test our COAs against a thinking adversary. The wargame ensures we bring viable COAs to the commander for his decision. The results of COA analysis are better COAs.
- The wargame confirms:
  - Paths to threat critical vulnerabilities
  - Mission analysis

l. A Synchronization Matrix is a decision-making tool and a method of recording the results of wargaming. Key results that should be recorded include decision points, potential governing factors, CCIR, COA adjustments, branches, and sequels. Using a synchronization matrix helps the staff visually synchronize the COA across time and space in relation to the adversary's possible COAs. The wargame and synchronization matrix efforts will be particularly useful in identifying cross-component support resource requirements. Figures XVII-6 through XVII-9 are examples of synchronization matrix frameworks and Figure XVII-10 through XVII-11 are examples of a time line matrix and matrix by phase.

	CRITICAL EVENT/PHASE/TIME:								
Sequence Number	Action	Reaction	Counter-action	Assets	Time	Decision Point	PIR	Procedural & Positive Controls	Remarks

Figure XVII-6. Example of Synchronization Matrix Framework.

These synchronization matrices might be combined into one that, for example, reflects the contributions that each component would provide, within each joint functional area, over time. The staff can adapt the synchronization matrix to fit the needs of the analysis. It should incorporate other operations, functions, and units that it wants to highlight.

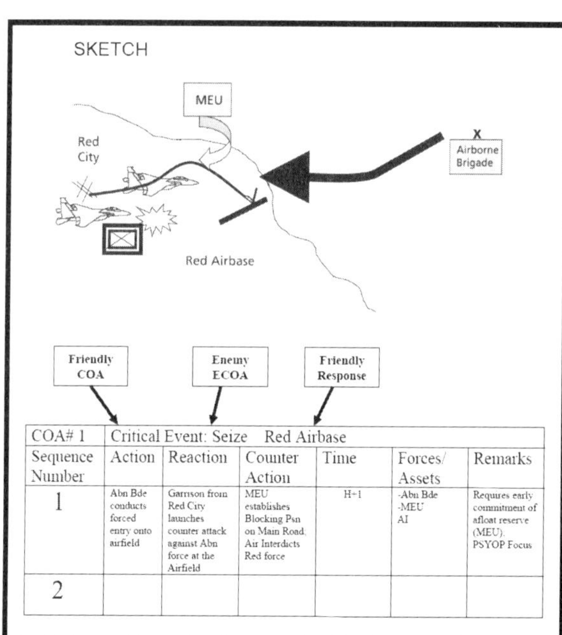

COA# 1	Critical Event: Seize Red Airbase					
Sequence Number	Action	Reaction	Counter Action	Time	Forces/ Assets	Remarks
1	Abn Bde conducts forced entry onto airfield	Garrison from Red City launches counter attack against Abn force at the Airfield	MEU establishes Blocking Psn on Main Road; Air Interdicts Red force	H+1	-Abn Bde -MEU AI	Requires early commitment of afloat reserve (MEU). PSYOP Focus
2						

Figure XVII-7. Example Wargame Synchronization Matrix.

Friendly COA #1 Short Name: ____________

Enemy COA – (Most Likely / Most Dangerous)

Time / Phase / Critical Event : __________

Situation:

Start:
End:

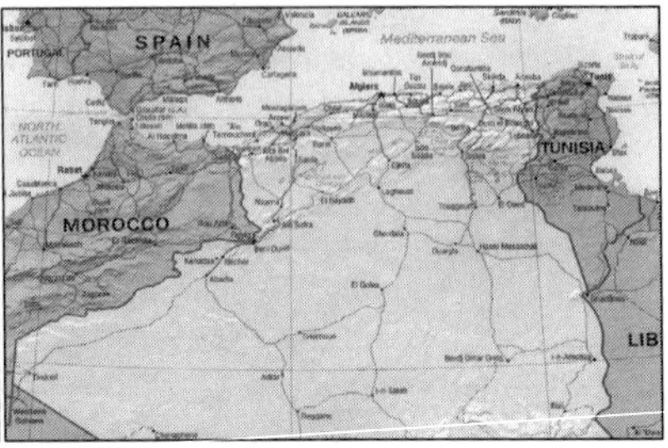

	Areas	Action	Enemy Reaction	Counteraction
COMPONENTS	ARFOR/Land Component			
	MARFOR/Land Component			
	NAVFOR/Maritime Component			
	ARFOR			
	JFACC			
	JSOTF			
	Others (e.g. USCG & rest of Interagency			
JOINT FUNCTIONAL AREAS	Movement & Maneuver			
	Intelligence			
	Firepower			
	Support			
	Command & Control			
	Protection			
OTHERS	Decision Points			
	CCIR			
	Branches			
	Risks			
	Issues			

Figure XVII-8. Example of Synchronization Matrix Framework.

Friendly COA #1 Short Name: Simultaneous NEO

Enemy COA – (Most Likely / Most Dangerous)

Time / Phase / Critical Event : Evacuations

Situation: The host government cannot guarantee the safety of American and third country nationals (TCN). The environment for a NEO is "uncertain." Rebel factions are becoming more violent.

Start: JTF 780 positioned for NEO
End: Successful NEO and forces return to safe havens.

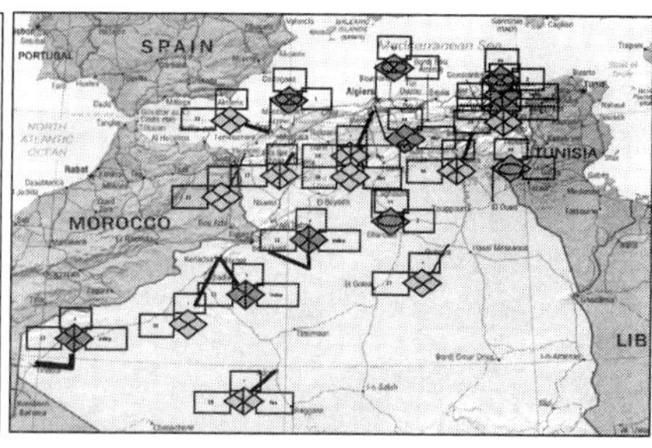

	Areas	Action	Enemy Reaction	Counteraction
COMPONENTS	ARFOR/Land Component	-Preparing to secure evac. site at #2 -Located with NAVFOR afloat	Rebel forces declares all AMCITS "enemies of the revolution" & US	-Launches and secures evac. Site #2 -Evac. AMCITs/TCN onto C-130
	MARFOR/ Land Component	-Preparing to secure evac. site at #3 -Located with NAVFOR (afloat)	military presence as evidence of plans to invade their country Several TCN's	-Launches and secures evac. Site #3 -Evac. AMCITs/TUN onto NAVFOR ships
	NAVFOR/Maritime Component	-Supporting MARFOR -Supporting JFACC -Supporting ARFOR (ISB TR)	killed in grenade attack near Site #2	-Supporting MARFOR -Supporting JFACC -Supporting ARFOR (ISB TR)
	ARFOR	-Supporting JFACC -Preparing evac. Control center at site A		-Supporting JFACC -Operates evac. Control center at site A
	JFACC	-Preparing to gain air superiority -Preparing to provide CAS/Coord. airspace		-Gains/maintains air superiority -Provide CAS /Coordinate airspace
	JSOTF	-Preparing to secure evac. site at #1 -Located at ISB B		-Launches and secures evac. Site #1 -Evac. AMCITs/TUN onto to ISB B
	Others (e.g. USCG & rest of Interagency	-DOS – Preparing to implement emergency evac. plans. -Pres/SECDEF prepare EXORD		-DOS – Implement emergency evac plan -Pres/SECDEF – Releases EXORD
JOINT FUNCTIONAL AREAS	Movement & Maneuver	-JTF must be in position NLT___ -JTF be prepared to execute EXORD +4hr		-NAVFOR & JFACC insures freedom of sea and air movements
	Intelligence	-NAIs – Sites 1, 2 & 3 _PIR – Threat against AMCITs, rebel troop movements, rebel attacks, etc.		-Collects info at NAIs -Collects and processes PIR
	Firepower	-Targeting effects continue		-AC-130 spt ARFOR/MARFOR/JSOTF (OPCON to JSOTF, TACON others)
	Support	-HN spt being provided by countries X&Y		-HN spt not available
	Command & Control	-IO/IW -PSYOP spt to ARFOR/MARFOR & JSOTF -JTF HQ located on USS MTW		-NAVFOR provides direct spt to MARFOR & ARFOR -Continues IW/C2W spt -ROE Changes
	Protection	-Weapons status___ -Deception theme___		-Weapons status___ -Deception theme___
OTHERS	Decision Points	-Recommendation for evac. EXORD		-When to declare "endstates" and mission accomplishment
	CCIR	-Taking of any AMCITs as hostages -Violence directed at AMCITs or TCNs		-# of AMCITs/TCNs evacuated -Any interference with the evacuation
	Branches	-Site #1, 2, or 3 is unusable -Hostages taken		-Implement branch plans
	Risks	-Military presence will provoke rebels		-Significant casualties -Usable to locate all AMCITs
	Issues	-Army helos on carriers Approval for use of evac site ISB in Cty Y		-Policy for evac. of unidentified personnel requesting help

Figure XVII-9. Example of Synchronization Matrix Framework.

Friendly COA#1 Short Name: Simultaneous NEO

Enemy COA – (Most Likely / Most Dangerous)

Time / Phase / Critical Event : Evacuations

Situation: The host government cannot guarantee the safety of American and third country nationals (TCN). The environment for a NEO is "uncertain." Rebel factions are becoming more violent.

Start: JTF 780 positioned for NEO
End: Successful NEO and forces return to safe havens.

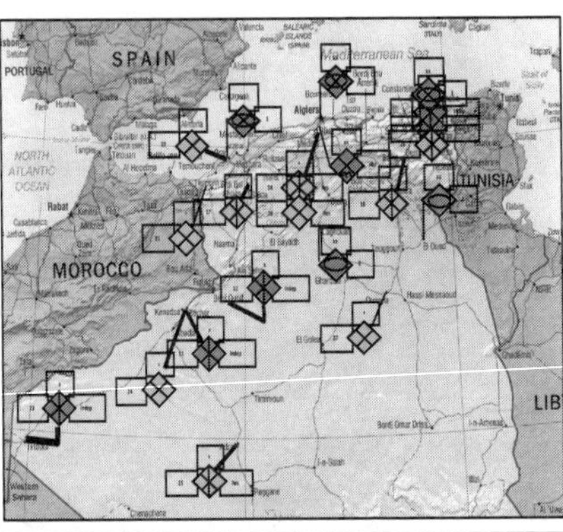

	Areas	Day D-2	D-Day
COMPONENTS	ARFOR/Land Component	-Preparing to secure evac. site at #2 -Located with NAVFOR afloat	-Launches and secures evac. Site #2 -Evac. AMCITs/TCN onto C-130
	MARFOR/ Land Component	-Preparing to secure evac. site at #3 -Located with NAVFOR (afloat)	-Launches and secures evac. Site #3 -Evac. AMCITs/TUN onto NAVFOR ships
	NAVFOR/Maritime Component	-Supporting MARFOR -Supporting JFACC -Supporting ARFOR (ISB TR)	-Supporting MARFOR -Supporting JFACC -Supporting ARFOR (ISB TR)
	ARFOR	-Supporting JFACC -Preparing evac. Control center at site A	-Supporting JFACC -Operates evac. Control center at site A
	JFACC	-Preparing to gain air superiority -Preparing to provide CAS/Coord. airspace	-Gains/maintains air superiority -Provide CAS /Coordinate airspace
	JSOTF	-Preparing to secure evac. site at #1 -Located at ISB B	-Launches and secures evac. Site #1 -Evac. AMCITs/TUN onto to ISB B
	Others (e.g. USCG & rest of Interagency	-DOS – Preparing to implement emergency evac. plans. -Pres/SECDEF prepare EXORD	-DOS – Implement emergency evac plan -Pres/SECDEF – Releases EXORD
JOINT FUNCTIONAL AREAS	Movement & Maneuver	-JTF must be in position NLT___ -JTF be prepared to execute EXORD +4hr	-NAVFOR & JFACC insures freedom of sea and air movements
	Intelligence	-NAIs – Sites 1, 2 & 3 _PIR – Threat against AMCITs, rebel troop movements, rebel attacks, etc.	-Collects info at NAIs -Collects and processes PIR
	Firepower	-Targeting effects continue	-AC-130 spt ARFOR/MARFOR/JSOTF (OPCON to JSOTF, TACON others)
	Support	-HN spt being provided by countries X&Y	-HN spt not available
	Command & Control	-IO/IW -PSYOP spt to ARFOR/MARFOR & JSOTF -JTF HQ located on USS MTW	-NAVFOR provides direct spt to MARFOR & ARFOR -Continues IW/C2W spt -ROE Changes
	Protection	-Weapons status___ -Deception theme___	-Weapons status___ -Deception theme___
OTHERS	Decision Points	-Recommendation for evac. EXORD	-When to declare "endstates" and mission accomplishment
	CCIR	-Taking of any AMCITs as hostages -Violence directed at AMCITs or TCNs	-# of AMCITs/TCNs evacuated -Any interference with the evacuation
	Branches	-Site #1, 2, or 3 is unusable -Hostages taken	-Implement branch plans
	Risks	-Military presence will provoke rebels	-Significant casualties -Usable to locate all AMCITs
	Issues	-Army helos on carriers Approval for use of evac site ISB in Cty Y	-Policy for evac. of unidentified personnel requesting help

Figure XVII-10. Sample Time Line Matrix for a JTF Operation.

	Time/Phase	Shaping	Deterr ence	Seize Initiative	Dominance	Stabilization	Enable CA
	Threat Enemy Action/Reaction	Component Tasks					
Functional UJTL Operational Level Tasks	Operational Intel/Surv. Recon Op Env. Aware	AR NAV MAR AF SOF					
	Operational Movement & Maneuver Force Application	AR NAV MAR AF SOF					
	Operational Firepower	AR NAV MAR AF SOF					
	Operational Protection Protection AT/NBC Def./ CBRNE/etc.	AR NAV MAR AF SOF					
	Operational Log & SPT Focused Logistics	AR NAV MAR AF SOF Host Nation SPT					
	Operational CMD & CTL ROE	Staff C4I Multinational Interagency					
Others/ Enablers	iNFO OPS JSOTF/etc.						
	Interagency						
	Multinational						
	ROE						
	Decision Pts. Link to CCIR						
	Risk						

XVII-11. Sample Matrix by Phase.

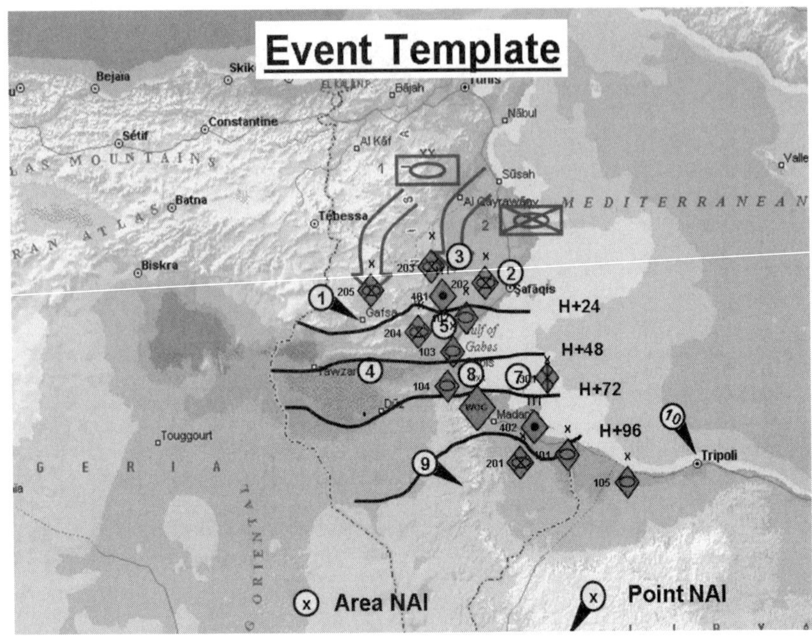

Figure XVII-12. Sample Event Template

Event Matrix

NAI	Event	NET/ NLT	DP & TAI	Tasked Collection Assets
3, 2	Orange WOG first echelon vic Mezzouna displaces from defensive positions	H + 6 H + 12	DP 3	RC-135, TR-1, F/A-18D, Predator Track 3
19, 20	Red SEC moving east across border toward Kasserine Pass	H +12 H +24	DP 1	National IMINT, National SIGINT U-2R, RC-135,
1	Orange 205th mech BDE moves north out of Gafsa to link up with lead BDE	H +6 H + 48	DP 1	National SIGINT, RC-135, Predator Track 6
8	LY 2d echelon forces move north to establish strongpoint defense vic Gabes	H Hour H + 60	DP 2	National SIGINT, Theater IMINT, RC-135, VMU Track 2
2, 5	Orange first echelon attacks Sfax	21 July / 26 July	DP 4	FA-18/FAC/FLIR National IMINT and SIGINT
10	Orange fire SCUDs at Blue ports and airfields	Phase 1 & 2	DP 5	National / Theater IMINT National / Theater SIGINT
7	Orange establish CSSC-3 seersucker battery vic Jerba Island	Phase 1 & 2	DP 6	JSTARS, Predator Track 1

Figure XVII-13. Sample Event Matrix.

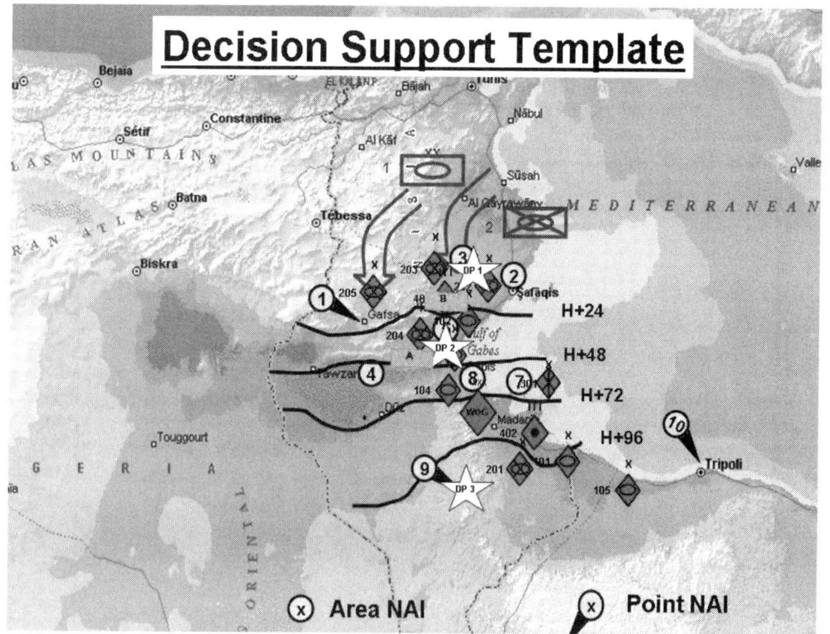

Figure XVII-14. Sample Decision Support Template.

DECISION SUPPORT MATRIX

DP	EVENTS & INDICATORS	NET/NLT	CDR'S OPTIONS
1	Whether or not enemy 1st echelon units are fixed; (NAI 1 - 3)	H+24/H+36	1 AD continues turning mvmnt or executes branch plan for envelopment of 1st echelon
2	1st echelon enemy forces withdrawing into/through Gabes; 103d Armor Bde covering withdrawal and as possible counterattk force; 204th, 201st, and 104th preparing for BHO vic Gabes; refugees being forced North; (NAI 4-5)	H+96/H+120	Bypass, isolate or clear Gabes; force options: 2d MarDiv or LF6F
3	Enemy delays 2dMarDiv and reorients on mountain passes IOT hold 1 AD and allow forces to withdraw to border; (NAI 9)	H+144/H+168	Options: defeat 2d echelon via encirclement; defeat 2d echelon by annihilation (1AD/2MAW) or allow enemy to withdraw

Figure XVII-15. Sample Decision Support Matrix.

7. Interpreting the Results of Analysis and Wargaming

Comparisons of advantages and disadvantages of each COA will be conducted during the next step of COA Determination. However, if the suitability, feasibility, or acceptability of any COA becomes questionable during the analysis, the CDR should modify or discard it and concentrate on other COAs. The need to create additional combinations of COAs may also be required (Figure XVII-16).

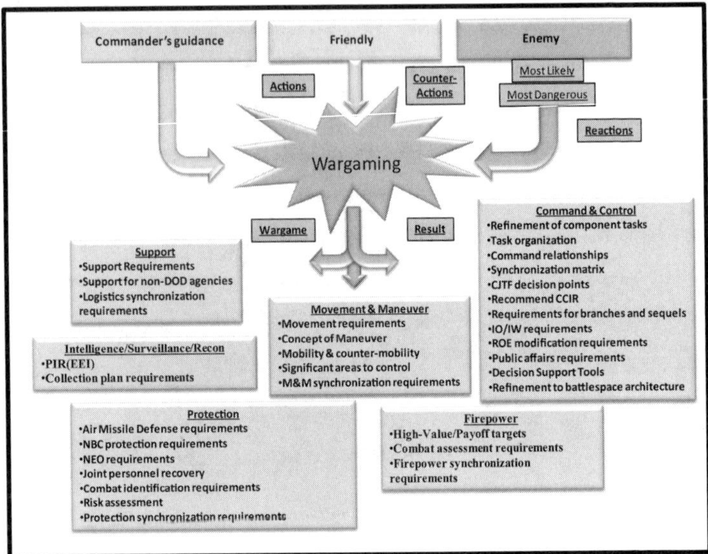

Figure XVII-16. Results of Analysis and Wargaming.

8. Flexible Plans

We have several techniques to help us develop adaptability. One of these is to make flexible plans. Flexible plans can enhance adaptability by establishing a COA that provides for multiple options. We can increase our flexibility by providing *branches* to deal with changing conditions on the battlefield that may affect the plan (e.g., changing dispositions, orientation, strength, movement) or by providing sequels for current and future operations. Sequels are courses of action to follow probable battle or engagement outcomes; victory, defeat, or stalemate.

> *"There were no decisions reached about how to exploit a victory in Sicily. It was an egregious error to leave the future unresolved."*
>
> *-General Omar Bradley on Operation Husky*

a. Branches and Sequels. Many operation plans require adjustment beyond the initial stages of the operation. Consequently, CDRs build flexibility into their plans by developing branches and sequels to preserve freedom of action in rapidly changing conditions. An effective plan places a premium on flexibility because operations never seem to proceed exactly as planned.

b. During the wargaming sequence we will identify operational branches and sequels giving our plans that flexibility and help focus our future planning efforts. Visualizing and planning branches and sequels are important because they involve transitions— changes in mission, type of operations, and often forces required for execution. Unless planned, prepared for, and executed efficiently, transitions can reduce the tempo of the operation, slow its momentum, and surrender the initiative to the adversary. Both branches and sequels should have execution criteria, carefully reviewed before their implementation and updated based on assessment of current operations. Both branches and sequels directly relate to the concept of phasing.

(1) *Branches are options built into the basic plan* and provide a framework for flexibility in execution. Branches may include shifting priorities, changing mission, unit organization and command relationships, or changing the very nature of the joint operation itself. Branches add flexibility to plans by anticipating situations that could alter the basic plan. Such situations could be a result of lingering planning assumptions not proven to be facts, anticipated events, opportunities, or disruptions caused by enemy actions, availability of friendly capabilities or resources, or even a change in the weather or season within the operational area. Commanders anticipate and devise counters to enemy actions to mitigate risk. Although anticipating every possible threat action is impossible, branches anticipate the most likely ones. Commanders execute branches to rapidly respond to changing conditions.

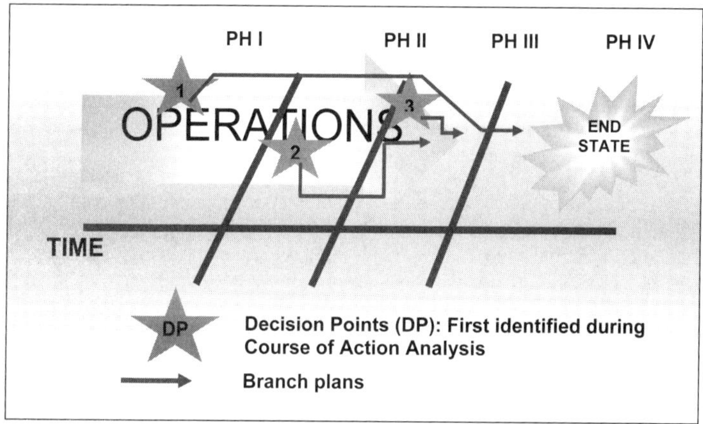

(2) *Sequels are subsequent operations* based on the possible outcomes of the current operation — victory, defeat, or stalemate. Sequel planning allows the CDR and staff to keep pace with a constantly evolving situation while staying focused on mission accomplishment. They are future operations that anticipate the possible outcomes —success, failure, or stalemate — of the current operations. A counteroffensive, for example, is a logical sequel to a defense; exploitation and pursuit follow successful attacks. Executing a sequel normally begins another phase of an operation, if not a new operation. In joint operations, phases can be viewed as the sequels to the basic plan. CDRs consider sequels early and revisit them throughout an operation. Without such planning, current operations leave forces poorly positioned for future opportunities, and leaders are unprepared to retain the initiative.

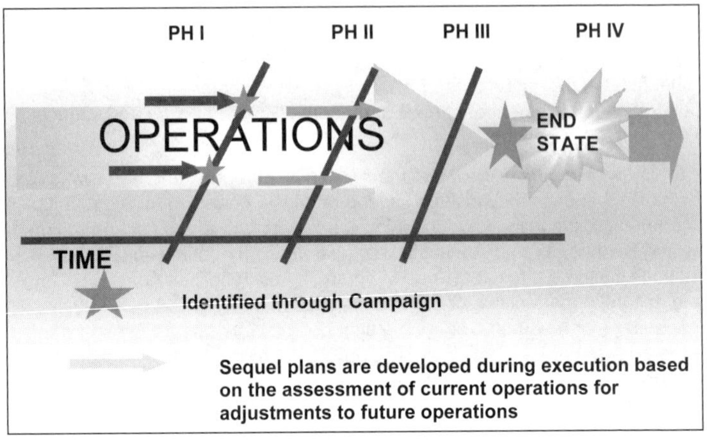

c. The value of *branches and sequels* is that they prepare us for several different actions. *We should keep the number of branches and sequels to a relative few.* We should not try to develop so many branches and sequels that we cannot adequately plan, train, or prepare for any of them. The skillful, well-thought-out use of branches and sequels becomes an important means of anticipating future courses of action. This anticipation helps accelerate the decision cycle and therefore increases tempo and flexibility.

d. Flexible plans avoid unnecessary detail that not only consumes time in their development, but has a tendency to restrict subordinates' latitude. Instead, flexible plans lay out what needs to be accomplished, but leave the manner of accomplishment to subordinates. This allows the subordinates the flexibility to deal with a broader range of circumstances.

e. Flexible plans are plans that can be easily changed. Plans that require coordination are said to be "coupled." If all the parts of a plan are too tightly coupled, the plan is harder to change because changing any one part of the plan means changing all the other parts. Instead, we should try to develop modular, loosely coupled plans. Then if we change or modify any one part of the plan, it does not directly affect all the other parts.

f. Finally, flexible plans should be simple plans. Simple plans are easier to adapt to the rapidly changing, complex, and fluid situations that we experience in combat

Course of Action (COA) Comparison 18

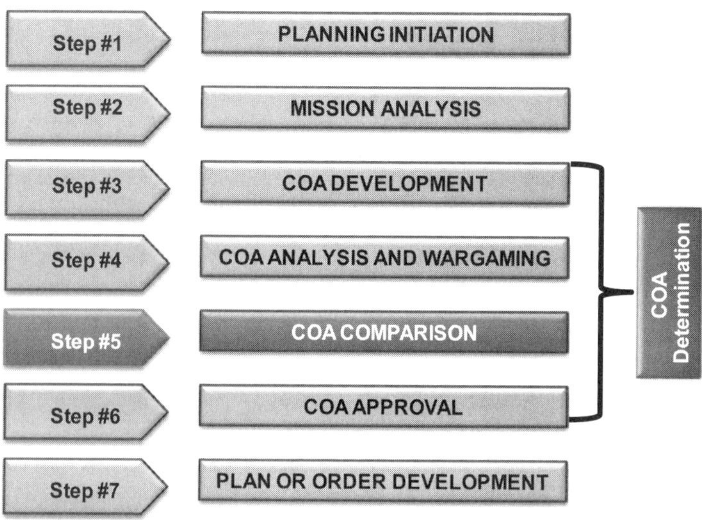

1. Step 5 to JOPP — COA Comparison

COA comparison is a *subjective* process whereby COAs are considered independently of each other and evaluated/compared against a set of criteria that are established by the staff and CDR. The goal is to identify and recommend the COA that has the highest probability of success against the enemy COA (ECOA) that is of the most concern to the CDR, so take some time and energy with this step. COA comparison facilitates the CDR's decision-making process by balancing the *ends, means, ways* and *risk* of each COA (Figure XVIII-1). The end product of this task is a briefing to the CDR on a COA recommendation and a decision by the CDR.

- *Determine comparison criteria*
- *Define and determine the standard for each criteria*
- *Assign weight or priority to comparison criteria*
- *Construct comparison method and record*
- *Conduct and record the comparison recommend COA*

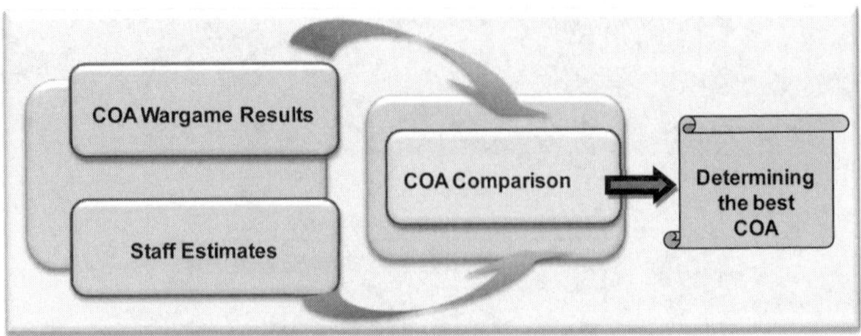

Figure XVIII -1. COA Comparison.

a. In COA comparison the CDR and staff evaluate all friendly COAs – against established criteria-governing factors (discussed in previous chapter), and select the COA which best accomplishes the mission. The CDR reviews the criteria list and adds or deletes as they see fit. The number of governing factors will vary, but there should be enough to differentiate COAs. Consequently, COAs are not compared to each other, but rather they are individually evaluated against the criteria that are established by the staff and CDR.

(1) COA comparison helps the CDR answer the following questions:

- What are the differences between each COA?
- What are the advantages and disadvantages?
- What are the risks?

Identify and select the COA that best accomplishes the mission.

b. Staff officers may each use their own matrix to compare COAs with respect to their functional areas. Matrices use the governing factors/evaluation criteria developed before the wargame. Decision matrices alone cannot provide decision solutions. Their greatest value is providing a method to compare COAs against criteria that, when met, produce mission success. They are analytical tools that staff officers use to prepare recommendations. CDRs provide the solution by applying their judgment to staff recommendations and making a decision.

Staff Estimate Matrix (Intel Estimate)

Evaluation Criteria	Frontal COA 1	Envelopment COA 2
Effects of Terrain		✓
Effects of Weather	✓	
Utilizes Surpise		✓
Attacks Critical Vulnerabilities		✓
Collection Support		✓
Counterintelligence	✓	
Totals	2	4

c. The staff helps the CDR identify and select the COA that best accomplishes the mission. The staff supports the CDR's decision-making process by clearly portraying his options and recording the results of the process. The staff compares feasible COAs to identify the one with the highest probability of success against the most likely enemy COA (MLCOA) and the most dangerous enemy COA (MDCOA).

COA Comparison

1. *Determine evaluation criteria*
2. *Develop a construct for making comparison*
3. *Perform COA comparison*
4. *Decide on COA and prepare to recommend to CDR*

2. Prepare for COA Comparison

The CDR and staff develop and evaluate a list of important criteria. Using the governing factors/evaluation criteria discussed during COA analysis and wargaming, the staff outlines each COA, highlighting advantages and disadvantages. Comparing the strengths and weaknesses of the COAs identifies their advantages and disadvantages relative to each other.

a. Determine/Define Comparison Criteria or Governing Factors. As discussed earlier, criteria are based on the particular circumstances and should be relative to the situation. There is no standard list of criteria, although the CDR may prescribe several core criteria that all staff directors will use (see Appendix G for examples). Individual staff sections, based on their estimate process, select the remainder of the criteria:

- Criteria are based on the particular circumstances and should be relative to the situation.
- Review CDRs guidance for relevant criteria.
- Identify implicit significant factors relating to the operation.
- Each staff identifies criteria relating to that staff function.
- Other criteria might include:
  - Political, social, and safety constraints; requirements for coordination with Embassy/Interagency personnel.
  - Fundamentals of Joint Warfare/SSTR.
  - Elements of Operational Art.
  - Mission accomplishment.
  - Risks.
  - Costs.

b. Define and Determine the Standard for each Criterion.

(1) Establish standard definitions for each governing factor. Define the criteria in precise terms to reduce subjectivity and ensure the interpretation of each governing factor remains constant between the various COAs.

(2) Establish definition prior to commencing COA comparison to avoid compromising the outcome.

(3) Apply standard for each criterion to each COA.

c. The staff evaluates feasible COAs using those governing factors most important to the CDR to identify the one COA with the highest probability of success. The selected COA should also:

(1) Mitigate risk to the force and mission to an acceptable level.

(2) Place the force in the best posture for future operations.

(3) Provide maximum latitude for initiative by subordinates.

(4) Provide the most flexibility to meet unexpected threats and opportunities.

3. Determine the Comparison Method and Record

Actual comparison of COAs is critical. The staff may use any technique that facilitates reaching the best recommendation and the CDR making the best decision. There are a number of techniques for comparing COAs. The most common technique is the decision matrix, which uses evaluation criteria to assess the effectiveness and efficiency of each COA. Here are examples of several decision matrices:

a. Weighted Numerical Comparison Technique. The example below provides a numerical aid for differentiating COAs. Values reflect the relative advantages or disadvantages of each COA for each of the criterion selected. Certain criteria have been weighted to reflect greater value (Figure XVIII-2 and XVIII-3 on the following page).

b. Determine the Weight of each Criterion Based on its Relative Importance and the CDR's Guidance. The CDR may give guidance that result in weighting certain criteria. The staff member responsible for a functional area scores each COA using those criteria. Multiplying the score by the weight yields the criterion's value. The staff member then totals all values. However, he must be careful not to portray subjective conclusions as the results of quantifiable analysis. Comparing COAs by category is more accurate than comparing total scores.

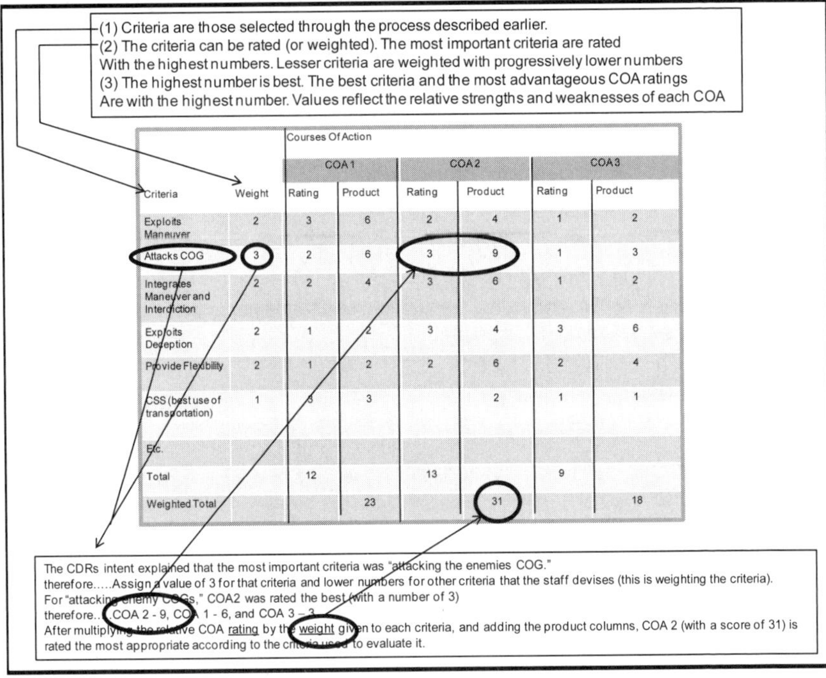
(1) Criteria are those selected through the process described earlier.
(2) The criteria can be rated (or weighted). The most important criteria are rated
With the highest numbers. Lesser criteria are weighted with progressively lower numbers
(3) The highest number is best. The best criteria and the most advantageous COA ratings
Are with the highest number. Values reflect the relative strengths and weaknesses of each COA
The CDRs intent explained that the most important criteria was "attacking the enemies COG."
therefore.....Assign a value of 3 for that criteria and lower numbers for other criteria that the staff devises (this is weighting the criteria).
For "attacking enemy COGs," COA2 was rated the best (with a number of 3)
therefore.....COA 2 - 9, COA 1 - 6, and COA 3 – 3
After multiplying the relative COA rating by the weight given to each criteria, and adding the product columns, COA 2 (with a score of 31) is rated the most appropriate according to the criteria used to evaluate it.

		Courses Of Action					
		COA 1		COA 2		COA 3	
Criteria	Weight	Rating	Product	Rating	Product	Rating	Product
Exploits Maneuver	2	3	6	2	4	1	2
Attacks COG	3	2	6	3	9	1	3
Integrates Maneuver and Interdiction	2	2	4	3	6	1	2
Exploits Deception	2	1	2	3	4	3	6
Provide Flexibility	2	1	2	2	6	2	4
CSS (best use of transportation)	1	3	3		2	1	1
Etc.							
Total		12		13		9	
Weighted Total			23		31		18

Figure XVIII-2. Example Numerical Comparison.

(4) Each staff section does this separately, perhaps using different criteria on which to base the COA comparison. The staff then assembles and arrives at a consensus for the criterion and weights. The Chief of Staff/DCJTF should approve the staff's recommendations concerning the criteria and weights to ensure completeness and consistency throughout the staff sections.

EXAMPLE #2 COA COMPARISON MATRIX FORMAT							
GOVERNING FACTOR	WEIGHT	COA #1		COA #2		COA #3	
		SCORE	WEIGHTED	SCORE	WEIGHTED	SCORE	WEIGHTED
SURPRISE	2	3	6	2	4	2	4
RISK	2	3	6	1	2	2	4
FLEXIBILITY	1	2	2	1	1	1	1
RETAILIATION	1	1	1	2	2	1	1
DAMAGE TO ALLIANCE	1	2	2	1	1	1	1
LEGAL BASIS	1	1	1	1	1	1	1
EXTERNAL SUPPORT	1	3	3	2	2	1	1
FORCE PROTECTION	1	3	3	3	3	1	1
OPSEC	1	3	3	2	2	2	2
TOTAL		27 - Green		18 - Yellow		16 - Red	

Figure XVIII-3. Comparison Matrix Format.

c. Non-Weighted Numerical Comparison Technique. The same as the previous method except the criteria are not weighted. Again, the highest number is best for each of the criteria.

d. Narrative or Bulleted Descriptive Comparison of Strengths and Weaknesses. Review criteria and describe each COA's strengths and weaknesses. See the example below, Figure XVIII-4.

Course of Action	Strengths	Weaknesses
COA 1	Narrative or bulletized discussion of strengths using the criteria	Narrative or bulletized discussion of weaknesses using the same criteria
COA 2	Same	Same
COA 3	Same	Same

Figure XVIII-4. Criteria for Strengths and Weaknesses.

e. Plus/Minus/Neutral Comparison. Base this comparison on the broad degree to which selected criteria support or are reflected in the COA. This is typically organized as a table showing (+) for a positive influence, (0) for a neutral influence, and (-) for a negative influence. Figure XVIII-5 is an example:

Criteria	COA 1	COA 2
Casualty estimate	--	--
Casualty evacuation routes	+	--
Suitable medical facilities	0	0
Flexibility	+	--

Figure XVIII-5. Plus/Minus/Neutral Comparison.

f. Stop Light Comparison. Criteria are judged to be acceptable or unacceptable with varying levels in between. Ensure you define each color in the stop light on a key along with corrective methods to elevate mid colors to green (Figure XVIII-6).

COA Criteria	COA 1	COA 2	COA 3
Speed	Red	Green	Yellow
Surprise	Red	Orange	Green
Risk	Orange	Yellow	Green
Total	Red	Orange	Green

Figure XVIII-6. Stop Light Comparison.

g. Descriptive Comparison. Simply a description of advantages and disadvantages of each COA (Figure XVIII-7).

COA	ADVANTAGES	DISADVANTAGES
COA 1	•Rapid Delivery •Meets critical needs	•Rough integration of forces •Rough transition •Complex organization •Not flexible at all •Adequate force protection
COA 2	•Rapid delivery •Meets critical needs •Smooth integration •Smooth transition	•Complex organization •Less flexible •Adequate force protection
COA 3	•Smooth integration •Smooth transition •Simplest organization •Adequate force protection •Best force protection	•Less rapid delivery •Does nor meet all critical needs

Figure XVIII-7. Desriptive Comparison.

4. COA comparison remains a subjective process and should not be turned into a mathematical equation. Using +,-, 0 or 1, 2, 3 are as appropriate as any other methods. The key element in this process is the ability to articulate to the CDR why one COA is preferred over another. Figure XVIII-8 is an example staff estimate comparison.

-Reviews the mission and situation from its own staff functional perspective.
-Examines the factors and assumptions for which it is the responsible staff.
-Analyzes each COA from its staff functional perspective.
-Concludes whether the mission can be supported and which COA can best be supported from its particular staff functional perspective

Staff	Decisive Air COA 1	Sequential Build COA 2
J2	✓	
J3		✓
J4	✓	
MARFOR	✓	
ARFOR		✓
AFFOR	✓	
NAVFOR	✓	
JFACC		✓
JFLCC		✓
JFMCC	✓	
TOTALS	6	4

The staff and components recommend COA 2

Figure XVIII-8. Staff Estimate Comparison Input.

5. Figure XVIII-9 depicts inputs and outputs for course-of-action comparison. Other products not graphically shown in the chart include updated JIPOE products, updated CCIR's, staff estimates, CDR's identification of branches for further planning and a Warning Order as appropriate. (following page)

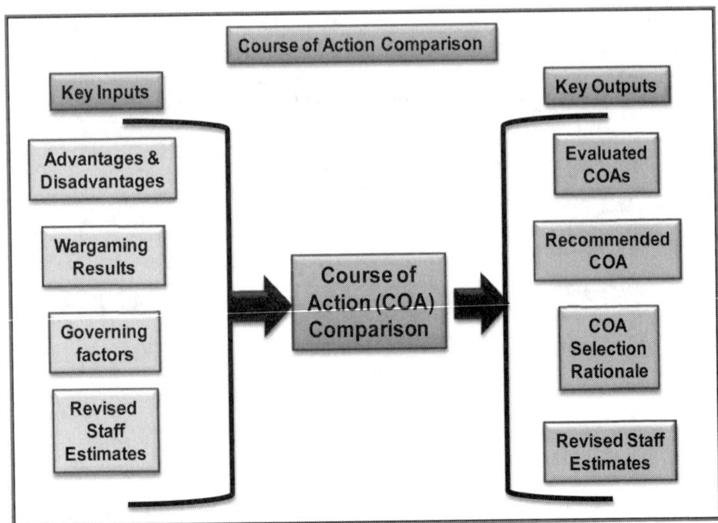

Figure XVIII-9. Course of Action Comparison.

6. The staff ensures the selected COA is faithfully captured as the concept of operations (CONOPS). The CONOPS – along with the joint functions (movement and maneuver, intelligence, firepower, support, C2, protection) – forms the basis for the operation plan (OPLAN) or order (OPORD).

Course of Action (COA) Selection & Approval

19

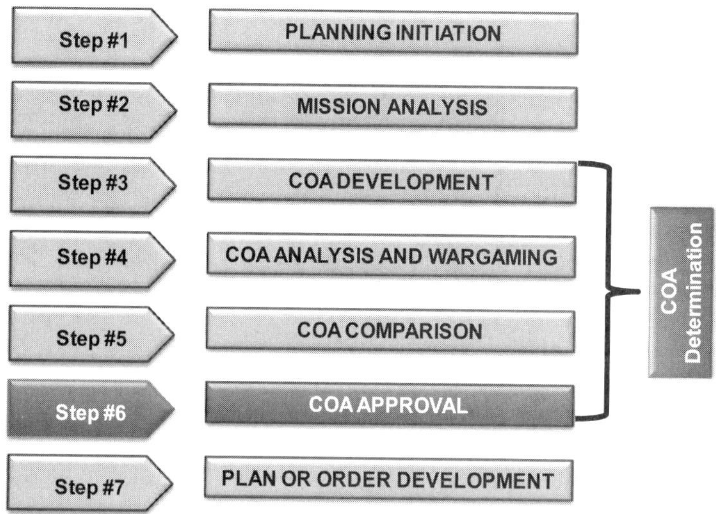

1. Step 6 to JOPP — COA Approval

a. COA Recommendation. Throughout the COA development process, the CCDR conducts an independent analysis of the mission, possible courses of action, and relative merits and risks associated with each COA. The CDR, upon receiving the staff's recommendation, combines his analysis with the staff recommendation resulting in a selected COA. The forum for presenting the results of COA comparison is the *CDR's Decision Brief.* Typically this briefing provides the CCDR with an update of the current situation, an overview of the COAs considered, and a discussion of the results of COA comparison. Figure XIX-1, on the following page, depicts the COA approval inputs and outputs.

2. Prepare the COA Decision Briefing

a. Prepare the COA Decision Briefing. This briefing often takes the form of a CDR's Estimate so take good notes of the CDR's comments. This information could include the current status of the joint force; the current JIPOE; and assumptions used in COA development. The CDR selects a COA based upon the staff recommendations and the CDR's personal estimate, experience, and judgment. See COA Decision Briefing Format on following page.

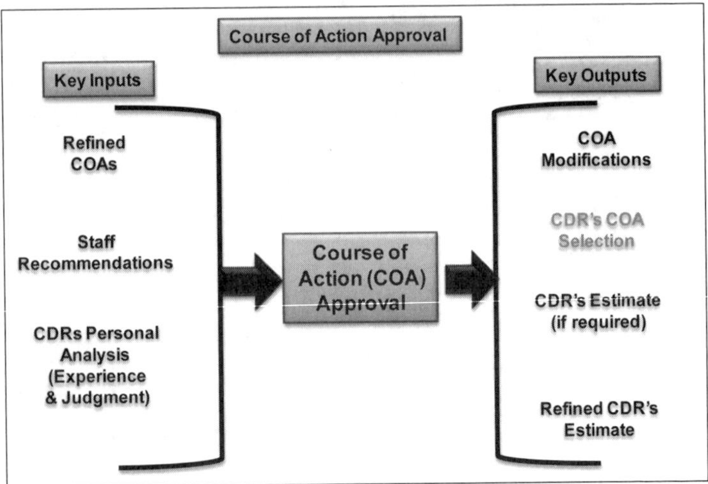

Figure XIX-1. Course of Action Approval.

3. Present the COA Decision Briefing

The staff briefs the CDR on the COA comparison and the analysis and wargaming results, including a review of important supporting information. All principal staff directors and the component CDR's should attend this briefing (physically present or linked by VTC).

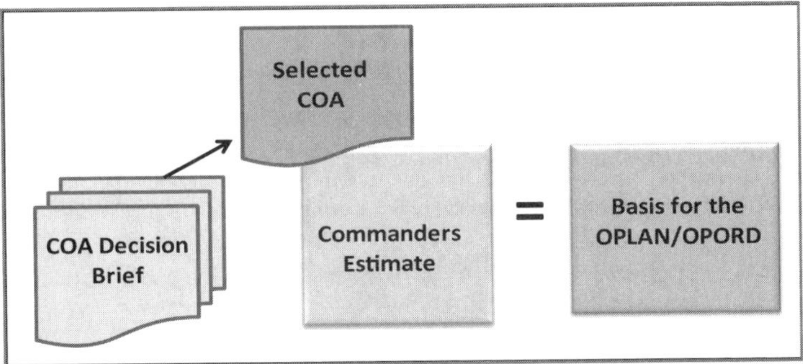

Figure XIX-2. COA Approval Inputs.

4. COA Selection/Modification by CDR

COA approval/selection is the end result of the COA comparison process. It gives the staff a concise statement of how the CDR intends to accomplish the mission, and provides the necessary focus for execution planning and OPLAN/OPORD development.

a. Review Staff Recommendations.

b. Apply Results of own COA Analysis and Comparison.

COA Decision Briefing Format

The JPG should prepare a briefing to provide the following to the CCDR:

a. The purpose of the briefing.

b. Enemy situation.

(1) Strength. A review of enemy forces, both committed/available for reinforcement.

(2) Composition. Order of battle, major weapons systems, and operational characteristics.

(3) Location and disposition. Ground combat and fire support forces, air, naval, missile forces, logistic forces and nodes, command and control (C2) facilities, and other combat power.

(4) Reinforcements. Land; air; naval; missile; nuclear, biological, and chemical (NBC), other advanced weapons systems; capacity for movement of these forces.

(5) Logistics. A summary of the enemy's ability to support combat operations.

(6) Time and space factors. The capability to move to and reinforce initial positions.

(7) Combat efficiency. The state of training, readiness, battle experience, physical condition, morale, leadership, motivation, tactical doctrine, discipline, and significant strengths and weaknesses.

c. Friendly Situation.

d. Mission Statements.

e. CDR's Intent Statement.

f. Operational Concepts and COAs Developed.

(1) Any changes from the mission analysis briefing in the following areas:

(a) Assumptions.

(b) Limitations.

(c) Enemy and friendly centers of gravity.

(d) Phasing of the operation (if phased).

(2) Present COA. As a minimum, discuss:

(a) COA # ___. (Short name, e.g., "Simultaneous Assault")

1 COA statement (brief concept of operations).

2 COA sketch.

3 COA architecture:

a Task organization.

b Command relationships.

c Organization of the operational area.

(b) Major differences between each COA.

(c) Summaries of COAs.

g. COA Analysis.

(1) Review of JPG's wargaming efforts.

(2) Add considerations from own experience.

h. COA Comparisons.

(1) Description of comparison criteria (e.g., governing factors) and comparison methodology.

(2) Weigh strengths/weaknesses with respect to comparison criteria.

i. COA Recommendations.

(1) Staff.

(2) Components.

c. <u>Consider Any Separate Recommendations from Supporting and Subordinate CDRs</u>.

d. <u>Review Guidance from the Higher Headquarters/Strategic Guidance</u>.

e. <u>The CDR May</u>:

(1) Concur with staff/component recommendations, as presented.

(2) Concur with staff/component recommended COAs, but with modifications.

(3) Select a different COA from the staff/component recommendations.

(4) Direct the use of a COA not formally considered.

(5) Defer the decision and consult with selected staff/CDRs prior to making a final decision.

5. Refine Selected COA

Once the CDR selects a COA, the staff will begin the refinement process of that COA into a clear decision statement to be used in the CDR's Estimate. At the same time, the staff will apply a final "acceptability" check.

a. <u>Staff Refines CDR's COA Selection into Clear Decision Statement</u>.

(1) Develop a brief statement that clearly and concisely sets forth the COA selected and provides only whatever information is necessary to develop a plan for the operation (no defined format).

(2) Describe what the force is to do as a whole, and as much of the elements of when, where, and how as may be appropriate.

(3) Express decision in terms of what is to be accomplished, if possible.

(4) Use simple language so the meaning is unmistakable.

(5) Include statement of what is acceptable risk.

b. <u>Apply Final "Acceptability" Check</u>.

(1) Apply experience and an understanding of situation.

(2) Consider factors of acceptable risk versus desired outcome consistent with higher CDR's intent and concept. Determine if gains are worth expenditures.

> ***Note**: The nature of a potential contingency could make it difficult to determine a specific endstate until the crisis actually occurs. In these cases, the CCDR may choose to present two or more valid COAs for approval by higher authority. A single COA can then be approved when the crisis occurs and specific circumstances become clear.*

6. Prepare the CDR's Estimate

a. Once the CCDR has made a decision on a selected COA, provides guidance, and updates his intent, the staff completes the CDR's Estimate. The CDR's Estimate provides a *concise narrative statement* of how the CCDR intends to accomplish the mission, and provides the necessary focus for campaign planning and OPLAN/OPORD development. Further, it responds to the establishing authority's requirement to develop a plan for execution. The CDR's Estimate[1] provides a continuously updated source of information from the perspective of the CCDR. CDRs at various levels use estimates during JOPP to support all

[1] Annex J of JOPES Volume I (CJCSI 3122.01) provides the format for a CDR's Estimate. See also Appendix C of this document.

aspects of COA determination and plan or order development. Outside of formal reporting requirements, a CDR may or may not use a CDR's Estimate as the situation dictates. The CDR's initial intent statement and planning guidance to the staff can provide sufficient information to guide the planning process. Although the CCDR will tailor the content of the CDR's Estimate based on the situation, a typical format for an estimate that a CCDR submits per APEX procedures is shown at Figure XIX-3.

Commander Estimate

Operational Description

- Purpose of the Operation
- References
- Description of Military Operations

Narrative – Five Paragraph Order

- Mission
- Situation and Courses of Action
- Analysis of Opposing Courses of Action (Adversary Capabilities and Intentions)
- Comparison of Friendly Courses of Action
- Recommendation or Decision

Remarks

- Remarks – Cite plan identification number of the file where detailed requirements have been loaded into the Joint Operation Planning and Execution System.

Figure XIX-3. Commander's Estimate.

(1) Precise contents may vary widely, depending on the nature of the crisis, time available to respond, and the applicability of prior planning. In a rapidly developing situation, the formal CDR's Estimate may be initially impractical, and the entire estimate process may be reduced to a CDRs' conference.

(2) In practice, with appropriate horizontal and vertical coordination, the CCDR's COA selection could already have been briefed to and approved by the CJCS and SECDEF. In the current global environment, where major military operations are both politically and strategically significant, even a CCDR's selected COA is normally briefed to and approved by the President or SECDEF. The CDR's Estimate then becomes a matter of formal record keeping and guidance for component and supporting forces.

b. The supported CDR may use simulation and analysis tools in the collaborative environment to assess a variety of options, and may also choose to convene a concept development conference involving representatives of subordinate and supporting commands, the Services, JS, and other interested parties. Review of the resulting CCDR's Estimate (also referred to as the strategic concept) requires maximum collaboration and coordination among all planning participants. The supported CDR may highlight issues for future interagency consultation, review or resolution to be presented to SECDEF during the IPR.[2]

[2] *CJCSI 3141.01, Management and review of Campaign and Contingency Plans*

c. CJCS Estimate Review. The Estimate Review determines whether the scope and concept of planned operations satisfy the tasking and will accomplish the mission; whether the assigned tasks can be accomplished using available resources in the time frames contemplated by the plan; and ensures the plan is proportional and worth the expected costs. Once approved for further planning by the SECDEF during an IPR discussion, the CCDR's Estimate becomes the CONOPS for the plan. A detailed description of the CONOPS will be included in Annex C of the plan.[3]

7. Concept Discussions (IPR)

Transition to concept development is marked by a decision to develop military options. During the concept development step, CCDRs develop, analyze, and compare viable COAs and develop staff estimates that are coordinated with the Military Departments when applicable. Analysis includes war gaming, operational modeling, and initial feasibility assessments. Concept discussions with the Joint Staff and senior leaders are important and will focus largely on issues the CCDR needs resolved with planning to this point.

a. The CCDR's estimate broadly outlines to the SECDEF, senior leaders, planners and subordinates how forces will conduct integrated, joint operations to accomplish the mission. Among other elements and as appropriate, the estimate communicates:

- Review of strategic guidance, assumptions, risk, termination criteria, and mission statement as well as any changes or modifications.
- Recommended COAs and supporting rationale.
- Descriptions and assessments of alternate COAs and the rationale for not recommending them.
- Review of the Operational Environment. Feasible enemy COAs and comparison of enemy and friendly COAs.
- Commander's intent and desired endstate.
- Assessed strategic and operational centers of gravity, as appropriate.
- Estimated level and duration of the operation.
- Nature, purpose, time-phasing, and interrelationship of operations, including specific relationships to strategic communication.
- Branches, sequels, or other options, including warning and response times, that involve scenarios likely to confront the command.
- Logistics feasibility.
- Gross transportation feasibility.
- Interagency and/or multinational involvement accomplished to date and future potential interagency and/or multinational involvement, any ally/partner nation support.
- The concept for sequencing the operation.

b. Transition to plan development is marked by approval of a COA and/or plan concept.

[3] *JOPES Volume I.*

Plan Development: Function III

20

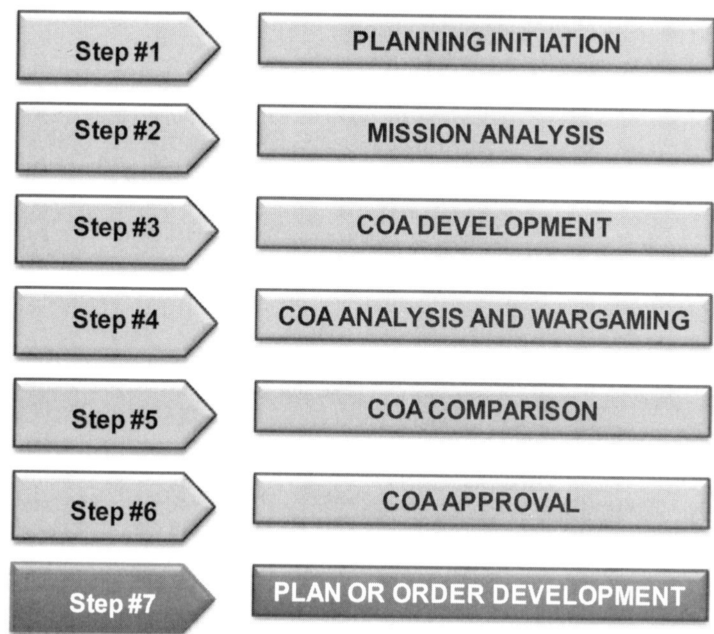

1. Function III — Plan/Order Development

During the Function of Plan or Order Development the CDR and staff, in collaboration with subordinate and supporting components and organizations, expand the approved COA into a detailed OPLAN or OPORD utilizing the CONPLAN, which is the centerpiece of the OPLAN or OPORD.

2. Step 7 to JOPP — Plan or Order Development

After completing Steps 1-6, we now have a document that's been well staffed which will aid us in developing our plan.

3. Format of Military Plans and Orders

Plans and orders can come in many varieties from the very detailed Campaign Plans and Operations Plans to simple verbal orders. They also include OPORDs, Warning Orders (WARNO), Planning Orders (PLANO), Alert Orders (ALERTO), Execute Orders (EXORD), and Fragmentary Orders (FRAGO). The more complex directives will contain much of the amplifying information in appropriate annexes and appendices. However, the directive should always contain the essential information in the main body. The form may depend on the time available, the complexity of the operation, and the levels of command involved. However, in most cases, the directive will be standardized in the five-paragraph format that was introduced back in step one. Following is a brief description of each of these paragraphs.

- *Paragraph 1 – Situation.* The commander's summary of the general situation that ensures subordinates understand the background of the planned operations. Paragraph 1 will often contain sub paragraphs describing the higher Commander's Intent, friendly forces, and enemy forces.
- *Paragraph 2 – Mission.* The commander inserts his restated mission (containing essential tasks) developed during the mission analysis.
- *Paragraph 3 – Execution.* This paragraph contains Commander's Intent, which will enable commanders two levels down to exercise initiative while keeping their actions aligned with the overall purpose of the mission. It also specifies objectives, tasks, and assignments for subordinates (by phase, as applicable—with clear criteria denoting phase completion).
- *Paragraph 4 – Administration and Logistics.* This paragraph describes the concept of support, logistics, personnel, public affairs, civil affairs, and medical services.
- *Paragraph 5 – Command and Control.* This paragraph specifies the command relationships, succession of command, and overall plan for communications.

4. Plan or Order Development

a. For plans and orders developed per APEX, the CJCS, in coordination with the supported and supporting CDRs and other members of the JCS, monitors planning activities, resolves shortfalls when required, and reviews the supported CDR's OPLAN for *adequacy, feasibility, acceptability, completeness, and compliance with policy and guidance.* The supported CDR will have IPR discussions as required (see Chapter VIII) with the SECDEF to confirm the plan's strategic guidance and receive approval of assumptions, the mission statement, the concept, the plan, and any further guidance required for plan refinement. Normally at this time the CJCS and USD(P) will include issues arising from, or resolved during, plan review (e.g., key risks, decision points). The intended result of these discussions is SECDEF approval of the basic plan and required annexes, the resolution of any remaining key issues, and approval to proceed with plan assessment (as applicable) with any amplifying guidance or direction.

If the President or SECDEF decides to execute the plan, all three joint operation planning elements — *situational awareness, planning, and execution* — continue in a complementary and iterative process.

b. The CCDR guides plan development by issuing a PLANORD or similar planning directive to coordinate the activities of the commands and agencies involved.

These planning activities typically will be accomplished in a parallel, collaborative, and iterative fashion rather than sequentially, depending largely on the planning time available. The same flexibility displayed in COA development is seen here again, as planners discover and eliminate shortfalls.

c. APEX provides specific guidance on these activities for organizations required to prepare a plan. However, these are typical types of activities that other organizations also will accomplish as they plan for joint operations. For example, a CCMD which is preparing a crisis-related OPLAN at the President's direction will follow specific procedures and milestones in force planning, TPFDD development, and shortfall identification. If required, a JTF subordinate to the CCDR will support this effort even as the CCDR and staff are planning for their specific mission and tasks.

5. Application of Forces and Capabilities

a. When planning the application of forces and capabilities, the CCDR should not be completely constrained by the strategic plan's force apportionment if additional resources are justifiable and no other COA within the allocation reasonably exists. The additional capability requirements will be coordinated with the JS through the development process. Risk assessments will include results using both allocated capabilities and additional capabilities. Operation planning is inherently an iterative process, with forces being requested and approved for certain early phases, while other forces may be needed or withdrawn for the later phases. This process is particularly complex when planning a campaign because of the potential magnitude of committed forces and length of the commitment. Finally, when making this determination the CCDR should also consider withholding some capability as an *operational reserve*.

b. When developing an OPLAN, the supported CCDR should designate the main effort and supporting efforts as soon as possible. This action is necessary for economy of effort and for allocating disparate forces, to include multinational forces. The main effort is based on the supported CCDR's prioritized objectives. It identifies where the supported CCDR will concentrate capabilities to achieve specific objectives. Designation of the main effort can be addressed in geographical (area) or functional terms. Area tasks and responsibilities focus on a specific area to control or conduct operations. Functional tasks and responsibilities focus on the performance of continuing efforts that involve the forces of two or more Military Departments operating in the same domain — air, land, sea, or space — or where there is a need to accomplish a distinct aspect of the assigned mission. In either case, designating the main effort will establish where or how a major portion of available friendly forces and assets are employed, often to attain the primary objective of a major operation or campaign.

c. Designating a main effort facilitates the synchronized and integrated employment of the joint force while preserving the initiative of subordinate CDRs. After the main effort is identified, joint force and component planners determine those tasks essential to accomplishing objectives. The supported CCDR assigns these tasks to subordinate CDRs along with the capabilities and support necessary to achieve them. As such, the CONOPS must clearly specify the nature of the main effort.

d. The main effort can change during the course of the operation based on numerous factors, including changes in the operational environment and how the adversary reacts to friendly operations. When the main effort changes, support priorities must change to ensure success. Both horizontal and vertical coordination within the joint force and with multinational and interagency partners is essential when shifting the main effort. Secondary efforts are important, but are ancillary to the main effort. They normally are designed to complement or enhance the success of the main effort (for example, by diverting enemy resources). Only necessary secondary efforts, whose potential value offsets or exceeds the resources required, should be undertaken, because these efforts divert resources from the main effort. Secondary efforts normally lack the operational depth of the main effort and have fewer forces and capabilities, smaller reserves, and more limited objectives.[1]

[1] *JP 5-0, Joint Operations Planning.*

6. Plan Development Activities

A number of activities are associated with plan development, as Figure XX-1 shows.

Plan Development Activities	
➢Force planning	➢Feasibility Analysis
➢Support Planning	➢Refinement
➢Nuclear Strike	➢Documentation
➢Deployment Planning	➢Plan Review and Approval
➢Shortfall Identification	➢Supporting Plan Development

Figure XX-1. Plan Development Activities.

a. Force Planning.

(1) The primary purposes of force planning are to (1) influence COA development and selection based on force allocations, availability, and readiness (2) identify all forces needed to accomplish the supported component CDR's CONOPS with some rigor and (3) *effectively phase the forces into the OE*. Force planning consists of determining the force requirements by operation phase, mission, mission priority, mission sequence, and operating area. It includes force allocation review, major force phasing; integration planning; force list structure development (TPFDD); followed by force list development. Force planning is the responsibility of the CCDR, supported by component CDRs in coordination with the GFM enterprise, including OSD, JS, JS Joint Force Coordinator (JS JFC), JFPs, the JFM, Services, Service components, Force Providers (FPs), and other supporting commands. Force planning begins early during CONOPS development and focuses on adaptability. The CDR determines force requirements; develops a letter of instruction or time phasing and force planning; and designs force modules to align and time-phase the forces in accordance with the CONOPS. Major forces and elements are selected from those apportioned or allocated for planning by the CJCS and included in the supported CDR's CONOPS by operation phase, mission and mission priority. Service components then collaboratively make tentative assessments of the specific sustainment capabilities required in accordance with the CONOPS. After the actual forces are identified (sourced), the CCDR refines the force plan to ensure it supports the CONOPS, provides force visibility, and enables flexibility. The CDR identifies and resolves or reports shortfalls with a risk assessment.

(2) In CAP, force planning focuses on the actual units designated to participate in the planned operation and their readiness for deployment. The supported CDR identifies force requirements as operational capabilities in the form of force packages to facilitate sourcing by the Services, JS JFC/JFPs and other FPs' supporting commands. A force package is a list (group of force capabilities) of the various forces (force requirements) that the supported CDR requires to conduct the operation described in the CONOPS. The supported CDR typically describes required force requirements in the form of capability descriptions or unit type codes, depending on the circumstances. The supported CDR submits the required force requests through the JS to the JS JFC/JFPs to develop sourcing recommendations. FPs review the readiness and deployability posture of their available units before recommending which units to allocate to the supported CDR's force requirements. Services and their component commands also determine mobilization requirements and plan for the provision of non-unit sustainment. The supported CDR will review the sourcing recommendations through the GFM process to ensure compatibility with capability requirements and concept of operations.[2] If the sourcing recommendations do not fully support the CONOP, the supported CCDR must provide the operational risk to support the SECDEF's decision.

[2] *JP 5-0, Joint Operations Planning.*

b. Support Planning. The purpose of support planning is to determine the sequence of the personnel, logistic, and other support required to provide distribution; maintenance; civil engineering, medical, and sustainment in accordance with the concept of operation. Support planning is conducted in parallel with other planning, and encompasses such essential factors as executive agent identification; assignment of responsibility for base operating support; airfield operations; management of non-unit replacements; health service support; personnel management; financial management; handling of prisoners of war and detainees; theater civil engineering policy; logistic-related environmental considerations; support of noncombatant evacuation operations and other retrograde operations; and nation assistance. Support planning is primarily the responsibility of the Service component CDRs and begins during CONOPS development. Service component CDRs identify and update support requirements in coordination with the Services, the Defense Logistics Agency, and USTRANSCOM. They initiate the procurement of critical and low-density inventory items; determine host nation support (HNS) availability; develop plans for total asset visibility; and establish phased delivery plans for sustainment in line with the phases and priorities of the CONOPS. They develop and train for battle damage repair; develop reparable retrograde plans; develop container management plans; develop force and line of communications protection plans; develop supporting phased transportation and support plans aligned to the CONOPS and report movement support requirements. Service component CDRs continue to refine their sustainment and transportation requirements as the FPs identify and source force requirements. During distribution planning, the supported CCDR and USTRANSCOM resolve gross distribution feasibility questions impacting inter-theater and intra-theater movement and sustainment delivery. USTRANSCOM and other transportation providers identify air, land, and sea transportation resources to support the approved CONOPS. These resources may include apportioned inter-theater transportation, GCC-controlled theater transportation, and transportation organic to the subordinate commands. USTRANSCOM and other transportation providers develop transportation schedules for movement requirements identified by the supported CDR. A transportation schedule does not necessarily mean that the supported CDR's CONOPS is transportation feasible; rather, the schedules provide the most effective and realistic use of available transportation resources in relation to the phased CONOPS.[3]

(1) *Support refinement* is conducted to confirm the sourcing of logistic requirements in accordance with strategic guidance and to assess the adequacy of resources provided through support planning. This refinement ensures support is phased in accordance with the CONOPS; refines support C2 planning; and integrates support plans across the supporting commands, Service components, and agencies. It ensures an effective but minimum logistics foot-print for each phase of the CONOPS.

(2) *Transportation refinement* simulates the planned movement of resources that require lift support to ensure that the plan is transportation feasible. The supported CDR evaluates and adjusts the concept of operation to achieve end-to-end transportation feasibility if possible, or requests additional resources if the level of risk is unacceptable. Transportation plans must be consistent and deconflicted with plans and timelines required by providers of Service-unique combat and support aircraft to the supported CCDR. Planning also must consider requirements of international law; commonly understood customs and practices; and agreements or arrangements with foreign nations with which the United States requires permission for overflight, access, and diplomatic clearance. If significant changes are made to the CONOPS, it should be assessed for feasibility and refined to ensure it is acceptable.

c. Nuclear Strike. CDRs must assess the military as well as political impact a nuclear strike would have on their operations. Nuclear planning guidance issued at the CCDR level is based on national-level political considerations and is influenced by the military mission.

[3] *JP 3-0, Doctrine for Joint Operation.*

Although USSTRATCOM conducts nuclear planning in coordination with the supported GCC and certain allied CDRs, the supported CDR does not effectively control the decision to use nuclear weapons.

d. Deployment Planning. Deployment planning is conducted on a continuous basis for all approved contingency plans and as required for specific crisis-action plans. In all cases, mission requirements of a specific operation define the scope, duration, and scale of both deployment and redeployment operation planning. Unity of effort is paramount, since both deployment and redeployment operations involve numerous commands, agencies, and functional processes. Because the ability to adapt to unforeseen conditions is essential, supported CCDRs must ensure their deployment plans for each contingency or crisis-action plan support global force visibility requirements.

(1) Operational Environment. For a given plan, deployment planning decisions are based on the anticipated operational environment, which may be permissive, uncertain, or hostile. The anticipated operational environment dictates the type of entry operations, deployment concept, mobility options, predeployment training, and force integration requirements. Normally, supported CCDRs, their subordinate CDRs, and their Service components are responsible for providing detailed situation information; mission statements by operation phase; theater support parameters; strategic and operational lift allocations by phase (for both force movements and sustainment); HNS information and environmental standards; and prepositioned equipment planning guidance.

(2) Deployment Concept. Supported CCDRs must develop a deployment concept and identify specific predeployment standards necessary to meet mission requirements. Supporting CCDRs provide trained and mission-ready forces to the supported CCMD deployment concept and predeployment standard. Services recruit, organize, train, and equip interoperable forces. The Services' predeployment planning and coordination with the supporting CCMD must ensure that predeployment standards specified by the supported CCDR are achieved, supporting personnel and forces arrive in the supported theater fully prepared to perform their mission, and deployment delays caused by duplication of predeployment efforts are eliminated. The Services and supporting CCDRs must ensure unit OPLANs are prepared; forces are tailored and echeloned; personnel and equipment movement plans are complete and accurate; command relationship and integration requirements are identified; mission-essential tasks are rehearsed; mission-specific training is conducted; force protection is planned and resourced; and sustainment requirements are identified. Careful and detailed planning ensures that only required personnel, equipment, and materiel deploy; unit training is exacting; missions are fully understood; deployment changes are minimized during execution; and the flow of personnel, equipment; and movement of materiel into theater aligns with the concept of operation.

(3) Movement Planning. Movement planning integrates the activities and requirements of units with partial or complete self-deployment capability, activities of units that require lift support, and the transportation of sustainment and retrogrades. Movement planning is highly collaborative and is enhanced by coordinated use of simulation and analysis tools.

(a) The supported command is responsible for movement control, including sequence of arrival, and exercises this authority through the CCDR TPFDD validation process. The supported CDR will use the organic lift and non-organic, common-user, strategic lift resources made available for planning by the CJCS. Competing requirements for limited strategic lift resources, support facilities, and intra-theater transportation assets will be assessed in terms of impact on mission accomplishment. If additional resources are required, the supported command will identify the requirements and provide rationale for those requirements. The supported CDR's operational priorities and any movement constraints (e.g., assumptions concerning the potential use of WMD) are used to prepare a movement plan. The plan will consider enroute staging locations to support the scheduled activity. This information, together with an estimate of required site augmentation, will be communicated to appropriate supporting CDR's. The global force manager and USTRANSCOM use the Joint Flow Analysis and Sustainment for Transportation model to assess transportation feasibility and develop recommendations on final port of embarkation selections for those units without organic lift capability. Movement feasibility requires current analysis and assessment of movement C2 structures and systems; available organic, strategic and theater lift assets; transportation infrastructure; and competing demands and restrictions.

(b) After coordinated review of the movement analysis by USTRANSCOM, the supported command, and the global FP may adjust the concept of operation to improve movement feasibility where operational requirements remain satisfied. CDR USTRANSCOM should adjust or reprioritize transportation assets to meet the supported CDR's operational requirements (fort to foxhole). If this is not an option due to requirements from other CDRs, then the supported CDR adjusts TPFDD requirements or is provided additional strategic and theater lift capabilities using (but not limited to) Civil Reserve Air Fleet and/or Voluntary Intermodal Sealift Agreement capabilities as necessary to achieve end-to-end transportation feasibility.

(c) Operational requirements may cause the supported CDR and/or subordinate CDRs to alter their plans, potentially impacting the deployment priorities or TPFDD requirements. Planners must understand and anticipate the impact of change. There is a high potential for a sequential pattern of disruption when changes are made to the TPFDD. A unit displaced by a change might not simply move on the next available lift, but may require reprogramming for movement at a later time. This may not only disrupt the flow, but may also interrupt the operation. Time is also a factor in TPFDD changes. Airlift can respond to short-notice changes, but at a cost in efficiency. Sealift, on the other hand, requires longer lead times, and cannot respond to change in a short period. These plan changes and the resulting modifications to the TPFDDs must be handled during the planning cycles.

(4) Joint Reception, Staging, Onward Movement, and Integration Planning. JRSOI planning is conducted to ensure an integrated joint force arrives and becomes operational in the OA as scheduled. Effective integration of the force into the joint operation is the primary objective of the deployment phase.

(5) TPFDD Letter of Instruction (LOI).[4] The supported CDR publishes supplemental instructions for time phasing force deployment data development in the TPFDD LOI. The LOI provides operation specific guidance for utilizing the processes and systems to provide force visibility and tracking; force mobility; and operational agility through the TPFDD and the validation process. It provides procedures for the deployment, redeployment, and rotations of the operation's forces. The LOI provides instructions on force planning sourcing, reporting, and validation. It defines planning and execution milestones and details movement control procedures and lift allocations to the CDR's components, supporting CDRs, and other members of the JPEC. A TPFDD must ensure force visibility, be tailored to the phases of the concept of operation, and be execution feasible.

(6) Deployment and JRSOI Refinement. Deployment and JRSOI refinement is conducted by the supported command in coordination with JS, USTRANSCOM, the Services, and supporting commands. The purpose of the deployment and JRSOI refinement is to ensure the force deployment plan maintains force mobility throughout any movements, provides for force visibility and tracking at all times, provides for effective force preparation, and fully integrates forces into a joint operation while enabling unity of effort. This refinement conference examines planned missions, the priority of the missions within the operation phases and the forces assigned to those missions. By mission, the refinement conference examines force capabilities, force size, support requirements, mission preparation, force positioning/basing, weapon systems, major equipment, force protection and sustainment requirements. It should assess the feasibility of force closure by the CDR's required delivery date and the feasibility of successful mission execution within the timeframe established by the CDR under the deployment concept. This refinement conference should assess potential success of all force integration requirements. Transition criteria for all phases should be evaluated for force redeployment or rotation requirements.

[4]*CJCSG 3122, Time-Phased Force and Deployment Data (TPFDD) Primer*

(7) For lesser-priority plans that may be executed simultaneously with higher-priority plans or on-going operations, CCMD and USTRANSCOM planners may develop several different deployment scenarios to provide the CCDR with a range of possible transportation conditions under which the plan may have to be executed based on risk to this plan and the other ongoing operations. This will help both the supported and supporting CCDRs identify risk associated with having to execute multiple operations in a transportation-constrained environment.

e. Shortfall Identification. Along with hazard and threat analysis, shortfall identification is performed throughout the plan development process. The supported CDR continuously identifies limiting factors and capabilities shortfalls and associated risks as plan development progresses. Where possible, the supported CDR resolves the shortfalls and required controls and countermeasures through planning adjustments and coordination with supporting and subordinate CDR's. If the shortfalls and necessary controls and countermeasures cannot be reconciled or the resources provided are inadequate to perform the assigned task, the supported CDR reports these limiting factors and assessment of the associated risk to the CJCS. The CJCS and the Service Chiefs consider shortfalls and limiting factors reported by the supported CDR and coordinate resolution. However, the completion of assigned plans is not delayed pending the resolution of shortfalls. If shortfalls cannot be resolved within the JSCP timeframe, the completed plan will include a consolidated summary and impact assessment of unresolved shortfalls and associated risks.

f. Feasibility Analysis. This Step in plan or order development is similar to determining the feasibility of a COA, except that it typically does not involve simulation-based wargaming. The focus in this Step is on ensuring the assigned mission can be accomplished using available resources within the time contemplated by the plan. The results of force planning, support planning, deployment planning, and shortfall identification will affect OPLAN or OPORD feasibility. The primary factors considered are whether the apportioned or allocated resources can be deployed to the joint operations area (JOA) when required, sustained throughout the operation, and employed effectively, or whether the scope of the plan exceeds the apportioned resources and supporting capabilities. Measures to enhance feasibility include adjusting the CONOPS, ensuring sufficiency of resources and capabilities, and maintaining options and reserves.

g. Refinement. During deliberate planning, plan refinement typically is an orderly process that follows plan development and is associated with plan assessment (Figure XX-2). Refinement then continues on a regular basis as circumstances related to the potential contingency change. In CAP, refinement is almost continuous throughout OPLAN or OPORD development. Planners frequently adjust the plan or order based on results of force planning, support planning, deployment planning, shortfall identification, revised JIPOE, and changes to strategic guidance. Refinement continues even after execution begins, with changes typically transmitted in the form of FRAGORDs rather than revised copies of the plan or order.

h. Documentation. When the TPFDD is complete and end-to-end transportation feasibility has been achieved and is acceptable to the CDR, the supported CDR completes the documentation of the final, transportation-feasible OPLAN or OPORD and coordinates distribution of the TPFDD within the JOPES database as appropriate.

i. Plan Review and Approval. When the final OPLAN or OPORD is complete, the supported CDR then submits it with the associated TPFDD file to the CJCS and SECDEF for review, approval, or modification. The JPEC reviews the supported CDR's OPLAN or OPORD and provides the results of the review to the CJCS. The CJCS reviews and recommends approval or disapproval of the OPLAN or OPORD to the SECDEF. After the CJCS's review, the SECDEF or President will review, approve, or modify the plan. The SECDEF may delegate the approval of contingency plans to the CJCS.

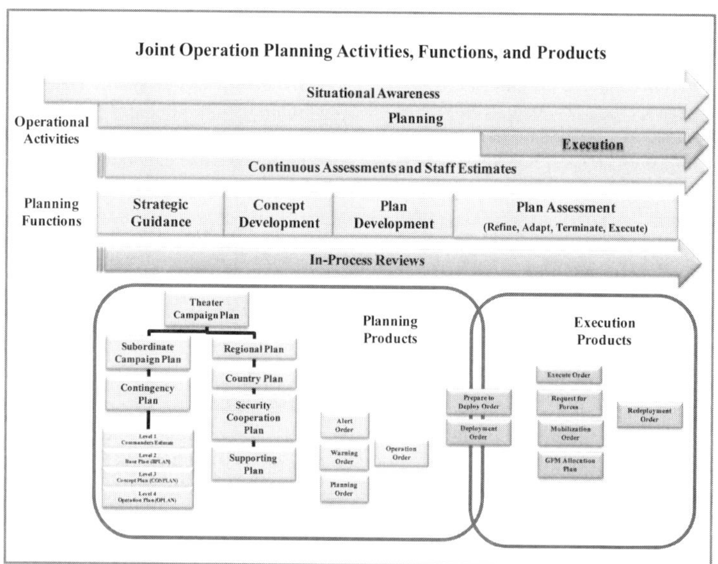

Figure XX-2. Joint Operations Planning Activities, Functions, and Products.

The President is the final approval authority for OPORDs. Plan review criteria are common to deliberate planning and CAP, as shown in Figure XX-3.

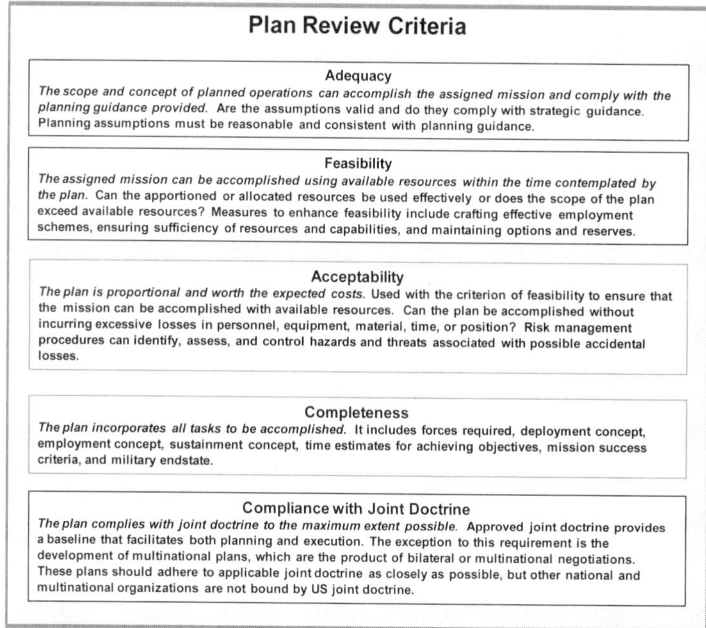

Plan Review Criteria

Adequacy
The scope and concept of planned operations can accomplish the assigned mission and comply with the planning guidance provided. Are the assumptions valid and do they comply with strategic guidance. Planning assumptions must be reasonable and consistent with planning guidance.

Feasibility
The assigned mission can be accomplished using available resources within the time contemplated by the plan. Can the apportioned or allocated resources be used effectively or does the scope of the plan exceed available resources? Measures to enhance feasibility include crafting effective employment schemes, ensuring sufficiency of resources and capabilities, and maintaining options and reserves.

Acceptability
The plan is proportional and worth the expected costs. Used with the criterion of feasibility to ensure that the mission can be accomplished with available resources. Can the plan be accomplished without incurring excessive losses in personnel, equipment, material, time, or position? Risk management procedures can identify, assess, and control hazards and threats associated with possible accidental losses.

Completeness
The plan incorporates all tasks to be accomplished. It includes forces required, deployment concept, employment concept, sustainment concept, time estimates for achieving objectives, mission success criteria, and military endstate.

Compliance with Joint Doctrine
The plan complies with joint doctrine to the maximum extent possible. Approved joint doctrine provides a baseline that facilitates both planning and execution. The exception to this requirement is the development of multinational plans, which are the product of bilateral or multinational negotiations. These plans should adhere to applicable joint doctrine as closely as possible, but other national and multinational organizations are not bound by US joint doctrine.

Figure XX-3. Plan Review Criteria.

j. Supporting Plan Development.

(1) Supporting CDRs prepare plans that encompass their role in the joint operation. Employment planning is normally accomplished by the JFC (CCDR or subordinate JFC) who will direct the forces if the plan is executed. Detailed employment planning may be delayed when the politico-military situation cannot be clearly forecast, or it may be excluded from supporting plans if employment is to be planned and executed within a multinational framework.

(2) The supported CDR normally reviews and approves supporting plans. However, the CJCS may be asked to resolve critical issues that arise during the review of supporting plans, and the JS may coordinate the review of any supporting plans should circumstances so warrant. Deliberate Planning does not conclude when the supported CDR approves the supporting plans. Planning refinement and maintenance continues until the operation terminates or the planning requirement is cancelled or superseded.

7. Transition

Transition is critical to the overall planning process. It is an orderly turnover of a plan or order as it is passed to those tasked with execution of the operation. It provides information, direction and guidance relative to the plan or order that will help to facilitate situational awareness. Additionally, it provides an understanding of the rationale for key decisions necessary to ensure there is a coherent shift from planning to execution. These factors coupled together are intended to maintain the intent of the concept of operations, promote unity of effort and generate tempo. Successful transition ensures that those charged with executing an order have a full understanding of the plan. Regardless of the level of command, such a transition ensures that those who execute the order understand the CDR's intent and concept of operations. Transition may be internal or external in the form of briefs or drills. Internally, transition occurs between future plans and future/current operations. Externally, transition occurs between the CDR and subordinate commands.

a. Transition Brief. At higher levels of command, transition may include a formal transition brief to subordinate or adjacent CDRs and to the staff supervising execution of the order. At lower levels, it might be less formal. The transition brief provides an overview of the mission, CDR's intent, task organization, and enemy and friendly situation. It is given to ensure all actions necessary to implement the order are known and understood by those executing the order. The brief should include items from the order or plan such as: higher headquarters mission (tasks and intent), mission, CDR's intent, CCIRs, task organization, situation (enemy and friendly), concept of operations, execution (including branches and sequels), and planning support tools (synchronization matrix, JIPOE products, etc.). The brief may include items from the order or plan such as:

- Higher headquarters mission (tasks and intent).
- Mission.
- Commander's Intent.
- CCIRs.
- Task organization.
- Situation (friendly and enemy).
- Concept of operations.
- Execution (including branches and potential sequels).
- Planning support tools (such as synchronization matrix, JIPOE products, etc.).

b. Confirmation Brief. A confirmation brief is given by a subordinate CDR after receiving the order or plan. Subordinate CDRs brief the higher CDR on their understanding of CDR's intent, their specific tasks and purpose, and the relationship between their unit's missions and the other units in the operation. The confirmation brief allows the higher CDR to identify potential gaps in the plan, as well as discrepancies with subordinate plans. It also gives the CDR insights into how subordinate CDRs intend to accomplish their missions.

c. Transition Drills. Transition drills increase the situational awareness of subordinate CDRs and the staff and instill confidence and familiarity with the plan. Sand tables, map exercises, rehearsals of concept (ROC) and rehearsals are examples of transition drills.

Plan Assessment 21

1. Function IV – Assessment

Assessment is a continuous activity of the operations process and a primary feedback mechanism that enables the command as a whole to learn and adapt. Effective assessment relies on an accurate understanding of the logic used to build the plan. Plans are based on imperfect understanding, assumptions and an operational approach on how the CDR expects a situation to evolve. The reasons or logic as to why the CDR believes the plan will produce the desired results are important considerations when determining how to assess the operation/plan. Continuous assessment helps CDRs recognize shortcomings in the plan and changes in the operational environment. In those instances when assessment reveals minor variances from the CDR's visualization, CDRs adjust plans as required. In those instances when assessment reveals a significant variance from the CDR's original visualization, CDRs reframe the problem and develop an entirely new plan as required.[1]

a. During plan assessment, the CCDR extends and refines planning while supporting and subordinate CDRs complete their plans for review and approval. The CCDR continues to develop branch plans and other options for the SECDEF and the President as required or directed. When required by the supported CCDR, supporting commands and agencies submit supporting plans within 60 days after SECDEF approval of the base plan. Service component supporting plans should be reviewed by the respective Military Department(s) for an assessment of Service Title 10 requirements. The CCDR and the JS continue to evaluate the situation for any changes that would "trigger" plan refinement, adaptation, termination, or execution (RATE). During the Plan Assessment stage, if the plan requires refinement or adaptation and depending on the amount of change required, the CCDR may direct his planning staff to reenter the planning process at any of the earlier stages.

b. Under APEX, "triggers" will alert the planning community to reassess and revise, if necessary, plans, thereby keeping them in a "living" state. These triggers include, but are not limited to, changes associated with:

(1) Implied or stated plan assumptions.

(2) Force or enemy military structure and/or capabilities.

(3) Readiness levels and availability of forces.

(4) COA timelines/concept/phases.

(5) Strategic guidance.

(6) Intentions (U.S. and enemy).

(7) Alliances.

(8) Key planning factors.

c. During assessment the CCDR will conduct as many IPRs as required with the SECDEF to maintain plans in a living state. Top-priority plans and CPG-directed plans unique to specific CDRs are required to be reviewed every six months at a minimum. Further, these plans require a review if there are significant changes in the following; strategy, risk and/or tolerance of risk, assumptions, U.S. capabilities, enemy and/or adversary intent or capabilities, resources, or alliances.

[1] *FM 5-0 Overview, Combined Arms Doctrine Directorate.*

Plan assessment encompasses four distinct evaluations of a plan; refinement (R), adaptation (A), termination (T), or execution (E).

d. Depending on the nature or significance of triggers, plans may only require refinement. Refining a plan to keep it in a living state does not require an additional IPR. However, an IPR is required if the plan requires a more complex *adaptation*, is recommended for *termination*, or is required for *execution*. The CCDR may conduct as many IPR(s) with the SECDEF during Plan Refinement (Figure XXI-1). These IPR(s) would likely focus on branches and sequels, updated intelligence, and changes in assumptions or the situation that requires major reassessment or significant plan modification. This meeting may include a variety of other pertinent issues (e.g., information operations, Special Access Programs (SAP), sensitive targets, nuclear escalation mitigation, and Annex V considerations).

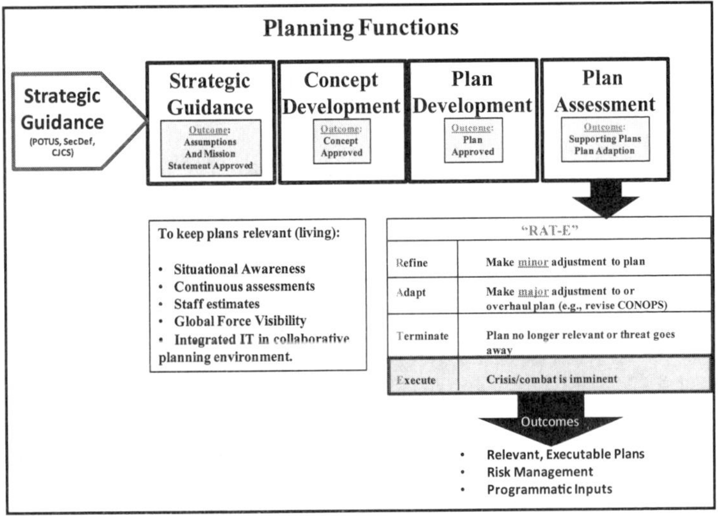

Figure XXI-1. Planning Functions .

2. Plan Assessment Overview

a. Plan assessment is part of planning and the plan review process. Effective plan assessment measures progress toward mission accomplishment (achieving intermediate military objectives (IMOs) as applicable and progress toward end states), identifies changes in the operational and strategic environment, and risk associated with the potential requirement to execute contingency plans. Accordingly, assessment considerations should:

(1) Be developed in concert with mission success criteria;

(2) Help guide operational design of campaign and contingency plans;

(3) Employ common methods that can be developed and applied across all planning and assessment requirements, and be briefed during IPR.

b. CCDRs are tasked to develop campaign plans that integrate security cooperation and other foundational activities with operations and contingency plans IAW the strategic policy guidance provided by the GEF and JSCP. Campaign plans also provide for conducting a comprehensive assessment of how the CCMDs activities are contributing to the achievement of IMOs theater objectives, and how those activities best deter, shape, or mitigate the potential to execute assigned plans. Accordingly, plan assessments should:

(1) Provide the basis for the RATE recommendation during all IPRs.

(2) Ensure that assessment of subordinate campaign and contingency plans nest under the assessment of the CCDR's TCP, as well as the FCP they support. This nesting provides the mechanism to synchronize assessment activities across the CCDR's planning requirements and eliminate redundant or contradictory activities (Figure XXI-2).

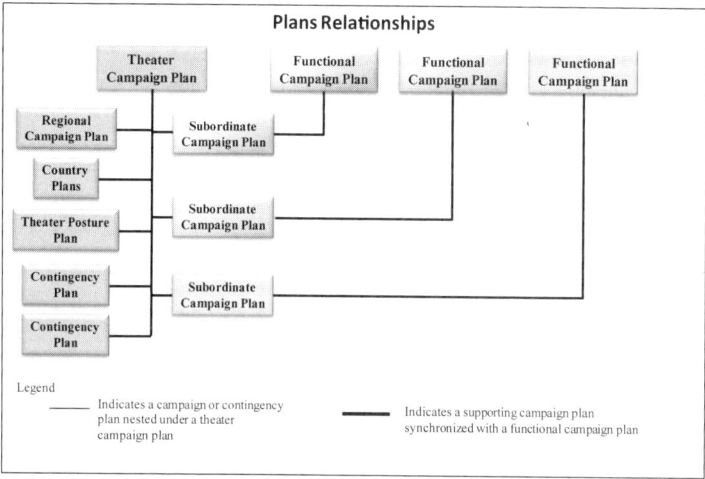

Figure XXI-2. Nesting of Plans in APEX System.

3. Plan Assessment Requirements

For the purposes of the management and review of GEF- and JSCP-tasked plans, plan assessment occurs through three main vehicles:

- Campaign plan assessments;
- Contingency plan assessments;
- CCDR's RATE assessment during IPR.

a. Campaign Plan Assessment.

(1) The CCDR assesses the campaign plan to determine his progress towards mission accomplishment and inform decision-making at the operational and strategic level through existing processes:

- IPRs.
- CCMD readiness reporting.
- GFM.
- Global Defense Posture.
- Program Review.
- Other APEX activities.

(2) CCDRs will annually assess their campaigns. CCDR assessments inform the IPR process, the Comprehensive Joint Assessment (CJA), and assumptions on the operational environment for their contingency plans.

(a) The campaign plan assessment portion of the CJA is formatted to address CCMD progress toward and the risks to accomplishment of theater or functional GEF objectives in relationship to their IMOs.

(b) Campaign plan IPRs includes assessment results and conclusions as detailed in CJCSI 3141.01 series. CCDRs may expand the scope during the IPR to facilitate strategic dialog with SECDEF.

(3) Campaign plan assessment should follow a common approach such as that suggested in JP 5-0 to meet CCDR and other senior leader information requirements.

(4) CCDR key recommendations shall be integrated into CCMD Comprehensive Joint Assessment (CJA) responses. Campaign plan IPRs will include campaign plan recommendations as deemed appropriate by CCDRs. Recommendations are a narrative text and;

(a) Provide a focused statement clearly defining the responsible DOD or IA office, action required, and timing.

(b) Categorize the recommendation within the Doctrine, Organization, Training, Materiel, Leadership and Education, Personnel, Facilities, and Policy framework.

(c) Include a POC at the submitting command to provide more information as required on the recommendation.

(d) Clearly identify key assessment findings that serve as evidence for the recommendation.

(e) Include narrative text identifying which IMO challenges will be overcome through recommendation implementation.

(f) No more than five, in CCDR determined priority order, recommendations will be submitted each assessment cycle.

(5) Joint Staff Campaign Plan Assessment Process, Responsibilities, and Output.

b. Process.

(1) A JS campaign plan assessment team will formally stand up at the start of every FY and meet as necessary until the JSCP directed output report is complete and necessary tasks are assigned to an OPR.

(2) The JS campaign plan assessment process will utilize a collaborative approach to optimize the review of multiple crosscutting strategic assessment inputs to the CJCS. To accomplish the assessment, the JS campaign plan assessment team will collaborate with the CCMDs, Services, and CSAs to collect, analyze, and clarify campaign assessment data. However, the JS campaign plan assessment process is not limited to assessing only conclusions and recommendations provided through any single assessment process, or to only input from the CCMDs, Services, or CSAs.

(3) The JS campaign plan assessment process will utilize existing assessment products and review and analyze results independently to capture and display data to integrate theater perspectives into a global common assessment picture for the CJCS, which will also inform OSD. The conclusions will be utilized to make recommendations regarding plan prioritization, resourcing, and planning requirements.

(4) Review of the JS campaign plan assessment report will require JS planner level coordination. In the event a JS directorate nonconcurs or concurs with a critical comment, the plan assessment team will utilize the directorate plan assessment POC to serve as the entry point for adjudication.

(5) The JS campaign plan assessment process and team are the primary vehicles for JS directorates to execute the JSCP direction for campaign plan assessment.

b. Responsibilities.

(1) The JS J5/JOWPD has overall responsibility for the conduct of each JS campaign plan assessment. JS J5/JOWPD is responsible for the planning, preparation, scheduling, execution, and coordination of each assessment and for the management, administration, and coordination necessary to ensure that the process outlined is thoroughly and efficiently executed.

(2) IAW the JSCP, JS directorates will participate on the JS campaign plan assessment team. JS directorates and OCJCS/Legal Counsel (LC) provide subject matter expertise and the JS perspective in their respective functional areas on issues raised during the

assessment. JS directorates shall provide an action officer to participate as a member of the campaign plan assessment team when formally requested by JS J5. POCs should immediately contact JS J5/JOWPD to exchange contact information.

(3) Directorate action officers assigned to the plan assessment team shall participate in all phases of the assessment and assist in deriving findings and developing feasible and actionable recommendations to correct issues.

c. Output. IAW the JSCP, the JS plan assessment team shall produce an assessment report of their findings.

4. Contingency Plan Assessments

a. Contingency plan assessments can be conducted on all JSCP tasked plans that have an associated TPFDD. Contingency plan assessments inform the Joint Combat Capabilities Assessment (JCCA) process. JCCA plan assessments measure the Department's ability to successfully execute contingency plans with the highest visibility or having the most severe consequences, as well as those most stressing to ground, maritime, air, and special operations forces. The Joint Combat Capability Assessment Group (JCCAG), through the JS J5, is responsible for proposing the plan assessment schedule and having it approved by the Global Force Management Board (GFMB). To inform the GFM process, the JCCAG, through the JS J5, will brief the insights from the latest contingency plan assessment during the GFMB. Plans may also be assessed in tandem with other related or supporting plans, or plans that if executed simultaneously would stress the force and pose a risk to the execution of the plan or plans in question. Force flow and associated timelines are a key metric for assessment. The expectation is selected plans will be assessed with fidelity and timeliness to allow flexibility for emerging assessment requirements due to a changing security environment without posing significant negative impact on the sourcing throughput of the JFPs.

(1) Contingency sourcing solutions and force flow timelines and feasibility data provided by the JFPs and USTRANSCOM are key components of contingency plan assessments. To focus on the sufficiency and executability of the most strategically important contingency plans, the plans assessment process typically follows the contingency sourcing and force flow analysis done by the supported CCMD as part of force planning to capitalize on the work already accomplished during development of contingency plans.

(2) To accomplish the assessment, JS J5/JOWPD collaborates with the supported CCMD, JFPs, USTRANSCOM, CSAs, and JS directorates to collect and analyze plan assessment data to finalize a briefing product through a series of Secure Video-teleconferences (SVTCs). JS directorates analyze their functional areas during the assessment.

b. The output of contingency plan assessment is an indicator of the general ability to execute a plan or group of plans. It is supported by an analysis of the impact of force sourcing and logistical shortfalls, readiness deficiencies, transportation feasibility, and military and strategic risk.

c. JCCA plan assessments are provided to the JCS Tank. All contingency plan assessments inform the Chairman's Annual Risk Assessment. The results of the plan assessments and the CCDR's own internal assessments should be used to inform the dialog during IPRs and provide support to the CCDR's RATE recommendation.

5. Plan Review Assessment

a. During Plan review Assessment, the CCDR extends and refines planning while considering branches and sequels, updated intelligence, and changes in assumptions or the situation that require major reassessment or significant plan modification. CCDRs will consider a number of factors as they assess and revise their plans. These key planning factors include, but are not limited to, changes associated with the following:

- Results of campaign plan and other relevant CCDR assessments that influence the RATE recommendation.
- Key planning factors (Strategic guidance, OE, facts, assumptions, and limitations; non-DOD USG department or agency or whole-of-government plans).
- Force or enemy military capabilities and intent, to include adversary non-state actors.
- USG, global, regional partner, and key supporting partner capabilities.
- TPFDD analysis versus force and transportation availability comparison.
- Readiness levels and availability of forces.
- COA timelines.
- Objectives (U.S. and enemy).
- Alliances.
- Input and feedback from socialization sessions with other USG departments and agencies.
- CCMD assessment and its impact on the direction of future planning in order to make a recommendation to RATE the plan.

(1) The execution of campaign plan tasks will be monitored and measured for progress toward achieving each IMO/endstate. It will incorporate new data as it is obtained to reframe the discussion regarding branches or sequels, or modifications.

(2) The RATE recommendation for TCPs and FCPs should focus on necessary plan revisions and or adaption's, since the plans are in execution and will not be terminated. Recommendations regarding RATE decisions of TCP and FCP branch plans may also be made based on plan assessments.

(3) Should a branch or sequel be considered for near-term implementation, sourcing, current readiness, and logistics impacts to the plan should be discussed as well as likely COAs.

(4) Ultimately, a recommendation should be made to the SecDef regarding the direction of future planning based on the totality of the CCDR's assessments. The way forward should include mitigation solutions if necessary to support the CCDR's continued planning or execution.

6. Interagency Collaboration

As part of Plan Assessment, and with approval of the SECDEF, the CCDR may present the plan's Annex V (Interagency Coordination) to the OSD/JS Annex V working group (JS/OSD working group will prep plans for IA collaboration) for transmittal to the NSC and/or the HSC, if the plan impacts U.S. territory, with the intent of obtaining managed interagency staffing and plan development. In advance of authorization for formal transmittal of Annex V to the NSC, the CCDR may request interagency consultation on approved Annex V elements by the JS/OSD working group. During this step, the CCDR may present his plan for multinational involvement.[2]

[2] *Adaptive Planning Roadmap II.*

Assessment Fundamentals 22

This chapter provides the fundamentals of assessment, including its definition, purpose, and process. It discusses how assessment works with the levels of war and offers considerations for effective assessment. This chapter also covers assessment working groups and assessment support with operations research/systems analysis.

1. Assessment Fundamentals

a. Assessment is the continuous monitoring and evaluation of the current situation and progress of the joint operation.[1] Commanders, assisted by their staffs and subordinate commanders, continuously assess the operational environment and the progress of the operation. Based on their assessment, commanders direct adjustments thus ensuring the operation remains focused on accomplishing the mission.

b. Assessment involves deliberately comparing forecasted outcomes with actual events to determine the overall effectiveness of force employment. More specifically, assessment helps the commander determine progress toward attaining the desired end state, achieving objectives, and performing tasks. It also involves continuously monitoring and evaluating the operational environment to determine what changes might affect the conduct of operations. Assessment helps commanders determine if they need to reframe the problem and develop an entirely new plan.

c. Throughout planning, preparation and execution commanders integrate their own assessments with those of the staff, subordinate commanders, and other partners in the AOR. Primary tools for assessing progress of the operation include the operation order, the common operational picture, personal observations, running estimates, and the assessment plan. The latter includes measures of effectiveness, measures of performance, and reframing criteria. The commander's visualization forms the basis for the commander's personal assessment of progress. Running estimates provide information, conclusions, and recommendations from the perspective of each staff section. They help to refine the common operational picture and supplement it with information not readily displayed.

d. The assessment process is continuous and directly tied to the commander's decisions throughout planning, preparation, and execution of operations. Staffs help the commander by monitoring the numerous aspects that can influence the outcome of operations and providing the commander timely information needed for decisions. The CCIR process is linked to the assessment process by the commander's need for timely information and recommendations to make decisions. The assessment process helps staffs by identifying key aspects of the operation that the commander is interested in closely monitoring and where the commander wants to make decisions. Examples of commander's critical decisions include when to transition to another phase of a campaign, what the priority of effort should be, or how to adjust command relationships between component commanders.

e. The assessment process begins during mission analysis when the commander and staff consider what to measure and how to measure it to determine progress toward accomplishing a task, creating an effect, or achieving an objective. During planning and preparation for an operation, for example, the staff assesses the joint force's ability to execute the plan based on available resources and changing conditions in the operational environment. However, the discussion in this section focuses on assessment for the purpose of determining the progress of the joint force toward mission accomplishment.

[1] *FM 3-0, Operations*

f. CDR's and their staffs determine relevant assessment actions and measures during planning. They consider assessment measures as early as mission analysis and include assessment measures and related guidance in commander and staff estimates. They use assessment considerations to help guide operational design because these considerations can affect the sequence and type of actions along LOOs/lines of effort. During execution, they continually monitor progress toward accomplishing tasks, creating effects, and achieving objectives. Assessment actions and measures help commanders adjust operations and resources as required, determine when to execute branches and sequels, and make other critical decisions to ensure current and future operations remain aligned with the mission and end state.

(1) Normally, the joint force J-3 or J-5, assisted by the J-2, is responsible for coordinating assessment activities. For subordinate commanders' staffs, this may be accomplished by equivalent elements within joint functional and/or Service components. The chief of staff facilitates the assessment process and determination of CCIRs by incorporating them into the headquarters' battle rhythm. Various elements of the CDR's staff use assessment results to adjust both current operations and future planning.

(2) Friendly, adversary, and neutral diplomatic, informational, and economic actions applied in the operational environment can impact military actions and objectives. When relevant to the mission, the commander also must anticipate using assessment to evaluate the results of these actions. This typically requires collaboration with other agencies and multinational partners—preferably within a common, accepted process—in the interest of unified action. For example, failure to coordinate overflight and access agreements with foreign governments in advance or to adhere to international law regarding sovereignty of foreign airspace could result in mission delay, failure to meet US objectives, and/or an international incident. Many of these organizations may be outside the CDR's authority. Accordingly, the CDR should grant some joint force organizations authority for direct coordination with key outside organizations—such as interagency elements from DOS or the Department of Homeland Security, national intelligence agencies, intelligence sources in other nations, and other CCMDs—to the extent necessary to ensure timely and accurate assessments.

2. Assessment Process

a. Assessment is continuous; it precedes and guides every operations process activity and concludes each operation or phase of an operation. The three activities that make up the assessment process are also continuous; they are logically sequential while constantly executed throughout the operations process. This process applies to assessments of every type and at every echelon. Broadly, assessment consists of the following activities:

- Monitoring the current situation to collect relevant information.
- Evaluating progress toward attaining end state conditions, achieving objectives, and performing tasks.
- Recommending or directing action for improvement.

(1) Monitoring.

(a) Monitoring is continuous observation of those conditions relevant to the current operation. Monitoring within the assessment process allows staffs to collect relevant information, specifically that information about the current situation that can be compared to the forecasted situation described in the commander's intent and concept of operations. Progress cannot be judged, nor effective decisions made, without an accurate understanding of the current situation.

(b) During planning, commanders monitor the situation to develop facts and assumptions that underlie the plan. During preparation and execution, commanders and staffs monitor the situation to determine if the facts are still relevant, if their assumptions remain valid, and if new conditions emerged that affect the operations.

Assessment Levels and Measures

Ref: JP 5-0, Joint Operation Planning (Dec '06), pp. D-6 to D-8.

Assessment occurs at all levels and across the entire range of military operations. Even in operations that do not include combat, assessment of progress is just as important and can be more complex than traditional combat assessment. As a general rule, the level at which a specific operation, task, or action is directed should be the level at which such activity is assessed. To do this, JFCs and their staffs consider assessment ways, means, and measures during planning, preparation, and execution.

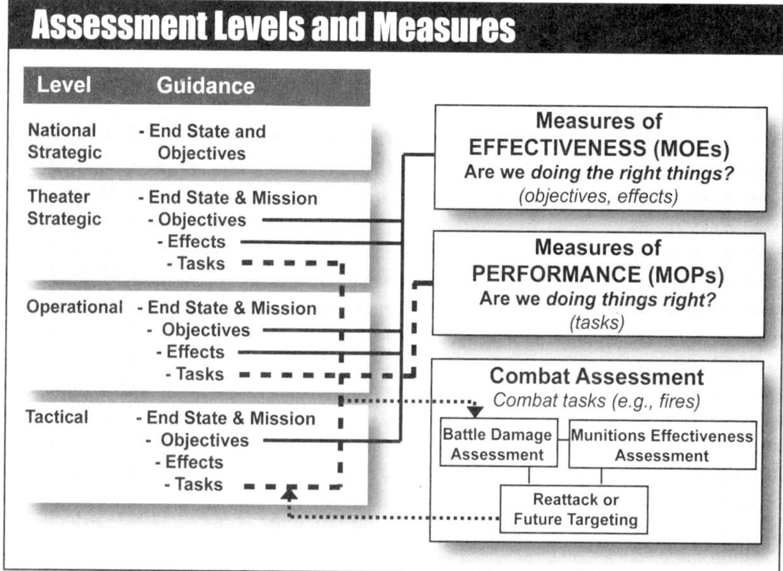

Ref: JP 5-0, Joint Operation Planning, fig. D-1, p. D-7.

Assessment at the Operational and Strategic Levels

Assessment at the operational and strategic levels typically is broader than at the tactical level (e.g., combat assessment) and uses MOEs that support strategic and operational mission accomplishment. Strategic- and operational-level assessment efforts concentrate on broader tasks, effects, objectives, and progress toward the end state.

Tactical-Level Assessment

Tactical-level assessment typically uses MOPs to evaluate task accomplishment. The results of tactical tasks are often physical in nature, but also can reflect the impact on specific functions and systems. Tactical-level assessment may include assessing progress by phase lines; neutralization of enemy forces; control of key terrain or resources; and security, relief, or reconstruction tasks.

Combat Assessment

Combat assessment is an example of a tactical-level assessment and is a term that can encompass many tactical-level assessment actions. Combat assessment typically focuses on determining the results of weapons engagement (with both lethal and nonlethal capabilities), and thus is an important component of joint fires and the joint targeting process. Combat assessment is composed of three related elements: **battle damage assessment (BDA), munitions effectiveness assessment (MEA),** and **future targeting** or **reattack recommendations.**

(c) Commander's critical information requirements and decision points focus the staff's monitoring activities and prioritize the unit's collection efforts. Information requirements concerning the enemy, terrain and weather, and civil considerations are identified and assigned priorities through intelligence, surveillance, and reconnaissance (ISR) synchronization. Operations officers use friendly reports to coordinate other assessment-related information requirements. To prevent duplicated collection efforts, information requirements associated with assessing the operation are integrated into both the ISR plan and friendly force information requirements.

(d) Staffs monitor and collect information from the common operational picture and friendly reports. This information includes operational and intelligence summaries from subordinate, higher, and adjacent headquarters and communications and reports from liaison teams. The staff also identifies information sources outside military channels and monitors their reports. These other channels might include products from civilian, host-nation, and other government agencies. Staffs apply information management and knowledge management principles to facilitate getting this information to the right people at the right time.

(e) Staff sections record relevant information in running estimates. Each staff section maintains a continuous assessment of current operations as a basis to determine if they are proceeding according to the commander's intent, mission, and concept of operations. In their running estimates, staff sections use this new information, updated facts, and assumptions as the basis for evaluation.

(2) Evaluating.

(a) The staff analyzes relevant information collected through monitoring to evaluate the operation's progress. Evaluating is using criteria to judge progress toward desired conditions and determining why the current degree of progress exists. Evaluation is the heart of the assessment process where most of the analysis occurs. Evaluation helps commanders determine what is working, determine what is not working, and gain insights into how to better accomplish the mission.

(b) Criteria in the forms of measures of effectiveness (MOEs) and measures of performance (MOPs) aid in determining progress toward attaining end state conditions, achieving objectives, and performing tasks. MOEs help determine if a task is achieving its intended results. MOPs help determine if a task is completed properly. MOEs and MOPs are simply criteria—they do not represent the assessment itself. MOEs and MOPs require relevant information in the form of indicators for evaluation.

(c) A measure of effectiveness is a criterion used to assess changes in system behavior, capability, or operational environment that is tied to measuring the attainment of an end state, achievement of an objective, or creation of an effect (JP 3-0/JP 5-0). MOEs help measure changes in conditions, both positive and negative. MOEs help to answer the question "Are we doing the right things?" MOEs are used at the strategic, operational, and tactical levels to assess the impact of military operations and measure changes in the operational environment, changes in system behavior, or changes to adversary capabilities. MOEs are based on observable or collectable indicators. Several indicators may make up an MOE, just like several MOEs may assist in assessing progress toward the achievement of an objective or moving toward a potential crisis or branch plan execution. Indicators provide evidence that a certain condition exists or certain results have or have not been attained, and enable decision makers to direct changes to ongoing operations to ensure the mission remains focused on the end state. MOE assessment is implicit in the continuous nature of the JIPOE process. Upon the collection of indicators, JIPOE analysts can compare the baseline intelligence estimate used to inform the plan against the current situation to measure changes. MOEs are commonly found and tracked in formal assessment plans.

Examples of MOEs for the objective provide a safe and secure environment may include:

- Increase/decrease in insurgent activity.
- Increase/decrease in reporting of insurgent activity to host-nation security forces.

MOE	MOP	Indicator
Answers the question: Are we doing the right things?	Answers the question: Are we doing things right?	Answers the question: What is the status of this MOE or MOP?
Measures purpose accomplishment	Measures task completion	Measures raw data inputs to inform MOEs and MOPs
Measures why in the mission statement	Measures what in the mission Statement	Information used to make measuring what or why possible
No hierarchical relationship to MOPs	No hierarchical relationship to MOEs	Subordinate to MOEs and MOPs
Often formally tracked in formal assessment plans	Often formally tracked in execution Matrixes	Often formally tracked in formal assessment plans
Typically challenging to choose the correct ones	Typically simple to choose the correct ones	Typically as challenging to select correctly as the supporting MOE or MOP

Figure XXII-1. Additional Information Concerning MOEs, MOPs, and Indicators

- Increase/decrease in civilian injuries involving mines and unexploded ordnance.
- Attitude/opinion/behavioral changes in selected populations.
- Changes in media portrayal of events.

(d) A measure of performance is a criterion used to assess friendly actions that is tied to measuring task accomplishment. MOPs help answer questions such as "Was the action taken?" or "Were the tasks completed to standard?" A MOP confirms or denies that a task has been properly performed. MOPs are commonly found and tracked at all levels in execution matrixes. MOPs are also heavily used to evaluate training. MOPs help to answer the question "Are we doing things right?"

(e) In general, operations consist of a series of collective tasks sequenced in time, space, and purpose to accomplish missions. The current operations cells use MOPs in execution matrixes and running estimates to track completed tasks. Evaluating task accomplishment using MOPs is relatively straightforward and often results in a yes or no answer. Examples of MOPs include:

- Route X cleared.
- Generators delivered, are operational, and secured at Villages A, B, and C.
- $15,000 spent for schoolhouse completion.

(f) In the context of assessment, an indicator is an item of information that provides insight into a measure of effectiveness or measure of performance. Staffs use indicators to shape their collection effort as part of ISR synchronization. Indicators take the form of reports from subordinates, surveys and polls, and information requirements. Indicators help to answer the question "What is the current status of this MOE or MOP?" A single indicator can inform multiple MOPs and MOEs. Examples of indicators for the MOE "Decrease in insurgent activity" are:

- Number of hostile actions per area each week.
- Number of munitions caches found per area each week.

g. Evaluation includes analysis of why progress is or is not being made according to the plan. Commanders and staffs propose and consider possible causes. In particular, the question of whether changes in the situation can be attributed to friendly actions is addressed. Commanders and staffs consult subject matter experts, both internal and external to the staff, on whether staffs have identified the correct underlying causes for specific changes in the situation. Assumptions identified in the planning process are challenged to determine if they are still valid.[2]

h. A key aspect of evaluation is determining variances—the difference between the actual situation and what the plan forecasted the situation would be at the time or event. Based on the significance of the variances, the staff makes recommendations to the commander on how to adjust operations to accomplish the mission more effectively.

i. Evaluating includes considering whether the desired conditions have changed, are no longer achievable, or are not achievable through the current operational approach. This is done by continually challenging the key assumptions made when framing the problem. When an assumption is invalidated, then reframing may be in order.

(3) Recommending Or Directing Action.

(a) Monitoring and evaluating are critical activities; however, assessment is incomplete without recommending or directing action. Assessment may diagnose problems, but unless it results in recommended adjustments, its use to the commander is limited.

(b) Based on the evaluation of progress, the staff brainstorms possible improvements to the plan and makes preliminary judgments about the relative merit of those changes. Staff members identify those changes possessing sufficient merit and provide them as recommendations to the commander or make adjustments within their delegated authority. Recommendations to the commander range from continuing the operation as planned, to executing a branch, or to making adjustments not anticipated. Making adjustments includes assigning new tasks to subordinates, reprioritizing support, adjusting the ISR synchronization plan, and significantly modifying the COA. Commanders integrate recommendations from the staff, subordinate commanders, and other partners with their personal assessment. From those recommendations, they decide if and how to modify the operation to better accomplish the mission. (See chapter 5 for a detailed discussion of decisions during execution.)

(c) Assessment diagnoses threats, suggests improvements to effectiveness, and reveals opportunities. The staff presents the results and conclusions of its assessments and recommendations to the commander as an operation develops. Just as the staff devotes time to analysis and evaluation, so too must it make timely, complete, and actionable recommendations. The chief of staff or executive officer ensures the staff completes its analyses and recommendations in time to affect the operation and for information to reach the commander when it is needed.

(d) When developing recommendations, the staff draws from many sources and considers its recommendations within the larger context of the operations. While several ways to improve a particular aspect of the operation might exist, some recommendations could impact other aspects of the operation. As with all recommendations, the staff should address any future implications.

3. Assessment and the Levels Of War

a. Assessment occurs at all levels of war and across the entire range of military operations. Even in operations that do not include combat, assessment of progress is just as important and can be more complex than traditional combat assessment. The situation and level dictate the focus and methods leaders use to assess. Normally, commanders assess those specific operations or tasks that they were directed to accomplish. This properly

[2] *JP 3-0, Joint Operations*

focuses collection and assessment at each level, reduces redundancy, and enhances the efficiency of the overall assessment process.

b. Assessment at the operational and strategic levels typically is broader than at the tactical level (e.g., combat assessment) and uses MOEs that support strategic and operational mission accomplishment. Strategic- and operational-level assessment efforts concentrate on broader tasks, effects, objectives, and progress toward the end state. Continuous assessment helps the JFC and joint force component commanders determine if the joint force is doing the right things to achieve objectives, not just doing things right. The JFC also can use MOEs to determine progress toward success in those operations for which tactical-level combat assessment ways, means, and measures do not apply.

c. Tactical-level assessment typically uses MOPs to evaluate task accomplishment. The results of tactical tasks are often physical in nature, but also can reflect the impact on specific functions and systems. Tactical-level assessment may include assessing progress by phase lines; neutralization of enemy forces; control of key terrain or resources; and security, relief, or reconstruction tasks. Assessment of results at the tactical level helps commanders determine operational and strategic progress, so JFCs must have a comprehensive, integrated assessment plan that links assessment activities and measures at all levels.

d. Combat assessment is an example of a tactical-level assessment and is a term that can encompass many tactical-level assessment actions. Combat assessment typically focuses on determining the results of weapons engagement (with both lethal and nonlethal capabilities), and thus is an important component of joint fires and the joint targeting process (see JP 3-60, Joint Targeting). Combat assessment is composed of three related elements: battle damage assessment (BDA), munitions effectiveness assessment (MEA), and future targeting or reattack recommendations. However, combat assessment methodology also can be applied by joint force functional and Service components to other tactical tasks not associated with joint fires (e.g., disaster relief delivery assessment, relief effectiveness assessment, and future relief recommendations).

4. Considerations for Effective Assessment

The following considerations help commanders and staffs develop assessment plans and conduct effective assessments:

- Assessment is continuous.
- Commanders drive assessment through prioritization.
- Assessment incorporates the logic behind the plan.
- Assessment facilitates learning and adapting.
- Commanders and staffs use caution when establishing cause and effect.
- Commanders and staffs combine quantitative and qualitative indicators.
- Assessment incorporates formal and informal methods.

a. Assessment is Continuous.

(1) Assessment is a continuous activity of the operations process. The focus of assessment, however, changes for each activity. During planning, assessment focuses on understanding current conditions of the operational environment and developing an assessment plan, including what and how to assess progress. Understanding the commander's intent and desired future conditions is key when building the assessment plan. During preparation, assessment focuses on determining the friendly force's readiness to execute the operation and on verifying the assumptions on which the plan is based. During execution, assessment focuses on evaluating progress of the operation. Based on their assess-

ment, commanders direct adjustments to the order, ensuring the operation stays focused on accomplishing the mission. They adjust their assessment plan as required.

(2) Assessment is continuous, even when the unit is not actively engaged in operations. At a minimum, staffs maintain running estimates of friendly force capabilities and readiness within their areas of expertise. Some running estimates, such as the intelligence estimate, also assess operational environments to which the unit is likely to deploy.

b. Commanders Drive Assessment Through Prioritization.

(1) The commander's role is central to the assessment process. Commanders establish priorities for assessment and discipline the staff to meet the requirements of time, simplicity, and level of detail based on the situation. While the staff does the detailed work, to include collecting and analyzing information, commanders ultimately make the assessment. Commanders are also responsible for decisions made based on their assessments.

(2) Assessment represents the commander's assessment of progress. During assessments, commanders consider information and recommendations by the staff, subordinate commanders, and other partners within and outside of their AOR. Commanders then apply their judgment to assess progress. As commanders monitor the situation, they compare the current situation to their initial commander's visualization and commander's intent. Based on their assessment of progress, commanders direct adjustments to the order—ensuring focus throughout planning, preparation, and execution of operations and on the end state—or reframe the problem and develop an entirely new plan.

(3) To assist commanders learning throughout the conduct of operations, they establish their commander's critical information requirements, set priorities for assessment in the form of MOEs and reframing criteria, and explicitly state assumptions. When results fail to meet expectations, commanders decide whether this is due to a failure in implementing the plan (execution) or if the plan and its underlying logic are flawed.

c. Assessment Incorporates the Logic Behind the Plan.

(1) Effective assessment relies on an accurate understanding of the logic used to build the plan. Each plan is built on assumptions and an operational approach—a broad conceptualization of the general actions that will produce the conditions that define the desire end state. The reasons or logic as to why the commander believes the plan will produce the desired results are important considerations when determining how to assess the operations. Recording and understanding this logic helps the staffs recommend the appropriate MOPs, MOEs and indicators for assessing the operation. It also helps the commander and staff to determine if they need to reframe the problem if assumptions prove false or the logic behind the plan appears flawed as operations progress.

(2) When conducting design, the logic used to drive more detailed planning is captured in the design concept that includes the mission narrative and problem statement. As planning continues, staff sections identify and record the logic behind the plan relating to their area of expertise in their running estimates. They also record assumptions and include key assumptions as part of the operation plan or order. An explicit record of this logic used in building the plan proves valuable to the commander and staff as well as to follow-on units and other civilian and military organizations in understanding the plan and assessing the progress of operations.

d. Assessment Facilitates Learning and Adapting.

(1) One of the most important questions when assessing the operation is whether the plan is still relevant. Assessment entails measuring progress according to the plan. It also includes periodically reexamining the logic and assumptions of the original plan to determine if the plan is still relevant. Throughout an operation, higher operational objectives may change and conditions may develop that did not exist during planning. These conditions may create a somewhat different situation from the one the commander originally visual-

ized. When this occurs, modifications to the plan may be in order, or it may be necessary to reframe the problem.

(2) The assessment process prompts the decision to reframe in several ways. Commanders and staffs continuously challenge the key assumptions in the plan. When an assumption is invalidated, reframing may be in order. Another sign of a requirement to reframe is when task completion measured by MOPs is high but purpose accomplishment measured by MOEs is low. That suggests that the wrong tasks have been assigned and reframing is needed.

(3) As commanders assess and learn throughout the operation, they determine if achieving their original objectives leads to the desired end state. Collaboration and dialog with higher, lower, and adjacent commanders and staffs, backed up by quantitative and qualitative assessments, contribute to this learning. Assessing helps commanders to update their commander's visualization (which may include a revised end state), direct changes to the order, and adapt the force to better accomplish the mission.

e. Commanders and Staffs Use Caution When Establishing Cause and Effect.

(1) Establishing cause and effect is sometimes difficult, but it is crucial to effective assessment. Sometimes, establishing causality between actions and their effects can be relatively straightforward, such as in observing a bomb destroy a bridge. In other instances, especially regarding changes in human behavior, attitudes, and perception, establishing links between cause and effect proves difficult. Commanders and staffs must guard against drawing erroneous conclusions in these instances.

(2) Understanding how cause and effect works requires careful consideration and shrewd judgment. Even when two variables seem to be correlated, commanders must still make assumptions to establish which one is cause and which one is effect. In fact, both may be caused by a third unnoticed variable. Commanders clearly acknowledge all assumptions made in establishing causes and effects. The payoff for correctly identifying the links between causes and effects is effective and smart recommendations. Commanders and staffs are well-advised to devote the time, effort, and energy needed to properly uncover connections between causes and effects. Assumptions made in establishing cause and effect must be recorded explicitly and challenged periodically to ensure they are still valid.

(3) In its simplest form, an effect is a result, outcome, or consequence of an action. Direct effects are the immediate, first-order consequences of a military action unaltered by intervening events. They are usually immediate and easily recognizable. For example, an enemy command and control center destroyed by friendly artillery or a terrorist network courier captured by a direct-action mission. Establishing the link between cause and effect in the physical domains is usually straightforward, as is assessing progress.

(4) It is often difficult to establish a link or correlation that clearly identifies actions that produce effects beyond the physical domains. The relationship between action taken (cause) and nonphysical effects may be coincidental. Then the occurrence of an effect is either purely accidental or perhaps caused by the correlation of two or more actions executed to achieve the effect. For example, friendly forces can successfully engage enemy formations with fire and maneuver at the same time as psychological operations. The psychological operations might urge enemy soldiers to surrender. If both these events occur at the same time, then correlating an increase in surrendering soldiers to psychological operations will be difficult. As another example, friendly forces may attempt to decrease population support for an insurgency in a particular city. To accomplish this task, the unit facilitates the reconstruction of the city's power grid, assists the local authorities in establishing a terrorist tips hotline, establishes a civil-military operations center, and conducts lethal operations against high-payoff targets within the insurgency. Identifying the relative impact of each of these activities is extremely challenging but is critical for allocating resources smartly to accomplish the mission. Unrecognized influences completely invisible to assessors can also cause changes unforeseen or attributed inaccurately to actions of the force.

f. Commanders and Staffs Combine Quantitative and Qualitative Indicators.

(1) Effective assessment incorporates both quantitative (observation based) and qualitative (opinion based) indicators. Human judgment is integral to assessment. A key aspect of any assessment is the degree to which it relies upon human judgment and the degree to which it relies upon direct observation and mathematical rigor. Rigor offsets the inevitable bias, while human judgment focuses rigor and processes on intangibles that are often key to success. The appropriate balance depends on the situation—particularly the nature of the operation and available resources for assessment—but rarely lies at the ends of the scale.

(2) A balanced judgment for any assessment identifies the information on which to concentrate. Amassing statistics is easy. Determining which actions imply success proves far more difficult due to dynamic interactions among friendly forces, adaptable enemies, populations, and other aspects of the operational environment such as economics and culture. This is especially true of operations that require assessing the actions intended to change human behavior, such as deception or stability operations. Using quantitative and qualitative indicators reduces the likelihood and impact of the skewed perspective that results from an overreliance on either expert opinion or direct observation.

(a) Quantitative. In the context of assessment, a quantitative indicator is an observation-based (objective) item of information that provides insight into a MOE or MOP. Little human judgment is involved in collecting a quantitative indicator. Someone observes an event and counts it. For example, that individual tally the monthly gallons of diesel provided to host-nation security forces by a unit or the monthly number of tips provided to a tips hotline. Then the commander or staff collects that number.

1 Some human judgment is inevitably a factor even when dealing with quantitative indicators. Choosing which quantitative indicators to collect requires significant human judgment prior to collection. During collection the choice of sources, methods, and standards for observing and reporting the events require judgment. After collection, the commander or staff decides whether to use the number as an indicator in a formal assessment plan and for which MOEs or MOPs.

2 Quantitative indicators prove less biased than qualitative indicators. In general, numbers based on observations are impartial (assuming that the events in question were observed and reported accurately). Often, however, these indicators are less readily available than qualitative indicators and more difficult to select correctly. This is because the judgment aspect of which indicators validly inform the MOE is already factored into qualitative indicators to a degree. Experts factor in all considerations they believe are relevant to answering questions. However, this does not occur inherently with quantitative indicators. The information in quantitative indicators is less refined and requires greater judgment to handle appropriately than information in qualitative indicators.

3 Public opinion polling can be easily miscategorized. It often provides an important source of information in prolonged stability operations. Results of a rigorously collected and statistically valid public opinion poll are quantitative, not qualitative. Polls take a mathematically rigorous approach to answering the question of what people really think; they do not offer opinions on whether the people are correct.

4 While the results of scientifically conducted polls are quantitative, human judgment is involved in designing a poll. Decisions must be made on what questions to ask, how to word the questions, how to translate the questions, how to select the sample, how to choose interviewers, what training to give interviewers, and what mathematical techniques to use for getting a sample of the population.

(b) Qualitative. In the context of assessment, a qualitative indicator is an opinion-based (subjective) item of information that provides insight into a MOE or MOP. A high degree of human judgment is involved when collecting qualitative indicators. Qualitative indicators are themselves opinions, not just observed opinions of others such as polls. For example, the division commander estimates the effectiveness of the host-nation forces on a scale of 1 to 5. Sources of qualitative indicators include subject matter experts' opinions and judgments as well as subordinate commanders' summaries of the situation.

1 Qualitative indicators can account for real-world complexities that cannot be feasibly measured using quantitative indicators. Qualitative indicators are also more readily available; commanders often have access to staff principals and other subject matter experts from whom to garner opinions. In some cases, the only available indicator for a particular MOE or MOP is an expert opinion. For example, determining changes in the size and number of enemy sanctuaries may prove impossible without asking local commanders. Without large amounts of objective data, subjective indicators can be used to give a relatively informed picture. However, subjective measures have a higher risk of bias. Human opinion is capable of spectacular insight but also vulnerable to hidden assumptions that may prove false.

2 Differentiating between quantitative and qualitative indicators is useful but signifies a major tendency rather than a sharp distinction in practice. Quantitative indicators often require a degree of judgment in their collection. For example, determining the number of mortar attacks in a given area over a given period requires judgment in categorizing attacks as mortar attacks. A different delivery system could have been used, or an improvised explosive device could have been mistaken for a mortar attack. The attack could also have landed on a boundary, requiring a decision on whether to count it. Similarly, qualitative indicators always have some basis in observed and counted events. The same indicator may be quantitative or qualitative depending on the collection mechanism. For example, the indicator may measure a change in market activity for village X. If a Soldier observes and tracks the number of exchanges, then the indicator is quantitative. If the battalion commander answers that question in a mandated monthly report based on a gut feel, then the indicator is qualitative.

(5) Furthermore, because commanders synchronize actions across the warfighting functions to achieve an objective or obtain an end state condition, the cumulative effect of these actions may make the impact of any individual task indistinguishable. Careful consideration and judgment are required, particularly when asserting cause-and-effect relationships in stability operations.

g. Assessment Incorporates Formal and Informal Methods.

(1) Assessment may be formal or informal; the appropriate level of formality depends entirely on the situation. As part of their planning guidance, commanders address the level of detail they desire for assessing an upcoming operation. In protracted stability operations, commanders may desire a formal assessment plan, an assessment working group, and standard reports. Subordinate units use these tools to assess local or provincial governance, economics, essential services, or the state of security. In fast-paced offensive or defensive operations or in an austere theater of operations, a formal assessment may prove impractical. To assess progress in those cases, commanders rely more on reports and assessments from subordinate commanders, the common operational picture, operation updates, assessment briefings from the staff, and their personal observations. The principles in this chapter apply to formal and informal assessment methods.

(2) A common informal assessment method is the after action review (AAR). Leaders use the AAR to assess unit performance in training and throughout an operation. Leaders at all echelons conduct AARs to generate candid, professional unit evaluations that include specific recommendations for improving unit performance.

(3) Collecting, assembling, and analyzing information takes time and resources. Commanders balance time and resources for assessment just as they do for planning, preparation, and execution. To help achieve this balance, commanders and staffs ask the following questions:

- What will be assessed and to what detail?
- How will a particular task, objective, end state condition, or assumption be assessed? What MOEs and MOPs will be used?

- What information requirements (indicators) are needed to support a particular assessment?
- Who on the staff has primary responsibility for assessing a particular area? What is the collection plan?

(4) Commanders must be careful, however, not to over assess. Staffs can easily get bogged down in developing formal assessment procedures for numerous tasks and objectives. Additional numerous reports, questions, and information requirements from higher headquarters can smother subordinate commanders and their staffs. Often standard reports, operational and intelligence summaries, and updates by subordinate commanders suffice. Higher echelons should never ask for something that the lower echelon does not need for its own purposes. The chief of staff or executive officer helps the commander achieve the right balance between formal and informal assessments.

h. The assessment process and related measures should be relevant, measurable, responsive, and resourced so there is no false impression of accomplishment. Quantitative measures can be helpful in this regard.

(1) Relevant. MOPs and MOEs should be relevant to the task, effect, operation, operational environment, end state, and commander's decisions. This criterion helps avoid collecting and analyzing information that is of no value to a specific operation. It also helps ensure efficiency by eliminating redundant efforts.

(2) Measurable. Assessment measures should have qualitative or quantitative standards they can be measured against. To effectively measure change, a baseline measurement should be established prior to execution to facilitate accurate assessment throughout the operation. Both MOPs and MOEs can be quantitative or qualitative in nature, but meaningful quantitative measures are preferred because they are less susceptible to subjective interpretation.

(3) Responsive. Assessment processes should detect situation changes quickly enough to enable effective response by the staff and timely decisions by the commander. The CDRs results within the operational environment and develop indicators that can respond accordingly. Many actions directed by the CDR require time to implement and may take even longer to produce a measurable result.

(4) Resourced. To be effective, assessment must be adequately resourced. Staffs should ensure that resource requirements for data collection efforts and analysis are built into plans and monitored. Effective assessment can help avoid both duplication of tasks and unnecessary actions, which in turn can help preserve combat power.

i. Commanders and staffs derive relevant assessment measures during the planning process and reevaluate them continuously throughout preparation and execution. They consider assessment measures during mission analysis, refine these measures in the commanders planning guidance and in commander's and staff's estimates, wargame the measures during COA development, and include MOEs and MOPs in the approved plan or order. An integrated data collection management plan is critical to the success of the assessment process and should encompass all available tactical, theater, and national intelligence sources.

j. Just as tactical tasks relate to operational- and strategic-level tasks, effects, and objectives, there is a relationship between assessment measures. By monitoring available information and using MOEs and MOPs as assessment tools during planning, preparation, and execution, commanders and staffs determine progress toward creating desired effects, achieving objectives, and attaining the military end state, and modify the plan as required. Well-devised MOPs and MOEs, supported by effective information management, help the commanders and staffs understand the linkage between specific tasks, the desired effects, and the commander's objectives and end state.

Execution 23

1. Execution

> *"In the military, as in any organization, giving the order might be the easiest part. Execution is the real game."*
>
> *LTG Russel Honore*

Planning and preparation accomplish nothing if not executed effectively. Execution is putting a plan into action by applying forces and capabilities to accomplish the mission and using situational awareness to assess progress and make execution and adjustment decisions.[1]

a. In execution the CCDR, staffs, components and supporting commanders focus their efforts on translating decisions into actions. During execution, the situation may change rapidly and operations the commander envisioned in the plan may bear little resemblance to actual events in execution.[2]

b. Execution begins when the President decides to use a military option to resolve a crisis. The EXORD directs the supported CDR to initiate military operations, defines the time to initiate operations, and conveys guidance not provided earlier. It will be issued by the Chairman to direct execution of an OPORD or other military operation to implement a Presidential or SecDef decision. The EXORD will be issued by authority and direction of the Secretary of Defense upon decision by the President to execute a military operation. Under the full CAP procedures, an EXORD would normally result from a Presidential decision, following execution planning initiated by a Planning or Alert Order. In a particularly time-sensitive situation requiring an immediate response, an EXORD may be issued without prior formal crisis planning, as would normally take place in Situation Awareness and Planning components of CAP.

c. When prior execution planning has been accomplished through adaptation of an existing plan or the development of an emergency OPORD, most of the guidance necessary for execution will already have been passed to the implementing commands, either through an existing plan or by a previously issued WARNORD, PLANORD, Alert Order, PTDO, DEPORD, or Redeployment Order. Under these circumstances, the EXORD need only contain the authority to execute the planned operation and any additional essential guidance, such as the date and time for execution. Reference to previous planning documents is sufficient for additional guidance.

d. In the no-prior-warning response situation where a crisis event or incident requires an immediate response without any prior formal planning, the EXORD must pass all essential guidance that would normally be issued in the WARNORD, PLANORD, and Alert Order. Under such rapid reaction conditions, the EXORD will generally follow the same paragraph headings as the Planning or Alert Order. If some information may be desirable but is not readily available, it can be provided in a subsequent message because the EXORD will normally be very time-sensitive.

[1] *FM 3-0, Operations*

[2] *ADRP 5-0, The Operations Process*

2. TPFDD and Execution

As discussed is Chapter 7 - GFM, the supported CCDR is responsible for the plan including the TPFDD. The TPFDD is a collection of data and information in a database that is built and refined incrementally and modified or updated throughout the operation's planning and execution phases. The TPFDD captures the who, what, when, where, and how for forces and equipment required based on the operational sequence required to support a joint operation. Supporting Services, CCDRs and Agencies support the development and data entry into the TPFDD. The TPFDD development process is collaborative, requiring support, cooperation and information sharing among widely dispersed members of the JPEC. The sourcing decisions resulting from the GFM processes are documented in the TPFDD. The TPFDD represents Annex A of an OPLAN or an OPORD.

a. Execution. The supported CCDR identifies a specific TPFDD for both deployment and redeployment to support the operation. The supported CCDR is responsible for deployment, employment, sustainment, rotational, and redeployment operations planned and executed during joint force missions in their AORs.

b. Force Definition. The supported CCDR, in coordination with his supported component commanders and supporting JFCs, determines the capabilities, quantity, and time phasing of forces required to accomplish assigned tasks. The supported CCDR's components translate specified and implied tasks to types of forces required to meet mission requirements.

c. Requirement Identification and Execution Sourcing. During execution sourcing, the supported CCDR can execution source assigned forces, but must request forces through the GFM allocation process if additional forces are required. The supported CCDR must request the force by transmitting a RFF (emergent force requirement discussed in Chapter 7). The GFM allocation process results in a decision by the SECDEF to direct a force provider (FP) to allocate a force to the supported CCDR to meet a specific requirement. The decision is communicated by the Chairman in the GFM Allocation Plan (GFMAP). The ordered FP is then responsible for sourcing the requirement and tailoring TPFDD personnel and cargo to accomplish specific mission requirements. Execution sourcing results in identifying, or sourcing, the actual unit(s) that will deploy to meet requirements.

d. TPFDD Sourcing. The CCDR TPFDD validation is the authorization to deploy/redeploy the force. Validation is the trigger for the execution sequence for force movements. Validation is the execution procedure used during both deployment and redeployment by the supported CCDR to confirm to the JPEC that all validated TPFDD records contain no fatal transportation errors and accurately reflect that current status. When a TPFDD is in execution and forces are being execution sourced, FPs identify units and tailor the TPFDD. This execution sourcing will replace previously entered information in the TPFDD that was assumed in planning.

> *"A good plan violently executed now is better than a perfect plan executed next week." Gen. George S. Patton*

3. Planning During Execution

Planning continues during execution, with an initial emphasis on refining the existing plan and producing the OPORD. As the operation progresses, planning generally occurs in three distinct but overlapping timeframes: future plans, future operations, and current operations as Figure XXIII-1 depicts.

Planning During Execution

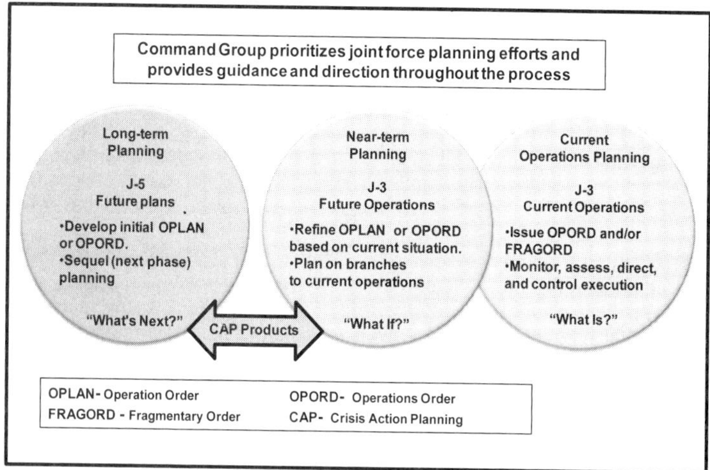
Command Group prioritizes joint force planning efforts and provides guidance and direction throughout the process
Long-term Planning
J-5
Future plans
•Develop initial OPLAN or OPORD.
•Sequel (next phase) planning
"What's Next?"
CAP Products
Near-term Planning
J-3
Future Operations
•Refine OPLAN or OPORD based on current situation.
•Plan on branches to current operations
"What If?"
Current Operations Planning
J-3
Current Operations
•Issue OPORD and/or FRAGORD
•Monitor, assess, direct, and control execution
"What Is?"
OPLAN- Operation Order
FRAGORD - Fragmentary Order
OPORD- Operations Order
CAP- Crisis Action Planning

Figure XXIII-1. Planning During Execution (JP 5-0)

a. The joint force J-5's effort focuses on future plans. The timeframe of focus for this effort varies according to the level of command, type of operation, CDR's desires, and other factors. Typically the emphasis of the future plans effort is on planning the next phase of operations or sequels to the current operation. In a campaign, this could be planning the next major operation (the next phase of the campaign).

b. Planning also occurs for branches to current operations (future operations planning). The timeframe of focus for future operations planning varies according to the factors listed for future plans, but the period typically is more near-term than the future plans timeframe. Future planning could occur in the J-5 or JPG, while future operations planning could occur in the joint operations center (JOC) or J-3.

c. Finally, current operations planning addresses the immediate or very near-term planning issues associated with ongoing operations. This occurs in the JOC or J-3.

d. During execution of a plan, accomplishment of the plan's tasks are monitored and measured for how successfully each objective was completed, along with the input of new data and information as it is obtained to allow selection of branches or sequels, if applicable, or the plan to be modified as necessary. During execution, planning will continue for future operations within the plan to include branches and sequels. The planning functions discussed in previous chapters will continue to be the basis for planning, although planners may reenter the planning process at any of the earlier functions.

> *"The most successful people are those who are good at plan B."*
> *James Yorke*

4. Plan Adjustment

Events that offer better ways to success are opportunities. Commanders recognize opportunities by continuously monitoring and evaluating the situation.[3]

a. Uncertainty and risk are inherent in all military operations. Recognizing and acting on opportunity means taking risks. Reasonably estimating and intentionally accepting risk is not gambling. Carefully determining the risks, analyzing and minimizing as many hazards as possible, and executing a supervised plan that accounts for those hazards contributes to successfully applying military force. Gambling, in contrast, is imprudently staking the success of an entire action on a single, improbable event (i.e., assuming away the threat).[4]

b. When commanders embrace opportunity, they accept risk. It is counterproductive to wait for perfect preparation and synchronization. The time taken to fully synchronize forces and warfighting functions in a detailed order could mean a lost opportunity. It is far better to quickly summarize the essentials, get things moving, and send the details later. Leaders optimize the use of time with warning orders, fragmentary orders, and verbal updates.

c. Assessment in execution identifies variances, their magnitude and significance, and the need for decisions and what type—whether execution or adjustment. The commander and staff assess the probable outcome of the operation to determine whether changes are necessary to accomplish the mission, take advantage of opportunities, or react to unexpected threats.

d. During execution, assessing allows the commander and staff to determine the existence and significance of variances from the operations as envisioned in the initial plan. The staff makes recommendations to the commander about what action to take concerning variances they identified.

5. A commander's visualization of the situation allows subordinate and supporting commanders—and in some cases higher headquarters—to adjust their actions rapidly and effectively. Throughout execution, the Chairman through his staff monitors movements, deployment and employment of forces, assesses achievement of tasks, and resolves shortfalls as necessary. The Chairman monitors and directs action needed to ensure successful completion of military operations. Execution continues until the operation is terminated or the mission is accomplished or revised.

> *"Determine the thing that can and shall be done, and then we shall find the way."*
> *Abraham Lincoln*

[3] *FM 3-0, Operations*

[4] *FM 3-0, Operations*

Summary 24

> *"Long range planning does not deal with future decisions, but with the future of present decisions."* *Peter F. Drucker*

The dawn of the 21st Century presents multiple, diverse and difficult strategic challenges for the United States. A national level of effort involving extensive interagency (and in most cases, alliance and coalition partners) integration and coordination is required not only to win the current fight, but to ensure a government capable of "defending the people" and our vital national interests abroad utilizing a "whole of government" approach to achieve enduring ends.

> *"Separate ground, sea, and air warfare is gone forever. If ever again we should be involved in war, we will fight in all elements, with all services, as one single concentrated effort. Peacetime preparatory and organizational activity must conform to this fact."*
>
> *President Dwight D. Eisenhower-Special Message to the Congress on Reorganization of the Defense Establishment, 3 April 1958*

Planning for Planners has outlined an approach to planning utilizing the construct of the Joint Operation Planning Process (JOPP). As this document has presented, the JOPP is key to making logical, sequential and learned decisions. It is a standardized planning process that is conceptually easy to understand and capable of being applied in contingency or crisis environments; however, it is only a guide. CDRs today operate in an environment of potentially overwhelming information. Our job as planners is to present that information in a logical flow to the CDR, using JOPP as a tool to assist us. Remember, the plan you're working on is probably not the CDR's only plan, nor concern. The CDR's time is valuable and your job is to get the point across in a relevant, well thought-out manner. Give the CDR solutions, not problems, or if you are lacking solutions, give the CDR options - well massaged, learned and vetted.

In today's global operating environment, we, as joint planners, must adjust our view of the adversary and the environment writ large. As the figure below visualizes, today's adversaries are dynamic, adaptive foes who operate within a complex, interconnected OE that we need to fully understand and penetrate. A logical way of gaining that increased understanding is to break the OE into its major parts, examine these parts individually, and then study the relationships and interaction between them to comprehend not only what is occurring, but why, and then plan to it, realizing that as the environment evolves, so will your plan!

> *"First comes thought; then organization of that thought, into ideas and plans; then transformation of those plans into reality. The beginning, as you will observe, is in your imagination."*
>
> *Napoleon Hill*

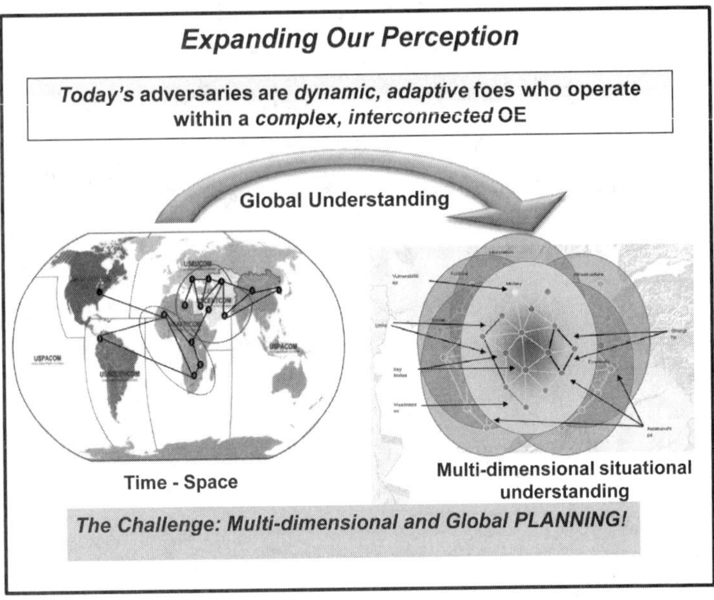

Expanding Our Perception
Today's adversaries are dynamic, adaptive foes who operate within a complex, interconnected OE
Global Understanding
USPACOM
Time - Space
Multi-dimensional situational understanding
The Challenge: Multi-dimensional and Global PLANNING!

General (ret) Luck in his paper titled "*Insights on Joint Operations: The Art and Science,*" says the challenge for us then is how to understand and visualize this new adversary so that we can effectively defend our national interests. The traditional military-centric single center of gravity [focus] approach that worked so well in the Cold War doesn't allow us to accurately analyze, describe, and visualize today's emerging networked, adaptable, asymmetric adversary. This adversary has no single identifiable 'source of all power.' Rather, because of globalization, the information revolution, and, in some cases, the non-state characteristic of our adversary, this form of adversary can only be described (and holistically attacked) as a system of systems. This environment requires astute multi-dimensional and adaptive planners. Our resources are limited, but our adversaries are not. Plan well, plan often, don't be married to your plan and never stop asking the most important questions in planning; "**WHAT IF**," and as General Zinni, USMC is often quoted as saying, "**THEN WHAT**?"

> *"And for the support of this declaration, with a firm reliance on the protection of Divine Providence, we mutually pledge to each other our Lives, our Fortunes, and our sacred Honor."*
>
> *Final lines of the Declaration of Independence, July 4, 1776*

References

Primary Sources

Adaptive Planning Roadmap I and II.

Budget of the U.S. Government.

CJCS Guide 3130, *Adaptive Planning and Execution (APEX) Overview and Policy Framework.*

CJCS Guide 3122, *Time-Phased Force and Deployment Data (TPFDD) Primer.*

CJCSI 1301.01, *Individual Augmentation Procedures.*

CJCSI 3110.01, *Joint Strategic Capabilities Plan.*

CJCSI 3141.01, *Management and Review of Campaign and Contingency Plans.*

CJCSI 3401.01, *Joint Combat Capability Assessment.*

CJCSI 3500.02, *Universal Joint Task List (UJTL) Policy and Guidance for the Armed Forces of the United States.*

CJCSI 4120.02, Change 1, *Assignment of Movement and Mobility Priority.*

CJCSI 5715.01, *Joint Staff Participation in Interagency Affairs.*

CJCSI 8501.01, *Chairman of the Joint Chiefs of Staff, Combatant Commanders, Chief, National Guard Bureau, and Joint Staff Participation in the Planning, Programming, and Budgeting System.*

CJCSM 3122.01, *Joint Operation Planning and Execution System (JOPES) Volume I.*

CJCSM 3122.03, *Joint Operation Planning and Execution System (JOPES) Volume II.*

CJCSM 3122.02, *Joint Operation Planning and Execution System (JOPES) Volume III.*

CJCSM 3130.00, *Crisis Action TPFDD Development and Deployment Execution.*

CJCSM 3130.06, *Global Force Management (GFM) Allocation Policies and Procedures.*

CJCSM 3500.04, *Universal Joint Task List.*

CJCSM 3500.05, *Joint Task Force Headquarters Master Training Guide.*

DODD, 1322.18, *Military Training.*

DODD 1400.31, *DOD Civilian Work Force Contingency and Emergency Planning and Execution.*

DODD 1404.10, *DOD Civilian Expeditionary Workforce.*

DODD 3000.05, *Military Support for Stability, Security, Transition and Reconstruction (SSTR).*

DODD 3000.06, *Combat Support Agencies.*

DODD 3020.42, *Defense Continuity Plan Development.*

DODD 3025.18, *Defense Support to Civil Authorities.*

DODD 5100.1, *Functions of the Department of Defense and its Major Components.*

DODD 5100.3, *Support of the Headquarters of Combatant and Subordinate Joint Commands.*

DODD 5143.01, *Under Secretary of Defense for Intelligence (USD(I)).*

DODI 3000.05, *Stability Operations.*

DODI 1100.22, *Policy and Procedures for Determining Workforce Mix.*

DODI 3020.41 *Contractor Personnel Authorized to Accompany the US Armed Forces.*

DODI 3025.21, *Defense Support of Civilian Law Enforcement Agencies*

DODI 5122.05, *Assistant Secretary of Defense for Public Affairs (ASD(PA)).*

DODI 8260.03, *The Global Force Management (GFM) Data Initiative (DI).*

Guidance for Development of the Force.

Guidance for Employment of the Force.

Global Force Management, *Business Rules.*

Global Force Management, *Initial Capabilities Document.*

Global Force Management Implementation Guidance.

Joint Doctrine Note 7/06, *Incorporating and Extending the UK Military Effects-Based Approach.*

Joint Force 2020: *Capstone Concept for Joint Operations (CCJO).*

Joint Operating Concept (JOC), *Deterrence Operations.*

Joint Operating Concept (JOC), *Irregular Warfare (IW).*

Joint Operating Concept (JOC), *Major Combat Operations.*

Joint Operating Concept (JOC), *Military Support to Stabilization, Security, Transition, and Reconstruction Operations.*

Joint Operating Concept (JOC), *Military Support to Shaping Operations.*

Joint Warfighting Center Pamphlet, *Design in Military Operations.*

Joint Pub 0-2, *Unified Action Armed Forces (UNAAF).*

Joint Pub 1, *Doctrine for the Armed Forces of the United States.*

Joint Pub 1-02, *Department of Defense Dictionary of Military and Associated Terms.*

Joint Pub 2-01, *Joint and National Intelligence Support to Military Operations.*

Joint Pub 2-01.3, *Intelligence Preparation of the Operational Environment.*

Joint Pub 2-03, *Geospatial Intelligence Support to Joint Operations.*

Joint Pub 3-0, *Joint Operations.*

Joint Pub 3-05, *Special Operations.*

Joint Pub 3-07, *Stability Operations.*

Joint Pub 3-07.3, *Peace Operations.*

Joint Pub 3-08, *Interorganizational Coordination during Joint Operations.*

Joint Pub 3-10, *Joint Security Operations in Theater.*

Joint Pub 3-13, *Information Operations.*

Joint Pub 3-13.2, *Military Information Support Operations.*

Joint Pub 3-16, *Multinational Operations.*

Joint Pub 3-17, *Air Mobility Operations.*

Joint Pub 3-22, *Foreign Internal Defense.*

Joint Pub 3-24, *Counterinsurgency Operations.*

Joint Pub 3-27, *Homeland Defense.*

Joint Pub 3-28, *Defense Support for Civil Authorities.*

Joint Pub 3-29, *Foreign Humanitarian Assistance.*

Joint Pub 3-30, *Command and Control for Joint Air Operations.*

Joint Pub 3-31, *Command and Control for Joint Land Operations.*

Joint Pub 3-32, *Command and Control for Joint Maritime Operations.*

Joint Pub 3-33, *Joint Task Force Headquarters.*

Joint Pub 3-35, *Deployment and Redeployment Operations.*

Joint Pub 3-57, *Civil-Military Operations,.*

Joint Pub 3-61, *Public Affairs.*

Joint Pub 4-0, *Joint Logistics.*

Joint Pub 4-01, *Joint Doctrine for the Defense Transportation System.*

Joint Pub 4-05, *Joint Mobilization Planning.*

Joint Pub 5-0, *Joint Operation Planning.*

Joint Pub 5-00.1, *Joint Doctrine for Campaign Planning. (superseded)*

Joint Pub 6-0, *Joint Communications System.*

Joint Doctrine Publication (JDP) 01, United Kingdom, *Campaigning.*

Joint Doctrine Publication (JDP) 3-07, United Kingdom, *Battlespace Management.*

Joint Doctrine Publication (JDP) 3-00, United Kingdom, *Campaign Execution.*

Joint Doctrine Publication 5-00 (JDP-5-00) United Kingdom, *Campaign Planning.*

Joint Warfare Publication (JWP) 3-00, United Kingdom, *Joint Operations Execution.*

National Framework for Strategic Communication, (Vice President of the United States Report)

National Defense Strategy of the United States of America.

National Response Framework.

National Strategy for Homeland Security.

The National Strategy to Secure Cyberspace.

National Military Strategy of the United States.

National Military Strategic Plan for the War on Terrorism.

National Security Strategy of the United States.

Deputy Secretary of Defense, *Quadrennial Defense Review (QDR) Report. Quadrennial Defense Review Building Partnership Capacity (BPC) Execution Roadmap.*

Quadrennial Defense Review, DOD.

Quadrennial Diplomacy and Development Review, DOS.

Implementation of the Adaptive Planning (AP) Roadmap II.

The Goldwater-Nichols Department of Defense Reorganization Act of 1986 (10 USC 161 et. seq. PL 99-433).

Title 10, US Code, as amended.

Unified Command Plan.

U.S. Department of State, U.S. Agency for International Development, Strategic Plan .

Strategic Communication and Public Diplomacy IPC, *U.S. National Strategy for Public Diplomacy and Strategic Communication.*

Books

Clausewitz, Carl von. *On War.* Edited and translated by Michael Howard and Peter Paret. Princeton, NJ: Princeton University Press, 1976.

Franks, Tommy with Malcom McConnell. *American Soldier.* Regan Books, 2004.

Gordon, Michael R. and Trainor, Bernard E. *Cobra II.* Pantheon Books, New York, 2006.

Koontz, Harold and *O'Donnell,* Cyril, *Principles of Management: An Analysis of Managerial Functions*, 5th ed. New York: McGraw-Hill,1972

Reynolds, Nicholas E. *U.S. Marines in Iraq, 2003 Basrah, Bagdad and Beyond: U.S. Marines in the Global War on Terrorism,* History Division, United States Marine Corps, Washington D.C. 2007.

Ricks, Thomas E., *Fiasco, The American Military Adventure in Iraq.* The Penguin Press, New York, 2006.

Strange, Dr. Joe, *Centers of Gravity and Critical Vulnerabilities: Building On The Clausewitzian Foundation So That We Can All Speak The Same Language.* Marine Corps University Foundation, 1996.

Strange, Dr.. Joe, Marine Corps University, *Perspectives on Warfighting,* Number 4, 1996.

Sun Tzu. *The Art of War.* Oxford, Oxford University Press, 1963.

Weigley, Russell F. *The American Way of War: A History of United States Military Strategy and Policy.* Bloomington, Indiana University Press, 1977.

Wright, Donald P and Reese, Timothy R. *On Point II. Transition to the New Campaign: The United States Army in Operation Iraqi Freedom May 2003-January 2005.*

Vego. Dr. Milan. *Joint Operational Warfare*, 20 September 2007.

Service Publications/War Colleges

AFDD 2-5, *Information Operations.*

AFDD 2-6, *Air Mobility Operations.*

AFDD 2-8, *Command and Control.*

Air Force War and Mobilization Plan.

Army Mobilization and Operations Planning and Execution System.

Coast Guard Capabilities Manual and Coast Guard Logistic Support and Mobilization Plan.

Field Manual (FM) 3-0, *Operations.*

FM 3-07, *Stability Operations.*

FM 3-24, MCWP 3-33.5, *Counterinsurgency.*

FM 3-93, *The Army in Theater Operations.*

FM 5-0, *The Operations Process.*

Joint Forces Staff College, Joint Command & Control and Information Operations School, *Joint Information Operations Planning Handbook.*

Marine Corps Capabilities Plan and Marine Corps Mobilization, Activation,

Integration, Deactivation Plan.

Marine Corps Doctrinal Publication-5 (MCDP-5), *Planning.*

Marine Corps Warfighting Publication (MCWP) 5-1, *Marine Corps Planning Process.*

Marine Corps Doctrinal Publication (MCDP) 1-2, *Campaigning.*

National Defense University. *Interagency Management of Complex Crisis Operations Handbook.* January 2003.

Navy Capabilities and Mobilization Plan.

Naval Doctrinal Publication 1, *Naval Warfare.*

Navy Warfare Publication 3-62M, *Seabasing.*

Naval War College, *CDR's Estimate of the Situation (CES).*

Naval War College, NWP 5-01, *Navy Planning.*

Sweeny, Dr. Patrick C., Naval War College, NWC 4111H, *Joint Operations Planning Process Workbook* and NWP 5-01, *Navy Planning.*

Naval War College, 4111H, *Joint Operation Planning Process (JOPP) Workbook.*

U.S. Army War College, *Campaign Planning Primer.*

U.S. Army War College, *Information Operations Primer.*

Articles/Monograph/Thesis

Betts, Richard K. *What Will it Take to Deter the United States?* Parameters, Winter 1995 pp. 70-79.

Burdon, John, Conway, Timothy, Santacroce, Michael A., *The Challenge of Achieving Adaptive Planning and Execution Objectives; Synchronizing with Global Force Management.* 15 September 2011.

Cole, Ronald H., Joint History Office, Office of the Chairman Joint Chiefs of Staff, *Operation Just Cause, Panama. The Planning and Execution of Joint Operations in Panama, February 1988-January 1990.* November 1995.

Cole, Ronald H., Joint History Office, Office of the Chairman Joint Chiefs of Staff, *Operation Urgent Fury, Grenada. The Planning and Execution of Joint Operations in Grenada, 12 October – 2 November 1983.* July 1997.

Conway, Timothy W., Santacroce, Michael A., *Utilizing the Current Allocation Process for Requesting Combat Support Agency (CSA) Support to CCDRs.* 03 November 2010.

Dickson, Keith. *Operational Design: A Methodology for Planners.*

Echevarria, Antulio J. *Clausewitz's Center of Gravity: It's Not What We Thought.* Strategic Studies Institute, U.S. Army War College, Carlisle Barracks, Pa. 2003.

Gardner, David W. *Clarifying Relationships Between Objectives, Effects and End States with Illustrations and Lessons from the Vietnam War.* Joint Advanced Warfighting School. 5 April 2007.

Goodwin, LCDR Ben B. *War Termination and the Gulf War: Can We Plan Better?* Naval War College. 9 February 2004.

Gray, Colin S. *Deterrence Resurrected: Revisiting Some Fundamentals.* Parameters, Summer 1991, pp. 13-21.

Guthrie, Colonel John W. U.S. Army War College, *The Theater Commander: Planning for Conflict Termination.* USMC. 15 March 2006.

Jaffe, Lorna S., Joint History Office, Office of the Chairman Joint Chiefs of Staff, *The Development of the Base Force 1989-1992.* July 1993

Janiczek, Rudolf M. *A Concept at the Crossroads: Rethinking Center of Gravity.* Strategic Studies Institute, U.S. Army War College, Carlisle Barracks, Pa. October 2007.

Kidder, Bruce L., *Center of Gravity: Dispelling the Myths* (Carlisle Barracks, PA. US Army War College, 1996), p.12.

Loney, Timothy J., U.S. Army War College, *Drafting a New Strategy for Public Diplomacy and Strategic Communication.* September 2009.

Luck, Gary, Gen (Ret), *Insights on Joint Operations: The Art and Science, Best Practices, The Move Toward Coherently Integrated Joint, Interagency, and Multinational Operations.* September 2006.

Melshen, Dr. Paul. *Mapping Out a Counterinsurgency Campaign Plan: Critical Considerations in Counterinsurgency Campaigning.* 17 March 2006.

Organizational Development of the Joint Chiefs of Staff 1942-1989, Historical Division, Joint Secretariat, Joint Chiefs of Staff. November 1989.

Poole, Walter S., Joint History Office, Office of the Chairman Joint Chiefs of Staff, *The Effort to Save Somalia August 1992-March 1994.* August 2005.

President of the United States, *A New Beginning, Talking Points.* Cairo, Egypt. June 2009.

Reed, James W., *Should Deterrence Fail: War Termination in Campaign Planning,* Parameters. Summer 1993 pp. 41-52.

Santacroce, Michael A., Conway, Timothy W., Eremita, Nicholas. *Moving Out With Global Force Management.* 23 March, 2010.

Sele, Richard K. *Engaging Civil Centers of Gravity and Vulnerabilities.* Military Review. September-October 2004.

Walsh, Mark R. and Harwood, Michael J., *Complex Emergencies: Under New Management,* Parameters. Winter 1998, pp. 39-50.

Warfare Studies Institute, Joint Air Operations Planning Course, *Joint Air Estimate Planning Handbook,* Version 5, 30 January 2005.

Webb, Willard J. and Poole, Walter S., Joint History Office, Office of the Chairman Joint Chiefs of Staff, The Joint Chiefs of Staff and The Vietnam War 1971-1973. 2007

National Publications/Official Reports/Policy/Web/Other

Adaptive Planning Combatant Commander Organizational Workload to Capacity Study, USJFCOM. July 2009.

"Al-Qa'ida (the Base)," *International Policy Institute for Counter-Terrorism,* on the World Wide Web at www.ict.org.il/inter_ter (accessed 3 April 2002).

Broadcasting Board of Governors Annual Report.

Broadcasting Board of Governors Budget Request, Executive Summary.

Conduct of the Persian Gulf War, Final Report to Congress. April 1992.

Conducting a DG Assessment: A Framework for Strategy Development, Office of Democracy and Governance, Bureau for Democracy, Conflict, and Humanitarian Assistance, U.S. Agency for International Development. November 2000.

Daggett, Stephen. CRS Report for Congress. *Quadrennial Defense Review 2010: Overview and Implications for National Security Planning.* May 17, 2010.

Drea, Edward J., Cole, Ronald H., Poole, Walter S., Schnabel, James F., Watson, Robert J.,

Webb, Willard J., Joint History Office, Office of the Chairman of the Joint Chiefs of Staff, *History of the Unified Command Plan 1946-2012*

Deputy Secretary of Defense Memorandum. *Implementation of the DOD Strategic Communication Plan for Afghanistan.* September 12, 2007.

Global Force Management Policy, Integration and Process Working Group (PIPWG) Session Summary. July 14, 2006.

JCS, J-7 Operational Plans and Joint Force Development Joint Education and Doctrine Division. *Irregular Warfare (IW) Execution Roadmap Task 2.6.8. DOD Cadre of Strategists and Campaign Planners.* 16 February 2007.

Katzman, Kenneth. CRS Report for Congress. *Iraq: Post-Saddam Governance and Security.* January 17, 2007.

Millennium Challenge Act of 2003. Title VI.

National Security Presidential Directive 1 (NSPD-1), *Organization of the National Security Council System.* February 13, 2001

PDD-56 *The Clinton Administration's Policy on Managing Complex Contingency Operations.* May 1997

PDD-1 Organization of the National Security Council System. February 2009.

Perl, Raphael F. CRS Report for Congress. *National Strategy for Combating Terrorism: Background and Issues for Congress.* November 1, 2007.

Report to Congress on the Implementation of DOD Directive 3000.05 Military Support for Stability, Security, Transition and Reconstruction (SSTR) Operations. April 1, 2007.

Secretary of Defense, Memorandum for the Secretaries of the Military Departments, Chairman of the Joint Chiefs of Staff, *Partial Mobilization (World Trade Center and Pentagon Attacks) and Redelegation of Authority under Title 10, USC, Section 123, 123a, 527, 12006, 12011, 12012, 12302, and 12305.* February 2003.

Statement of Michael S. Doran, Deputy Assistant Secretary of Defense, *Support to Public Diplomacy,* Before the Committee on Homeland Security and Governmental Affairs, United States Senate. 3 May 2007.

Strategic Communication and Public Diplomacy Policy Coordinating Committee (PCC). *U.S. National Strategy for Public Diplomacy and Strategic Communication.* December 14, 2006.

The National Security Policy Process: The National Security Council and Interagency System. Alan G. Whittaker, PhD., Frederick C. Smith, and Ambassador Elizabeth McKune. Annual Update: April, 2007.

The Atlantic Council of the United States. Nelson, Richard C., *How Should NATO Handle Stabilization Operations and Reconstruction Efforts?* September 2006

United States Department of State, Counterinsurgency for U.S. Government Policy Makers, A Work in Progress. U.S.G. Interagency Counterinsurgency Initiative. October 2007.

United States Government Draft Planning Framework for Reconstruction, Stabilization, and Conflict Transformation. November 1, 2007.

United States Department of State. *The Budget In Brief.*

United States Department of State. *Congressional Budget Justification Foreign Operations.*

United States Department of State. *Summary and Highlights International Affairs Function 150. Budget Request.*

United States Department of State/United States Agency for International Development, *Revised Strategic Plan.*

United States Government Accountability Office. Report. *U.S. Public Diplomacy. Interagency Coordination Efforts Hampered by the Lack of a National Communication Strategy.* April 2005.

United States Government Accountability Office. Report. *U.S. Public Diplomacy. Actions Needed to Improve Strategic Use and Coordination of Research. July 2007.*

United States Government Accountability Office. Testimony. Ford, Jess T. Director International Affairs and Trade. *U.S. Public Diplomacy. State Department Efforts Lack Certain Communication Elements and Face Persistent Challenges. May 3, 2006.*

U.S. Agency for International Development. *Operating in High Threat Environments.* Spring 2005.

U.S. Agency for International Development. *At Freedom's Frontiers. A Democracy and Governance Strategic Framework.* December 2005.

U.S. Agency for International Development. *Fragile States Strategy.* January 2005.

Working with the Office of U.S. Foreign Disaster Assistance, U.S. Army GTA 41-01-006, October 2007.

Planning Timelines & Dates

1. Times

(C-, D-, M-days end at 2400 hours Universal Time (Zulu time) and are assumed to be 24 hours long for planning.) The Chairman of the Joint Chiefs of Staff normally coordinates the proposed date with the CDRs of the appropriate unified and specified commands, as well as any recommended changes to C-day. L-hour will be established per plan, crisis, or theater of operations and will apply to both air and surface movements. Normally, L hour will be established to allow C-day to be a 24-hour day.

a. C-Day. The unnamed day on which a deployment operation commences or is to commence. The deployment may be movement of troops, cargo, weapon systems, or a combination of these elements using any or all types of transport. The letter "C" will be the only one used to denote the above. The highest command or headquarters responsible for coordinating the planning will specify the exact meaning of C-day within the aforementioned definition. The command or headquarters directly responsible for the execution of the operation, if other than the one coordinating the planning, will do so in light of the meaning specified by the highest command or headquarters coordinating the planning.

b. D-Day. The unnamed day on which a particular operation commences or is to commence.

c. F-Hour. The effective time of announcement by the Secretary of Defense to the Military Departments of a decision to mobilize Reserve units.

d. H-Hour. The specific hour on D-day at which a particular operation commences.

e. H-Hour (amphibious operations). For amphibious operations, the time the first assault elements are scheduled to touch down on the beach, or a landing zone, and in some cases the commencement of countermine breaching operations.

f. I-Day. (CJCSM 3110.01/JSCP). The day on which the Intelligence Community determines that within a potential crisis situation, a development occurs that may signal a heightened threat to U.S. interests. Although the scope and direction of the threat is ambiguous, the Intelligence Community responds by focusing collection and other resources to monitor and report on the situation as it evolves.

g. L-Hour. The specific hour on C-day at which a deployment operation commences or is to commence.

h. L-Hour (Amphibious Operations). In amphibious operations, the time at which the first helicopter of the helicopter-borne assault wave touches down in the landing zone.

i. M-Day. The term used to designate the unnamed day on which full mobilization commences or is due to commence.

j. N-Day. The unnamed day an active duty unit is notified for deployment or redeployment.

k. R-Day. Redeployment day. The day on which redeployment of major combat, combat support, and combat service support forces begins in an operation.

l. S-Day. The day the President authorizes Selective Reserve call-up (not more than 200,000).

m. T-Day. The effective day coincident with Presidential declaration of national emergency and authorization of partial mobilization (not more than 1,000,000 personnel exclusive of the 200,000 call-up).

n. W-Day. Declared by the President, W-day is associated with an adversary decision to prepare for war (unambiguous strategic warning).

2. Indications and Warning

Those intelligence activities intended to detect and report time-sensitive intelligence on foreign developments that could involve a threat to the United States or allied and/or coalition military, political, or economic interests or to U.S. citizens abroad. It includes forewarning of enemy actions or intentions; the imminence of hostilities; insurgency; nuclear/non-nuclear attack on the United States, its overseas forces, or allied and/or coalition nations; hostile reactions to U.S. reconnaissance activities; terrorists' attacks; and other similar events. Also called *I&W*. See also *information*; *intelligence* in JP 2-01, *DOD Dictionary of Military and Associated Terms.*

Commander's Estimate

1. Purpose

a. The commander's estimate, submitted by the supported commander in response to a CJCS WARNORD, provides the CJCS with time-sensitive information for consideration by the NCA in meeting a crisis situation. Essentially, it reflects the supported commander's analysis of the various COAs that may be used to accomplish the assigned mission and contains recommendations as to the best COA (recommended COAs submitted for President, SECDEF approval may be contained in current OPLANs or CONPLANs or may be developed to meet situations not addressed by current plans. Regardless of origin, these COAs will be specifically identified when they involve military operations against a potential enemy). Although the estimative process at the supported commander's level may involve a complete, detailed estimate by the supported commander, the estimate submitted to the CJCS will normally be a greatly abbreviated version providing only that information essential to the President, SECDEF and the CJCS for arriving at a decision to meet a crisis.

b. Supporting commanders normally will not submit a commander's estimate to the CJCS; however, they may be requested to do so by the supported commander. They may also be requested to provide other information that could assist the supported commander in formulating and evaluating the various COAs.

2. When Submitted

a. The Commander's Estimate will be submitted as soon as possible after receipt of the CJCS WARNORD, but no later than the deadline established by the CJCS in the WARNORD. Although submission time is normally 72 hours, extremely time-sensitive situations may require that the supported commander respond in 4 to 8 hours.

b. Follow-on information or revisions to the Commander's Estimate should be submitted as necessary to complete, update, or refine information included in the initial estimate.

c. The supported commander may submit a Commander's Estimate at the commander's own discretion, without a CJCS WARNORD, to advise the SECDEF and CJCS of the commander's evaluation of a potential crisis situation within the AOR. This situation may be handled by a SITREP instead of a Commander's Estimate.

3. How Submitted

The Commander's Estimate is submitted by record communication, normally with a precedence of IMMEDIATE or FLASH, as appropriate. GCCS Newsgroup should be used initially to pass the commander's estimate but must be followed by immediate record communication to keep all crisis participants informed.

4. Addressees

The message is sent to the CJCS with information copies to the Services, components, supporting commands and combat support agencies, USTRANSCOM, and other appropriate commands and agencies.

5. Contents

a. TheCommander's Estimate will follow the major headings of a commander's estimate of the situation as outlined in Appendix A to Enclosure J but will normally be substantially abbreviated in content. As with the WARNORD, the precise contents may vary widely, depending on the nature of the crisis, time available to respond, and the applicability of prior planning. In a rapidly developing situation, a formal Commander's Estimate may be initially impractical, and the entire estimative process may be reduced to a commander's conference, with corresponding brevity reflected in the estimate when submitted by record communications to the CJCS. Also, the existence of an applicable OPLAN may already reflect most of the necessary analysis.

b. The essential requirement of the Commander's Estimate submitted to the CJCS is to provide the SECDEF in a timely manner, with viable military COAs to meet a crisis. Normally, these will center on military capabilities in terms of forces available, response time, and significant logistic considerations. In the estimate, one COA will be recommended. If the supported commander desires to submit alternative COAs, an order of priority will be established. All COAs in the WARNORD will be addressed.

c. The estimate of the supported commander will include specific information to the extent applicable. The following estimate format is desirable but not mandatory and may be abbreviated where appropriate.

(1) Mission. State the assigned or deduced mission and purpose. List any intermediate tasks, prescribed or deduced, that the supported commander considers necessary to accomplish the mission.

(2) Situation and COA. This paragraph is the foundation of the estimate and may encompass considerable detail. Because the CJCS is concerned primarily with the results of the estimate rather than the analysis, for purposes of the estimate submitted, include only the minimum information necessary to support the recommendation.

(a) Considerations Affecting the Possible COA. Include only a brief summary, if applicable, of the major factors pertaining to the characteristics of the area and relative combat power that have a significant impact on the alternative COAs.

(b) Enemy Capability. Highlight, if applicable, the enemy capabilities and psychological vulnerabilities that can seriously affect the accomplishment of the mission, giving information that would be useful to the President, SECDEF, and the CJCS in evaluating various COAs.

(c) Terrorist Threat. Describe potential terrorist threat capabilities to include force protection requirements (prior, during, and post mission) that can affect the accomplishment of the mission.

(d) Own COA. List COAs that offer suitable, feasible, and acceptable means of accomplishing the mission. If specific COAs were prescribed in the WARNORD, they must be included. For each COA, the following specific information should be addressed:

1. Combat forces required; e.g., 2 FS, 1 airborne brigade. List actual units if known.

2. FP.

3. Destination.

4. Required delivery dates.

5. Coordinated deployment estimate.

6. Employment estimate.

7. Strategic lift requirements, if appropriate.

(3) Analysis of Opposing COA. Highlight enemy capabilities that may have significant impact on U.S. COAs.

(4) Comparison of Own COA. For the submission to the CJCS, include only the final statement of conclusions and provide a brief rationale for the favored COA. Discuss the advantages and disadvantages of the alternative COAs, if significant, in assisting the President, SECDEF, and the CJCS in arriving at a decision.

(5) Recommended COA. State the supported commander's recommended COA (recommended COA should include any recommended changes to the ROE in effect at that time) (CJCSM 3122.01)).

SAMPLE COMMANDER'S ESTIMATE

IMMEDIATE (OR FLASH AS APPROPRIATE)
FROM: COMUSCENTCOM MACDILL AFB FL
TO: CJCS WASHINGTON DC
INFO: CSA WASHINGTON DC
CNO WASHINGTON DC
CSAF WASHINGTON DC
CMC WASHINGTON DC
COMUSELEMNORAD PETERSON AFB CO
COMUSEUCOM VAIHINGEN GE
HQ AMC SCOTT AFB IL//CC//

COMUSPACOM HONOLULU HI
COMUSNORTHCOM PETERSON AFB CO
COMUSSOUTHCOM MIAMI FL
DIRNSA FT GEORGE G MEADE MD

DISTR: COMBATANT COMMANDER/DCOM/CCJ1/CCJ2/CCJ3/CCJ4/7/CCJ5/CCJ6
DRAFTER: LTC CHUCK SWANSON, USA CCJ7, EXT 53046
COMUSSTRATCOM OFFUTT AFB NE
COMUSSTRATCOM OFFUTT AFB NE
COMUSSOCOM MACDILL AFB FL
COMUSTRANSCOM SCOTT AFB IL
DISA WASHINGTON DC
DIA WASHINGTON DC
DLA FT BELVOIR VA
DIRECTOR DTRA FAIRFAX VA
CIA WASHINGTON DC
NGA HQ BETHESDA MD
COMSDDC FALLS CHURCH VA
COMSC WASHINGTON DC
COMDT COGARD WASHINGTON DC//G-OPF/G-OPD//
COMUSARCENT FT MCPHERSON GA
USCENTAF SHAW AFB SC//CC//
COMUSNAVCENT
COMLANTFLT NORFOLK VA
CORMARFORLANT
COMPACFLT PEARL HARBOR HI
COMPACAF HICKAM AFB HI
CORMARFORPAC
COMUSNAVEUR LONDON UK
C L A S S I F I C A T I O N
OPER/BLUENOSE//
MSGID/COMESTIMATE/COMUSCENTCOM//
REF/A/ORDER/CJCS/211742ZNOV _____ /____/NOTAL//
AMPN/CJCS Warning Order//
REF/B/DOC/USCENTCOM OPLAN XXXX//
AMPN/USCENTCOM OPLAN FOR CONTINGENCY OPERATIONS XXXX.//GENTEXT/
MISSION/

1. ( ) MISSION. WHEN DIRECTED BY THE SECDEF, USCENTCOM COMMANDER WILL CONDUCT MILITARY OPERATIONS IN SUPPORT OF THE GOVERNMENT OF BLUELAND (GOB) TO PROTECT AND DEFEND BLUELAND STRONG POINTS AND LINES OF COMMUNICATION (LOCS).//
GENTEXT/SITUATION/
2. ( ) SITUATION

A. ( ) THE INTERNAL STABILITY AND SECURITY OF BLUELAND AND ORANGELAND HAVE DETERIORATED BECAUSE OF CONTINUED YELLOWLAND SUPPORT OF THE REBEL FORCES SEEKING THE OVERTHROW OF THE GOVERNMENT. TENSIONS BETWEEN YELLOWLAND, BLUELAND, AND ORANGELAND HAVE BEEN HIGH BECAUSE OF OVERT YELLOWLAND SUPPORT OF THE COUP ATTEMPT, YELLOWLAND ARMS SHIPMENTS TO THE REBELS, AND A RECENT ALLIANCE OF HERETOFORE ANTAGONISTIC REBEL FORCES. ALL OF THESE ACTIONS AGAINST BLUELAND AND ORANGELAND BY YELLOWLAND REQUIRE PRUDENT CONSIDERATION OF POSSIBLE IMPLEMENTATION OF USCENTCOM OPLAN XXXX.

B. ( ) ASSIGNED AND SUPPORTING FORCES ARE IN ACCORDANCE WITH CURRENT USCENTCOM OPLAN XXXX.//

GENTEXT/ENEMY CAPABILITIES/
3. <statement on enemy capabilities>//

GENTEXT/OPERATIONAL CONSTRAINTS/
4. <list operational constraints>

GENTEXT/CONCEPT OF OPERATIONS/
5. <summary of concept of operations>

HEADING/COURSES OF ACTION/
GENTEXT/OWN COURSES OF ACTION/
6. ( ) USCENTCOM COMMANDER HAS DEVELOPED THE FOLLOWING COURSES OF ACTIONS (COAs):

A. ( ) COA 1. DEPLOY AND EMPLOY FORCES IN ACCORDANCE WITH USCENTCOM OPLAN XXXX TPFDD. TACTICAL FIGHTER AND RECONNAISSANCE WING TO USE BABA AFB AS MAIN OPERATING BASE. I MEF TO DEPLOY VIA STRATEGIC AIR TO JOIN WITH MPS EQUIPMENT. CVBG TO OPERATE MODLOC VIA SOUTHEASTERN SEA. TWO ARMY BDES DEPLOY TO PORT WASI VIA STRATEGIC AIR TO JOIN WITH EQUIPMENT SHIPPED BY SEA. SUBSEQUENT MILITARY ASSISTANCE OPERATIONS TO BE CONDUCTED AS REQUESTED BY GOB TO INCLUDE, BUT NOT BE LIMITED TO, NONCOMBATANT EVACUATION OPERATIONS (NEO), SHOW OF FORCE, AND PROTECTION AND DEFENSE OF BLUELAND STRONG POINTS AND LOCS.

B. ( ) COA 2. DEPLOY AND EMPLOY AIR FORCE AND NAVAL FORCES IN ACCORDANCE WITH USCENTCOM OPLAN XXXX. HOLD MEF AND ARMY BDES ON CALL. SUBSEQUENT MILITARY OPERATIONS TO BE CONDUCTED AS REQUESTED BY GOB.
GENTEXT/OPPOSING COURSES OF ACTION/

7. ( ) ANALYSIS OF OPPOSING COA. ENEMY CAPABILITIES CANNOT SIGNIFICANTLY DELAY SUCCESSFUL EXECUTION OF US MILITARY OPERATIONS UNDER EITHER COA. UNDER COA 2, HOWEVER, THERE IS AN INCREASED POSSIBILITY OF TERRORIST VIOLENCE AGAINST ISOLATED AMERICANS IN RETALIATION FOR US FORCE ARRIVAL. ARRIVAL OF SMALL AIR FORCE AND NAVAL FORCE PACKAGES

FOR SHOW OF FORCE RESTRICTS COMMANDERS POTENTIAL TO CONDUCT NEOS OR DEFENSIVE OPERATIONS WITHOUT GROUND FORCES.

GENTEXT/COMPARISON OF OWN COURSES OF ACTION/
8. ( ) COMPARISON OF OWN COAs

A. ( ) COA 1 PROVIDES FOR SIMULTANEOUS EMPLOYMENT OF THE ENTIRE TASK FORCE AND IS THE MOST DESIRABLE FOR TACTICAL EXECUTION. THE INITIAL PRESENCE OF AIR FORCE AND NAVAL FORCES COUPLED WITH THE ARRIVAL OF THE CG, I MEF (FORWARD) AND ASSOCIATED EQUIPMENT ABOARD MPS, PROVIDES CONSIDERABLE FLEXIBILITY FOR RAPID INSERTION OF SECURITY FORCES AS REQUIRED BY GOB. THIS COA REQUIRES THE LONGEST RESPONSE TIME (__ DAYS AIRLIFT AND __ DAYS SEALIFT (DEPLOYMENT ESTIMATE) FOR CLOSURE OF THE ENTIRE TASK FORCE. EMPLOYMENT COULD BEGIN IMMEDIATELY.

B. ( ) COA 2 HAS ADVANTAGE OF MOST RAPID RESPONSE (____ DAYS AIRLIFT AND ____ DAYS SEALIFT (DEPLOYMENT ESTIMATE)) FOR AIR FORCE AND NAVAL FORCES. IT PROVIDES FOR A REPRESENTATIVE FORCE TO BE ABLE TO RESPOND TO GOB AND DEMONSTRATE US RESOLVE IN AREA. ITS PRIMARY DISADVANTAGE IS THAT ALL GROUND FORCES ARE ON CALL. HOWEVER, RESPONSE TIME FOR MEF AND ARMY BDES COULD BE MINIMAL AS MPS AND MSC SHIPS COULD BE IN MODLOC POSITION OFF COAST OF PORT WASI PRIOR TO DEPLOYMENT OF PERSONNEL.
COAIDENT/

9. ( ) DECISION. RECOMMEND COA 1.

10. ( ) GENTEXT/OPERATIONAL OBJECTIVE/
GENTEXT/ADDITIONAL INFORMATION/ FORCE, LOGISTIC, AND TRANSPORTATION REQUIREMENT DETAILS HAVE BEEN LOADED INTO THE JOINT OPERATION PLANNING AND EXECUTION SYSTEM (JOPES) AND ARE AVAILABLE UNDER PLAN IDENTIFICATION NUMBER (PID) XXXXT (COA 1) AND PID XXXXU (COA 2).//

DECL/<source for classification>/<reason for classification>/<downgrade instructions or date>/<downgrading or declassification exemption code>//

(CJCSM 3122.01A, JOPES Volume I, Enclosure J)

Operational Plan Annexes

A Task Organization
B Intelligence
C Operations
D Logistics
E Personnel
F Public Affairs
G Civil-Military Affairs
H Meteorological and Oceanographic Services
J Command Relationships
K Communications System Support
L Environmental Considerations
M Geospatial Information and Services
N Space Operations
P Host Nation Support
Q Medical Service
R Reports
S Special Technical Operations
T Consequence Management
U Notional OPLAN Decision Guide
V Interagency Coordination
X Execution Checklist
Y Strategic Communication
Z Distribution
AA Religious Support

Annexes A-D, K, and Y are required annexes for a CAP OPORD per JOPES. All others may either be required by the JSCP or deemed necessary by the supported CCDR.

Command Relationships

I. Authorities

1. Levels of Authority. The specific command relationship (combatant command (command authority) (COCOM), operational control (OPCON), tactical control (TACON), and support) will define the level of authority a commander has over assigned or attached forces. A commander can also have authority when coordinating authority, administrative control (ADCON), and direct liaison authorized (DIRLAUTH) relationships have been specified. An overview of command relationships is shown in Figure E-1.[1]

Categories of Support

General Support
That support which is given to the supported force as a whole rather then to a particular subdivision thereof.

Mutual Support
That support which units render each other against an enemy because of their assigned tasks, their position relative to each other and to the enemy, and their inherent capabilities.

Direct Support
A mission requiring a force to support another specific force and authorizing it to answer directly to the supported force's request for assistance.

Close Support
That action of the supporting force against targets or objectives that are sufficiently near the supported force as to require detailed integration or coordination of the supporting action with the fire, movement, or other actions of the supported force.

Figure E-1. Command Relationships.

2. All forces under the jurisdiction of the Secretaries of the Military Departments (except those forces necessary to carry out the functions of the Military Departments) are assigned to CCMDs or CDR, U.S. Element North American Aerospace Defense

Command (NORAD) (USELEMNORAD) by the SECDEF in the "Forces for Unified Commands" memorandum. A force assigned or attached to a CCMD may be transferred from that command to another CCDR only when directed by the SECDEF and under procedures prescribed by the SECDEF and approved by the President. The command relationship the gaining commander will exercise (and the losing commander will relinquish) will be

[1] Extracted from JP 1, FM 3-31 and NWC 4111H.

specified by the SECDEF. Establishing authorities for subordinate unified commands and JTFs may direct the assignment or attachment of their forces to those subordinate commands and delegate the command relationship as appropriate (see Figure E-2).

Command Relationships Overview

- *Forces, not command relationships, are transferred between commands. When forces are transferred, the command relationship the gaining commander will exercise (and the losing commander will relinquish) over those forces must be specified.*
- *When transfer of forces to a joint force will be permanent (or for an unknown but long period of time) the forces should be reassigned. CCDRs will exercise COCOM (command authority) and subordinate JFCs will exercise operational control (OPCON) over reassigned forces.*
- *When transfer of forces to a joint force will be temporary, the forces will be attached to the gaining command and JFCs, normally through the Service component commander, will exercise OPCON over the attached forces.*
- *Establishing authorities for subordinate unified commands and JTFs direct the assignment or attachment of their forces to those subordinate commands as appropriate.*

Figure E-2. Command Relationships Overview.

3. The CCDR exercises combatant command (command authority) (COCOM) over forces assigned or reassigned by the President or SECDEF. Forces are assigned or reassigned when the transfer of forces will be permanent or for an unknown period of time, or when the broadest level of command and control (C2) is required or desired. Operational control (OPCON) of assigned forces is inherent in COCOM and may be delegated within the CCMD by the CCDR. Subordinate JFCs will exercise OPCON over assigned or reassigned forces.

4. The CCDR normally exercises OPCON over forces attached by the SECDEF. Forces are attached when the transfer of forces will be temporary. Establishing authorities for subordinate unified commands and JTFs normally will direct the delegation of OPCON over forces attached to those subordinate commands.

5. In accordance with the "Forces for Unified Commands" memorandum and the UCP, except as otherwise directed by the President or the SECDEF, all forces operating within the geographic area assigned to a specific CCDR shall be assigned or attached to, and under the command of, that CCDR. Transient forces do not come under the chain of command of the area commander solely by their movement across operational area boundaries, except when the CCDR is exercising tactical control (TACON) for the purpose of force protection. Unless otherwise specified by the SECDEF, and with the exception of the United States Northern Command (USNORTHCOM) AOR, a CCDR has TACON for exercise purposes whenever forces not assigned to that CCDR undertake exercises in that CCDR's AOR.

6. Brief Summary of U.S. Command Relationships.

a. Combatant Command (Command Authority).

(1) COCOM is the authority of a CCDR to perform those functions of command over assigned forces to include:

- Organizing and employing commands and forces.

- Assigning tasks.
- Designating objectives.
- Giving authoritative direction over all aspects of military operations, joint training.
- Logistics.

(2) COCOM should be exercised through the commanders of subordinate organizations. Normally, this authority is exercised through subordinate JFCs and Service and/or functional component commanders; however, it cannot be delegated to subordinate commanders. COCOM provides full authority to organize and employ commands and forces as the CCDR considers necessary to accomplish assigned missions.

b. Operational Control (OPCON).

(1) OPCON is the command authority exercised by commanders at any echelon at or below the level of COCOM and can be delegated or transferred.

(2) OPCON is inherent within COCOM and is the authority to perform those functions of command over subordinate forces involving:

- Organizing and employing commands and forces.
- Assigning tasks.
- Designating objectives.
- Giving authoritative direction necessary to accomplish the mission.

(3) OPCON includes authoritative direction over all aspects of military operations and joint training necessary to accomplish missions assigned to the command. It should be exercised through the commanders of subordinate organizations; normally, this authority is exercised through subordinate JFCs and Service and/or functional component commanders. OPCON normally provides full authority to organize commands and forces and employ those forces necessary to accomplish assigned missions. It does not include authoritative direction for logistics or matters of administration, discipline, internal organization, or unit training. The CCDR delegates these elements. OPCON does include the authority to delineate functional responsibilities and geographic JOAs of subordinate JFCs.

(4) The superior commander gives commanders of subordinate commands and JTFs OPCON of assigned or attached forces.

c. Tactical Control (TACON).

(1) TACON is the command authority over assigned or attached forces or commands, or military capability or forces made available for tasking. It is limited to the detailed and usually local direction and control of movements or maneuvers necessary to accomplish assigned missions or tasks.

(2) TACON may be delegated to and exercised by commanders at any echelon at or below the level of COCOM. TACON is inherent in OPCON.

d. Support.

(1) Support is a command authority. A support relationship is established by a superior commander between subordinate commanders when one organization should aid, protect, complement, or sustain another force.

(2) Support may be exercised by commanders at any echelon at or below the level of COCOM. This includes the President/SECDEF designating a support relationship between CCDRs as well as within a COCOM. The designation of supporting relationships is important as it conveys priorities to commanders and staffs who are planning or executing joint operations. The support command relationship is a flexible arrangement. The establishing authority is responsible for ensuring that both the supported and supporting commanders understand the degree of authority granted the supported commander.

(3) The supported commander should ensure that the supporting commander understands the assistance required. The supporting commander provides the assistance needed, subject to the supporting commander's existing capabilities and other assigned tasks. When the supporting commander cannot fulfill the needs of the supported commander, the establishing authority is notified by either the supported or supporting commander. The establishing authority is responsible for determining a solution.

(4) An establishing directive is normally issued to specify the purpose of the support relationship, the effect desired, and the action to be taken.

e. Direct Liaison Authorized (DIRLAUTH). Direct liaison authorized (DIRLAUTH) is that authority granted by a commander (any level) to a subordinate to directly consult or coordinate an action with a command or agency within or outside of the granting command. DIRLAUTH is more applicable to planning than operations and always carries with it the requirement of keeping the commander granting DIRLAUTH informed. DIRLAUTH is a coordination relationship, not an authority through which command may be exercised.

f. Functional Component Support Relationships.

(1) The Joint Force Land Component Commander (JFLCC) can be in either a supporting or supported relationship or both. The JFC's needs for unity of command and unity of effort dictate these relationships. Support relationships will be established by the JFC in appropriate campaign plans and orders. Similar relationships can be established among all functional and Service component commanders, such as the coordination of deep operations involving the JFLCC and the joint force air component commander (JFACC). Close coordination is necessary when the JFLCC provides joint suppression of enemy air defenses in support of JFACC operations. Examples are attack helicopters or multiple-launched rocket systems in Operation DESERT STORM as well as seizing and holding ports and airbases for friendly air and sea forces (such as in Operation JUST CAUSE). The JFLCC can also expect support to include airlift, close air support (CAS), and interdiction strikes from the JFACC.

(2) The JFC may task the JFLCC to conduct operations outside of the land AO. Land-based elements may conduct air and missile defense operations to protect the force and critical assets from air and missile attack and surveillance. These may include operational maneuver and/or operational fires against enemy ports and airbases outside of the land area of operations (AO). Similarly, the JFLCC can request from the JFC air support from other components to attack or isolate enemy land forces in the land AO. Figure E-3 illustrates a simultaneous support relationship scenario between the JFLCC and JFACC.

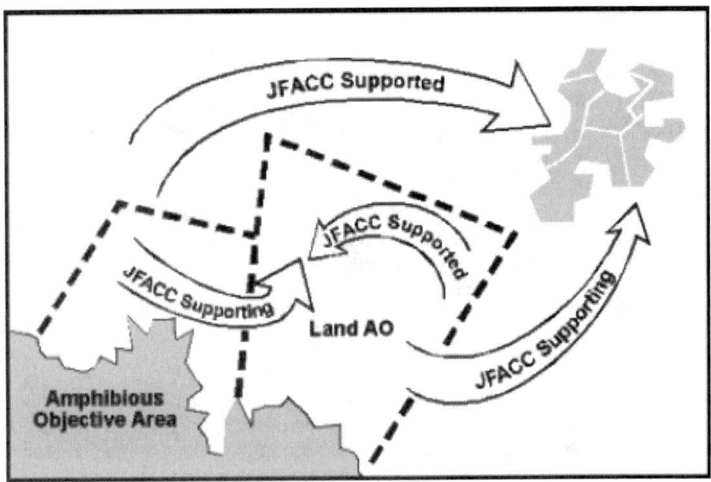

Figure E-3. JFLCC and JFACC Support Relationships.

g. Command Relationships with Service Components.

(1) The JFLCC functional component responsibility is normally assigned to a commander already serving as a Service component (e.g., ARFOR, MARFOR) to a JTF or subordinate unified command. Additionally, the JFC may use one of his Service components (e.g., Army Service component or Marine Service component) as the JFLCC reporting to him directly. The JFLCC retains Service component responsibility for assigned or attached forces but does not assume Service component responsibility for forces made available by other Service components. TACON is the normal relationship with these Service forces. In those cases in which the JFLC command is not formed from a Service component headquarters, the JFLCC has no Service component responsibilities (see Figure E-4).

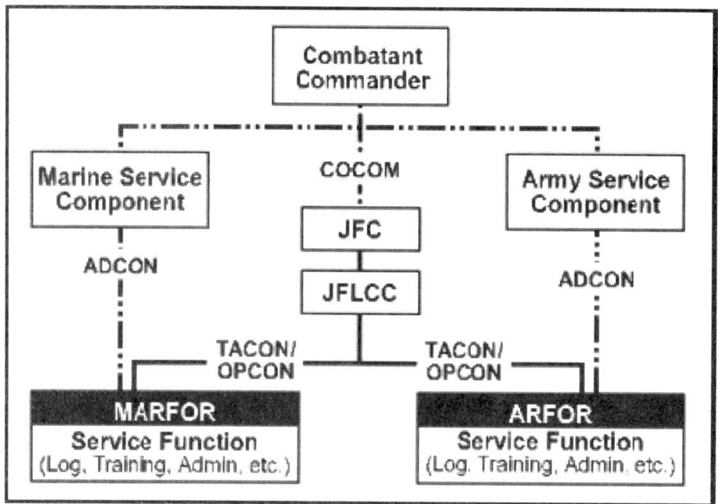

Figure E-4. Service Functions.

(2) Once the JFLC command is established, the operational requirements of the JFLCC subordinate commands are prioritized and presented to the joint force headquarters by the JFLCC. However, Service component commanders remain responsible for their military department Title 10 responsibilities, such as logistics and personnel support.

II. Detailed Description of Command Relationships

1. COCOM

COCOM is the command authority over assigned forces vested only in the commanders of CCMDs by Title 10, United States Code (USC), Section 164 (or as directed by the President in the Unified Command Plan [UCP]) and cannot be delegated or transferred.

a. Basic Authority. COCOM is the authority of a CCDR to perform those functions of command over assigned forces involving organizing and employing commands and forces, assigning tasks, designating objectives, and giving authoritative direction over all aspects of military operations, joint training, and logistics necessary to accomplish the missions assigned to the command. COCOM should be exercised through the commanders of subordinate organizations. Normally, this authority is exercised through subordinate JFCs and Service and/or Functional CCDRs (Functional Component Commander (FCCs)). COCOM provides full authority to organize and employ commands and forces as the CCDR considers necessary to accomplish assigned missions. b. Unless otherwise directed by the President or the SECDEF, the authority, direction, and control of the CCDR with respect to the command of forces assigned to that command includes the following:

(1) Exercise or delegate operational control (OPCON), tactical control (TACON), and establish support relationships among subordinate commanders over assigned or attached forces, and designate coordinating authorities, as described in subparagraphs (8), (9), and (10) below.

(2) Exercise directive authority for logistic matters (or delegate directive authority for a common support capability).

(3) Prescribe the chain of command to the commands and forces within the command.

(4) Employ forces within that command as necessary to carry out missions assigned to the command.

(5) Assign command functions to subordinate commanders.

(6) Coordinate and approve those aspects of administration and support, and discipline necessary to carry out missions assigned to the command.

(7) Give authoritative direction to subordinate commands and forces necessary to carry out missions assigned to the command, including authoritative direction over all aspects of military operations, joint training, and logistics.

(8) Coordinate with other CCDRs, United States Government (USG) agencies, and organizations of other countries regarding matters that cross the boundaries of geographic areas specified in the Unified Command Plan (UCP) and inform USG agencies or organizations of other countries in the AOR, as necessary, to prevent both duplication of effort and lack of adequate control of operations in the delineated areas.

(9) Unless otherwise directed by the SECDEF, function as the U.S. military single point of contact and exercise directive authority over all elements of the command in relationships with other CCMDs, DOD elements, U.S. diplomatic missions, other U.S. agencies, and organizations of other countries in the AOR. Whenever a CCDR conducts exercises, operations, or other activities with the military forces of nations in another CCDR's AOR, those exercises, operations, and activities and their attendant command relationships will be mutually agreed to between the CCDRs.

(10) Determine those matters relating to the exercise of COCOM in which subordinates must communicate with agencies external to the CCMD through the CCDR.

(11) Establish personnel policies to ensure proper and uniform standards of military conduct.

(12) Submit recommendations through the CJCS to the SECDEF concerning the content of guidance affecting the strategy and/or fielding of joint forces.

(13) Participate in the Planning, Programming, Budgeting, and Execution process.

(14) Participate in the Joint Strategic Planning System and the Adaptive Planning and Execution System (APEX).

(15) Concur in the assignment (or recommendation for assignment) of officers as commanders directly subordinate to the CCDR and to positions on the CCMD staff. Suspend from duty and recommend reassignment, when appropriate, of any subordinate officer assigned to the CCMD.

(16) Convene general courts-martial in accordance with the Uniform Code of Military Justice (UCMJ).

(17) In accordance with laws and national and DOD policies, establish plans, policies, programs, priorities, and overall requirements for the command and control (C2), communications system, and intelligence, surveillance, and reconnaissance (ISR) activities of the command.

c. Directive Authority for Logistics. CCDRs exercise directive authority for logistics and may delegate directive authority for a common support capability. The CCDR may delegate directive authority for as many common support capabilities to a subordinate JFC as required to accomplish the subordinate JFC's assigned mission. For some commodities or support services common to two or more Services, one Service may be given responsibility for management based on Department of Defense (DOD) executive agent (EA) designations or inter-Service support agreements. However, the CCDR must formally delineate this delegated directive authority by function and scope to the subordinate JFC or Service component commander. The exercise of directive authority for logistics by a CCDR includes the authority to issue directives to subordinate commanders, including peacetime measures necessary to ensure the following: effective execution of approved OPLANs; effectiveness and economy of operation; and prevention or elimination of unnecessary duplication of facilities and overlapping of functions among the Service component commands. CCDRs will coordinate with appropriate Services before exercising directive authority for logistics or delegate authority for subordinate commanders to exercise common support capabilities to one of their components.

(1) A CCDR's directive authority does not:

(a) Discontinue Service responsibility for logistic support;

(b) Discourage coordination by consultation and agreement; or

(c) Disrupt effective procedures or efficient use of facilities or organizations.

(2) Unless otherwise directed by the SECDEF, the Military Departments and Services continue to have responsibility for the logistic support of their forces assigned or attached to joint commands, subject to the following guidance.

(a) Under peacetime conditions, the scope of the logistic authority exercised by the commander of a CCMD will be consistent with the peacetime limitations imposed by legislation, DOD policy or regulations, budgetary considerations, local conditions, and other specific conditions prescribed by the SECDEF or the CJCS. Where these factors preclude execution of a CCDR's directive by component CDRs, the comments and recommendations of the CCDR, together with the comments of the component commander concerned, normally will be referred to the appropriate Military Department for consideration. If the matter is not resolved in a timely manner with the appropriate Military Department, it will be referred by the CCDR, through the CJCS, to the SECDEF.

(b) Under crisis action, wartime conditions, or where critical situations make diversion of the normal logistic process necessary, the logistic authority of CCDRs enables them to use all facilities and supplies of all forces assigned to their commands as necessary for the accomplishment of their missions. The President or SECDEF may extend this authority to attached forces when transferring those forces for a specific mission and should specify this authority in the establishing directive or order. Joint logistic doctrine and policy developed by the CJCS establishes wartime logistic support guidance to assist the CCDR in conducting successful joint operations.

2. Operational Control (OPCON)

OPCON is the command authority that may be exercised by commanders at any echelon at or below the level of CCMD and may be delegated within the command. When forces are transferred between CCMDs, the command relationship the gaining commander will exercise (and the losing commander will relinquish) over these forces must be specified by the SECDEF.

a. Basic Authority. Operational control (OPCON) is inherent in COCOM and is the authority to perform those functions of command over subordinate forces involving organizing and employing commands and forces, assigning tasks, designating objectives, and giving authoritative direction necessary to accomplish the mission. OPCON includes authoritative direction over all aspects of military operations and joint training necessary to accomplish missions assigned to the command. It should be exercised through the commanders of subordinate organizations; normally, this authority is exercised through subordinate JFCs and Service and/or functional CCDRs or functional component commanders. OPCON normally provides full authority to organize commands and forces and employ those forces as the commander considers necessary to accomplish assigned missions. It does not include authoritative direction for logistics or matters of administration, discipline, internal organization, or unit training. These elements of COCOM must be specifically delegated by the CCDR. OPCON does include the authority to delineate functional responsibilities and operational areas of subordinate JFCs.

b. Commanders of subordinate commands, including JTFs, normally will be given OPCON of assigned or attached forces by the superior commander.

c. OPCON conveys the authority for the following:

(1) Exercise or delegate OPCON and tactical control (TACON), establish support relationships among subordinates, and designate coordinating authorities.

(2) Give direction to subordinate commands and forces necessary to carry out missions assigned to the command, including authoritative direction over all aspects of military operations and joint training.

(3) Prescribe the chain of command to the commands and forces within the command.

(4) Organize subordinate commands and forces within the command as necessary to carry out missions assigned to the command.

(5) Employ forces within the command, as necessary, to carry out missions assigned to the command.

(6) Assign command functions to subordinate commanders.

(7) Plan for, deploy, direct, control, and coordinate the actions of subordinate forces.

(8) Establish plans, policies, priorities, and overall requirements for the intelligence, surveillance, and reconnaissance (ISR) activities of the command.

(9) Conduct joint training and joint training exercises required to achieve effective employment of the forces of the command, in accordance with joint doctrine established by the CJCS, and establish training policies for joint operations required to accomplish the mission. This authority also applies to forces attached for purposes of joint exercises and training.

(10) Suspend from duty and recommend reassignment of any officer assigned to the command.

(11) Assign responsibilities to subordinate commanders for certain routine operational matters that require coordination of effort of two or more commanders.

(12) Establish an adequate system of control for local defense and delineate such operational areas for subordinate commanders as deemed desirable.

(13) Delineate functional responsibilities and geographic operational areas of subordinate commanders.

d. The SECDEF may specify adjustments to accommodate authorities beyond OPCON in an establishing directive when forces are transferred between CCDRs or when members and/or organizations are transferred from the Military Departments to a CCMD. Adjustments will be coordinated with the participating CCDRs.

3. Tactical Control (TACON)

TACON is the command authority over assigned or attached forces or commands, or military capability or forces made available for tasking, that is limited to the detailed direction and control of movements or maneuvers within the operational area necessary to accomplish assigned missions or tasks.

a. Basic Authority. TACON is inherent in OPCON and may be delegated to and exercised by commanders at any echelon at or below the level of CCMD. When forces are transferred between CCDRs, the command relationship the gaining commander will exercise (and the losing commander will relinquish) over those forces must be specified by the SECDEF.

b. TACON provides the authority to:

(1) Give direction for military operations; and

(2) Control designated forces (e.g., ground forces, aircraft sorties, missile launches, or satellite payload management).

c. TACON provides sufficient authority for controlling and directing the application of force or tactical use of combat support assets within the assigned mission or task. TACON does not provide organizational authority or authoritative direction for administrative and logistic support; the commander of the parent unit continues to exercise these authorities unless otherwise specified in the establishing directive.

d. Functional component commanders typically exercise TACON over military capability or forces made available to the functional component for tasking.

4. Support

Support is a command authority. A support relationship is established by a superior commander between subordinate commanders when one organization should aid, protect, complement, or sustain another force.

a. Basic Authority. Support may be exercised by commanders at any echelon at or below the CCMD level. This includes the SECDEF designating a support relationship between CCDRs as well as within a CCMD. The designation of supporting relationships is important as it conveys priorities to commanders and staffs that are planning or executing joint operations. The support command relationship is, by design, a somewhat vague but very flexible arrangement. The establishing authority (the common superior commander) is responsible for ensuring that both the supported commander and supporting commanders understand the degree of authority that the supported commander is granted.

b. The supported commander should ensure that the supporting commanders understand the assistance required. The supporting commanders will then provide the assistance needed, subject to a supporting commander's existing capabilities and other assigned tasks. When a supporting commander cannot fulfill the needs of the supported commander, the establishing authority will be notified by either the supported commander or a supporting commander. The establishing authority is responsible for determining a solution.

c. An establishing directive normally is issued to specify the purpose of the support relationship, the effect desired, and the scope of the action to be taken. It also should include:

(1) The forces and resources allocated to the supporting effort;

(2) The time, place, level, and duration of the supporting effort;

(3) The relative priority of the supporting effort;

(4) The authority, if any, of the supporting commander to modify the supporting effort in the event of exceptional opportunity or an emergency; and

(5) The degree of authority granted to the supported commander over the supporting effort.

d. Unless limited by the establishing directive, the supported commander will have the authority to exercise general direction of the supporting effort. General direction includes the designation and prioritization of targets or objectives, timing and duration of the supporting action, and other instructions necessary for coordination and efficiency.

e. The supporting commander determines the forces, tactics, methods, procedures, and communications to be employed in providing this support. The supporting commander will advise and coordinate with the supported commander on matters concerning the employment and limitations (e.g., logistics) of such support, assist in planning for the integration of such support into the supported commander's effort as a whole, and ensure that support requirements are appropriately communicated within the supporting commander's organization.

f. The supporting commander has the responsibility to ascertain the needs of the supported force and take action to fulfill them within existing capabilities, consistent with priorities and requirements of other assigned tasks.

g. Several categories of support have been defined to better characterize the support that should be given. For example, land forces that provide fires normally are tasked in a direct support role.

h. There are four defined categories of support that a CCDR may direct over assigned or attached forces to ensure the appropriate level of support is provided to accomplish mission objectives. These include general support, mutual support, direct support, and close support. Figure E-5 summarizes each of the categories of support. The establishing directive will specify the type and extent of support the specified forces are to provide.

Categories of Support

General Support
That support which is given to the supported force as a whole rather then to a particular subdivision thereof.

Mutual Support
That support which units render each other against an enemy because of their assigned tasks, their position relative to each other and to the enemy, and their inherent capabilities.

Direct Support
A mission requiring a force to support another specific force and authorizing it to answer directly to the supported force's request for assistance.

Close Support
That action of the supporting force against targets or objectives that are sufficiently near the supported force as to require detailed integration or coordination of the supporting action with the fire, movement, or other actions of the supported force.

Figure E-5. Categories of Support.

III. Support Relationships Between Combatant Commands

1. The SECDEF establishes support relationships between the CCDRs for the planning and execution of joint operations. This ensures that the tasked CCDR(s) receives the necessary support. A supported CCDR requests capabilities, tasks supporting DOD components, coordinates with the appropriate Federal agencies (where agreements have been established), and develops a plan to achieve the common goal. As part of the team effort, supporting CCDRs provide the requested capabilities, as available, to assist the supported CCDR to accomplish missions requiring additional resources.

2. The CJCS organizes the joint planning and execution community for joint operation planning to carry out support relationships between the CCMDs. The supported CCDR has primary responsibility for all aspects of an assigned task. Supporting CCDRs provide forces, assistance, or other resources to a supported CCDR. Supporting CCDRs prepare supporting plans as required. Under some circumstances, a CCDR may be a supporting CCDR for one operation while being a supported CCDR for another.

IV. Support Relationships Between Component Commands

1. The JFC may establish support relationships between component commanders to facilitate operations. Support relationships afford an effective means to prioritize and ensure unity of effort for various operations. Component commanders should establish liaison with other component commanders to facilitate the support relationship and to coordinate the planning and execution of pertinent operations. Support relationships may change across phases of an operation as directed by the establishing authority.

2. The operational requirements of the FCCs subordinate forces are prioritized and presented to the JFC by the FCC, relieving the affected Service component commanders of this responsibility, but the affected Service component commanders are not relieved of their administrative and support responsibilities.

3. In rare situations, a supporting component commander may be supporting two or more supported commanders. In these situations, there must be clear understanding among all parties and a specification in the establishing directive, as to who supports whom, when, and with what prioritization. When there is a conflict over prioritization between component commanders, the CCDR having COCOM of the component commanders will have final adjudication.

V. Other Authorities Outside the Command Relationships

1. Administrative Control

Administrative control (ADCON) is the direction or exercise of authority over subordinate or other organizations with respect to administration and support, including organization of Service forces, control of resources and equipment, personnel management, logistics, individual and unit training, readiness, mobilization, demobilization, discipline, and other matters not included in the operational missions of the subordinate or other organizations. ADCON is synonymous with administration and support responsibilities identified in Title 10, USC. This is the authority necessary to fulfill Military Department statutory responsibilities for administration and support. ADCON may be delegated to and exercised by commanders of Service forces assigned to a CCDR at any echelon at or below the level of Service component command. ADCON is subject to the command authority of CCDRs. ADCON may be delegated to and exercised by commanders of Service commands assigned within Service authorities. Service commanders exercising ADCON will not usurp the authorities assigned by a CCDR having COCOM over commanders of assigned Service forces.

2. Coordinating Authority

Commanders or individuals may exercise coordinating authority at any echelon at or below the level of CCMD. Coordinating authority is the authority delegated to a commander or individual for coordinating specific functions and activities involving forces of two or more Military Departments, two or more joint force components, or two or more forces of the same Service (e.g., joint security coordinator exercises coordinating authority for joint security area operations among the component commanders). Coordinating authority may be granted and modified through a memorandum of agreement to provide unity of command and unity of effort for operations involving, RC, and AC forces engaged in interagency activities. The commander or individual has the authority to require consultation between the agencies involved but does not have the authority to compel agreement. The common task to be coordinated will be specified in the establishing directive without disturbing the normal organizational relationships in other matters. Coordinating authority is a consultation relationship between commanders, not an authority by which command may be exercised. It is more applicable to planning and similar activities than to operations. Coordinating authority is not in any way tied to force assignment. Assignment of coordinating authority is based on the missions and capabilities of the commands or organizations involved.

3. Direct Liaison Authorized (DIRLAUTH)

Direct liaison authorized (DIRLAUTH) is that authority granted by a commander (any level) to a subordinate to directly consult or coordinate an action with a command or agency within or outside of the granting command. DIRLAUTH is more applicable to planning than operations and always carries with it the requirement of keeping the commander granting DIRLAUTH informed. DIRLAUTH is a coordination relationship, not an authority through which command may be exercised.

VI. Command of National Guard and Reserve Units

All National Guard and reserve forces (except those forces specifically exempted) are assigned by the SECDEF to the CCMDs under the authority provided in Title 10, USC, Sections 162 and 167, as indicated in the "Forces for Unified Commands" memorandum. However, those forces are available for operational missions only when mobilized for specific periods in accordance with the law, or when ordered to active duty and after being validated for employment by their parent Service.

a. The authority CCDRs may exercise over assigned Reserve Component (RC) forces when not on active duty or when on active duty for training is Training and Readiness Oversight (TRO). CCDRs normally will exercise TRO over assigned forces through the Service component commanders. TRO includes the authority to:

(1) Provide guidance to Service component commanders on operational requirements and priorities to be addressed in Military Department training and readiness programs;

(2) Comment on Service component program recommendations and budget requests;

(3) Coordinate and approve participation by assigned Reserve Component (RC) forces in joint exercises and other joint training when on active duty for training or performing inactive duty for training; (4) Obtain and review readiness and inspection reports on assigned Reserve Component (RC) forces; and

(5) Coordinate and review mobilization plans (including post-mobilization training activities and deployability validation procedures) developed for assigned Reserve Component (RC) forces.

b. Unless otherwise directed by the SECDEF, the following applies:

(1) Assigned Reserve Component (RC) forces on active duty (other than for training) may not be deployed until validated by the parent Service for deployment.

(2) CCDRs may employ Reserve Component (RC) forces assigned to their subordinate component commanders in contingency operations only when the forces have been mobilized for specific periods in accordance with the law, or when ordered to active duty and after being validated for employment by their parent Service.

(3) Reserve Component (RC) forces on active duty for training or performing inactive-duty training may be employed in connection with contingency operations only as provided by law, and when the primary purpose is for training consistent with their mission or specialty.

c. CCDRs will communicate with assigned Reserve Component (RC) forces through the Military Departments when the RC forces are not on active duty or when on active duty for training.

d. CCDRs may inspect assigned Reserve Component (RC) forces in accordance with Department of Defense directive (DODD) 5106.4, Combatant Command Inspectors Genera, when such forces are mobilized or ordered to active duty (other than for training).

VII. U.S. vs. Alliance Command Relationships

Figure E-6, on the following page, offers a comparison between U.S. command relationships and the two alliance command relationships of NATO and CFC/USFK.

U.S. vs. Alliance Command Relationships

Figure E-6 offers a comparison between U.S. command relationships and the two alliance command relationships of NATO and CFC/USFK.

Authority	US COCOM	US OPCON	NATO OPCOM	NATO OPCON	CFC/USFK COMBINED OPCON	NATO TACOM	US & NATO TACON
	Most control						Least control
Direct authority to deal with DOD, US diplomatic missions, agencies	X						
Coordinate CINC boundary	X						
Granted to a command	X		X				
Delegated to a command		X		X	X	X	X
Set chain of command to forces	X	X					
Assign mission/designate objective	X	X	X				
Assign tasks	X	X	X			X	
Direct/employ forces	X	X	X	X	X		
Establish maneuver control measures	X	X	X	X	X	X	X
Reassign forces	X						
Retain OPCON	X	X	X				
Delegate OPCON	X	X	X	X with approval			
Assign TACOM	X	X					
Delegate TACON	X	X	X	X	X		
Retain TACON	X	X	X	X			
Deploy forces (information/within theater)	X	X	X	X			
Local direction/control designated forces	X	X					X
Assign separate employment of unit components	X	X	X				
Directive authority for logistics	X						
Direct joint training	X	X					
Exercise command of US forces in MNF	X	X					
Assign/reassign subordinate commanders/officers	X	May suspend or recommend reassignment					
Conduct internal discipline/training	X						

NATO *Full Command* and CFC/USFK *Command less OPCON* are basically equivalent to US COCOM, but only for internal matters

X – has this authority

[shaded] – denied this authority, or not specifically granted it

LEGEND

COCOM – Combatant command
OPCON – Operational control
OPCOM – Operational command
TACOM – Tactical command
TACON – Tactical control

Figure E-6. U.S. vs. Alliance Command Relationships.

Chain of Command

NCA

President

Secretary of Defense

CJCS

Military Department

Unified Commands

Service Forces (not assigned by "Forces For")

COCOM

Joint Task Force

Functional Component

Service Component Commands

OPCON

Service Component/ Forces

Subordinate Unified Command

OPCON

Functional Component

Service Component/ Forces

Joint Task Force

OPCON/TACON

Service Component/ Forces

OPCON/TACON

Service Component/ Forces

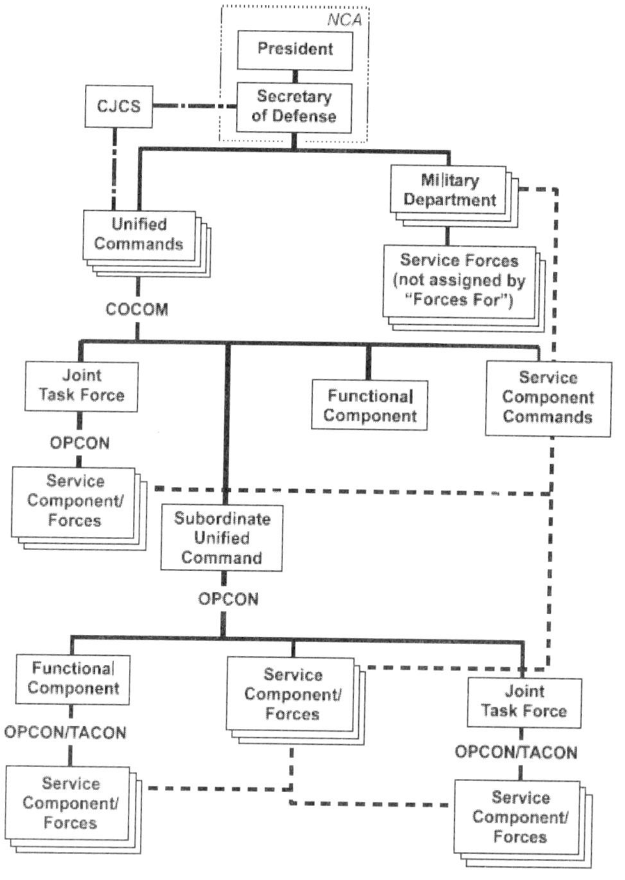

Note:
This diagram is only an example. It does not prescribe joint force organization. Service components at lower echelons may only contain service forces.

LEGEND
- Chain of Command
- Administrative Control
- Channel of Communication

The President exercises authority and control of the Armed Forces through a single chain of command with two branches. One branch goes from the SECDEF through CCDRs to Service component commands and subordinate forces. It is for conducting operations and support. The other branch comes from the SECDEF, through the military departments, to their respective major commands.[1]

1 FM 3-0, Operations

National Strategic Direction

Ref: JP 1, Doctrine for the Armed Forces of the United States (May '07), chap. II.

National strategic direction is governed by the Constitution, federal law, USG policy regarding internationally-recognized law and the national interest. This direction leads to unified action. At the strategic level, unity of effort requires coordination among government departments and agencies within the executive branch, between the executive and legislative branches, with NGOs, IGOs, the private sector, and among nations in any alliance or coalition.

The President of the United States

The President of the United States, advised by the NSC, is responsible to the American people for national strategic direction. When the United States undertakes military operations, the Armed Forces of the United States are only one component of a national-level effort involving all instruments of national power. Instilling unity of effort at the national level is necessarily a cooperative endeavor involving a number of Federal departments and agencies. In certain operations, agencies of states, localities, or foreign countries may also be involved.

The Secretary of Defense (SecDef)

The SecDef is responsible to the President for creating, supporting, and employing military capabilities. The SecDef provides authoritative direction, and control over the Services through the Secretaries of the Military Departments. SecDef exercises control of and authority over those forces not specifically assigned to the combatant commands and administers this authority through the Military Departments, the Service Chiefs, and applicable chains of command. The Secretaries of the Military Departments organize, train, and equip forces to operate across the range of military operations and provide for the administration and support of all those forces within their department, including those assigned or attached to the CCDRs.

The Chairman of the Joint Chiefs of Staff (CJCS)

The CJCS is the principal military adviser to the President, the NSC, and SecDef and functions under the authority of the President and the direction and control of the President and SecDef, and oversees the activities of the CCDRs as directed by SecDef. Communications between the President or the SecDef and the CCDRs are normally transmitted through the CJCS.

The Commanders of Combatant Commands (CCDRs)

Commanders of combatant commands exercise combatant command (COCOM) over assigned forces and are responsible to the President and SecDef for the performance of assigned missions and the preparedness of their commands to perform assigned missions.

The US Ambassador

In a foreign country, the US ambassador is responsible to the President for directing, coordinating, and supervising all USG elements in the HN, except those under the command of a CCDR. GCCs are responsible for coordinating with US ambassadors in their geographic AOR across the range of military operations, and for negotiating memoranda of agreement (MOAs) with the chiefs of mission in designated countries to support military operations.

Refer to The Joint Forces Operations & Doctrine SMARTbook (Guide to Joint, Multinational & Interagency Operations) for complete discussion of joint operations, unified action, and the range of military operations. Additional topics include joint doctrine fundamentals, joint operation planning, joint logistics, joint task forces, information operations, multinational operations, and IGO/NGO coordination.

Governing Factors & Evaluation Critieria

The following menu provides a good starting point for developing a COA comparison criteria list. Some possible sources for determining criteria are:

a. CCDR's intent statement.

b. CJTF's intent statement.

c. CJTF's subsequent guidance.

d. Implicit significant factors relating to the operation (e.g., need for speed, security).

e. Each staff member may identify factors relating to that staff function.

f. Other factors such as:

(1) Principles of war. COAs should provide for:

- Mass.
  - Masses friendly strengths against enemy weaknesses.
  - Synchronizes fires, maneuver, protections and support.
  - Creates combat asymmetries.
  - Concentrates results (not necessarily forces).
- Objective.
  - Directly, quickly, and economically contributes to the purpose of the operation.
  - Clearly defined, decisive, and attainable.
- Offensive.
  - Seizes, retains, exploits the initiative.
  - Attacks enemy center(s) of gravity.
  - Maintains freedom of action.
  - Deters aggression.
  - Robust and versatile reserves.
- Simplicity.
  - Easily integrated with adjacent operations.
  - Does not require extensive preconditions prior to execution.
  - Simplified combat identification procedures.
- Economy of force.
  - Minimizes forces assigned to secondary efforts.
  - Minimizes force support and sustainment.
- Maneuver.
  - Gains positional advantage over the enemy.
  - Positional advantage usually provides positions to deliver, or threaten to deliver, direct or indirect fires.
  - Keeps the enemy off balance and protects the friendly forces.
  - Preserves freedom of action.

- Unity of command.
  - Command relationships are well defined.
  - Minimizes requirements for real-time coordination.
  - Forces and operations are under a single commander with requisite authority.
- Security (and force protection).
  - Accomplishes the mission without provocation.
  - Reduces friendly vulnerabilities, enhances freedom of action.
  - Provides for prudent risk management.
- Surprise.
  - Effective IO/IW (especially deception and OPSEC).
  - Avoids stereotypical operations.
  - Applies unexpected combat power (lethal or nonlethal).

(2) Elements of operational art (JP 3-0). COAs should provide for:

- Synergy. Integration and synchronization of operations in a manner that applies force from different dimensions to shock, disrupt, and defeat opponents.
- Simultaneity and depth. Bring forces to bear on the opponent's entire structure in a near simultaneous manner to overwhelm and cripple enemy capabilities and the will to resist.
- Anticipation. The unexpected/opportunities to exploit the situation.
- Balance. The appropriate mix of forces/capabilities within the JTF, and the nature and timing of operations to disrupt an enemy's balance.
- Leverage. COAs gain, maintain, and exploit advantages in combat power across all dimensions.
- Timing and tempo. COAs conduct operations at a tempo and point in time that best exploit friendly capabilities and inhibit the enemy.
- Operational reach and approach. COAs provide basing, whether from overseas locations, sea-based platforms, or the continental United States, which directly affect operational reach. In particular, advanced bases underwrite the progressive ability of the JTF force to shield its components from enemy action and deliver symmetric and asymmetric blows with increasing power and ferocity.
- Forces and functions. COAs provide campaigns/operations that focus on defeating enemy forces or functions, or combination of both.
- Arranging operations. COAs provide a combination of simultaneous and sequential operations to achieve the desired endstate conditions quickly and at the least cost in personnel and other resources.
- Centers of gravity. COAs provide the ability to mass results against the enemy's sources of power in order to destroy or neutralize them.
- Direct versus indirect. COAs, to the extent possible, attack enemy centers of gravity directly. Where direct attack means attacking into an opponent's strength, seek an indirect approach.
- Decisive points. COAs should correctly identify and control decisive points that can gain a marked advantage over the enemy, and greatly influence the outcome of an action.
- Culmination. COAs synchronize logistics with combat operations.
- Termination. COAs account for the endstate of the operation.

(3) Other factors, e.g., political constraints, risk, financial costs, flexibility.

Acronyms & Abbreviations

A

AADC Area Air Defense Commander
AAR After-Action Report/Review
AAW Antiair Warfare
AC Active Component
ACA Airspace Control Authority
ACT Advanced Civilian Teams
ADMIN Administration
ADVON Advanced Echelon
ADP Automated Data Processing
AFIS Armed Forces Information Service
AFFOR Air Force Forces
AI Air Interdiction
ALCON All Concerned
ALERTORD Alert Order
AMC Air Mobility Command; Army Materiel Command
AMCIT American Citizen
AMEMB American Embassy
AMHS Automated Message Handling System
AMPE Automated Message Processing Exchange
AO Area of Operations
AOA Amphibious Objective Area
AOC Air Operations Center (USAF)
AOI Area of Interest
AOR Area of Responsibility
AP Adaptive Planning
APEX Adaptive Planning and Execution
APMT Automated Planning and Management Tools
APOD Aerial Port of Debarkation
APOE Aerial Port of Embarkation
ARFOR Army Forces
ASD Assistant Secretary of Defense
ASD (PA) Assistant Secretary of Defense (Public Affairs)
ASOC Air Support Operations Center
ATDS Airborne Tactical Data System
ATO Air Tasking Order
AUTODIN Automatic Digital Network

B

BBG Broadcast Board of Governors
BCE Battlefield Coordination Element
BDE Brigade
BMD Ballistic Missile Defense

C

C2 Command and Control
C2W Command and Control Warfare
C2WC Command and Control Warfare Commander
C3 Command, Control, and Communications
C3I Command, Control, Communications, and Intelligence
C3IC Coalition, Coordination, Communications, and Integration Center
C4 Command, Control, Communications, and Computers
C4I Command, Control, Communications, Computers, and Intelligence
C4S Command, Control, Communications, and Computer Systems
C-day Unnamed day on which a deployment operation begins
CA Civil Affairs
CAP Crisis Action Planning
CAS Close Air Support
CAT Crisis Action Team
CATF Commander Amphibious Task Force
CCDR Combatant Commander
CCMD Combatant Command
CCIR Commander's Critical Information Requirements
CC Critical Capabilities
CCO Complex Contingency Operation
CDR Commander
C-E Communications-Electronics
CE Communications-Electronics; Command Element (MAGTF)
CEOI Communications-Electronics Operating Instructions
CESP Civil Engineer Support Plan
CF Critical Factors

CHAP	Chaplain
CIA	Central Intelligence Agency
CJA	Comprehensive Joint Assessment
CJCS	Chairman of the Joint Chiefs of Staff
CJCSI	Chairman of the Joint Chiefs of Staff Instruction
CJCSM	Chairman of the Joint Chiefs of Staff Manual
CJTF	Commander Joint Task Force
CJTMP	CJCS Joint Training Master Plan
CLF	Commander Landing Forces; Combat Logistics Force
CMBT	Combatant
CMCS	Civil Military Coordination Section
CMCoord	Civil Military Coordination, UN Humanitarian
CMD	Command
CMDT	Commandant
CMO	Civil-Military Operations
CMOC	Civil-Military Operations Center
CMST	Collection Management Support Tools
CNA	Computer Network Attack
CNO	Computer Network Operations
COA	Course of Action
COCOM	Combatant Command (Command Authority)
COE	Common Operating Environment
COG	Center of Gravity
COIN	Counterinsurgency
COLISEUM	Community On-line Intelligence System for End Users Managers
COMARFOR	Commander of Army Forces
COMINT	Communications Intelligence
COMMARFOR	Commander of Marine Forces
COMMZ	Communication Zone
COMPT	Comptroller
COMPUSEC	Computer Security
COMSEC	Communications Security
CONOPS	Concept of Operations
CONPLAN	Operation Plan in Concept Format
CONUS	Continental United States
COS	Chief of Staff
CPG	Contingency Planning Guidance
CPX	Command Post Exercise
CR	Critical Requirements
CSA	Combat Support Agencies
CS	Civil Support, Combat Support
CSS	Combat Service Support
CSSA	Combat Service Support Area
CSSE	Combat Service Support Element (MAGTF)
CT	Counter Terrorism
CTAPS	Contingency Theater Automated Planning System
CV	Critical Vulnerabilities

D

D-day	Unnamed day on which operations commence or are scheduled to commence
D3A	Decide, Detect, Deliver and Assess
DASD(P)	Deputy Assistant Secretary of Defense for Plans
DCA	Defensive Counterair
DCM	Deputy Chief of Mission
DCJTF	Deputy Commander JTF
DCS	Defense Communications System
DCTN	Defense Commercial Telecommunication Network
DDN	Defense Data Network
DEPORD	Deployment Order
DHL	Dalsey, Hillblom and Lynn
DHS	Department of Homeland Security
DIA	Defense Intelligence Agency
DIME	Diplomatic, Informational, Military, Economic
D-IO	Defensive Information Operations
DIR	Director
DIRLAUTH	Direct Liaison Authorized
DIRMOBFOR	Director of Mobility Forces
DISA	Defense Information Systems Agency
DISN	Defense Information Systems Network
DISUM	Defense Intelligence Summary; Daily Intelligence Summary
DJIOC	Defense Joint Intelligence Operations Center
DJ-3	Joint Staff Director of Operations
DJ-7	Joint Staff Director for Operational Plans and Joint Force Development
DLA	Defense Logistics Agency
DMS	Defense Message System
DNSO	Defense Network Systems Organization
DOD	Department of Defense
DODD	Department of Defense Directive
DODIIS	Department of Defense Intelligence Information System
DODI	Department of Defense Instruction

DOJ	Department of Justice
DOS	Department of State
DOT	Department of Transportation
DP	Decision Point
DPG	Defense Planning Guidance
DSB	Defense Science Board
DSCA	Defense Support of Civil Authorities
DSN	Defense Switched Network
DTRA	Defense Threat Reduction Agency

E

EAP	Emergency Action Procedures
EBO	Effects based operations
ECOA	Enemy Course of Action
EEFI	Essential Elements of Friendly Information
EEI	Essential Elements of Information
eJMAPS	Electronic Joint Manpower and Personnel System
ELINT	Electronics Intelligence
EMC	Electromagnetic Compatibility
EMP	Electromagnetic Pulse
EMPINT	Electromagnetic Pulse Intelligence
EMS	Electromagnetic Spectrum
ENDEX	Exercise Termination
ENGR	Engineer
EO	Electro-Optical
EOD	Explosive Ordnance Disposal
ES	Electronic Warfare Support
ESF	Economic Support Fund
EW	Electronic Warfare
EWO	Electronic Warfare Officer
EXORD	Execute Order

F

F-hour	Time of announcement by DECDEF to mobilize Reserve units
FA	Feasibility Assessment
FAA	Federal Aviation Administration; Foreign Assistance Act
FAD	Feasible Arrival Date; Force Activity Designator
FCC	Functional Component Command
FDO	Flexible Deterrent Option
FEMA	Federal Emergency Management Agency
FFIR	Friendly Force Information Requirements
FHA	Foreign Humanitarian Assistance
FID	Foreign Internal Defense
FISINT	Foreign Instrumentation Signals Intelligence
FM	Field Manual
FMFM	Fleet Marine Force Manual
FOB	Forward Operations Base
FRAG	Fragmentation
FRAGO	Fragmentation Order
FSA	Forward Support Area
FSCL	Fire Support Coordination Line
FSCOORD	Fire Support Coordinator
FSE	Fire Support Element
FSN	Foreign Service National
FSSG	Force Service Support Group (Marine Air-Ground Task Force)
FTN	Force Tracking Number

G

GAO	General Accounting Office
GCE	Ground Combat Element (MAGTF)
GCC	Geographic Combatant Commander
GCCS	Global Command and Control System
GCSS	Global Combat Support System
GDP	Global Defense Posture
GDF	Guidance for Development of the Force
GEF	Guidance for Employment of the Force
GEOINT	Geospatial Intelligence
GENSER	General Service (message)
GENTEXT	General Text
GFM	Global Force Management
GFMAP	Global Force Management Allocation Plan
GFMB	Global Force Management Board
GFMIG	Global Force Management Implementation Guidance
GIRH	Generic Intelligence Requirements Handbook (USMC)
GI&S	Geospatial Information and Services
GOH	Government of Haiti
GTN	Global Transportation Network
GWOT	Global War on Terrorism

H

H-hour	Specific time an operation or exercise begins
HA	Humanitarian Assistance
HCA	Humanitarian and Civic Assistance

HN	Host Nation
HNG	Host-Nation Government
HNS	Host-Nation Support
HOC	Humanitarian Operations Center
HPT	High-Priority/Payoff Target(s)
HQ	Headquarters
HSS	Health Service Support
HUMINT	Human Intelligence
HVT	High-Value Target(s)

I

I&W	Indication and Warning
IA	Interagency; Information Assurance
IAW	In Accordance With
ICN	Interface Control Network
ICRC	International Committee of the Red Cross
IDAD	Internal Defense and Development
IDHS	Intelligence Data Handling System
IED	Improved Explosives Devices
IES	Imagery Exploitation System
IEW	Intelligence and Electronic Warfare
IFF	Identification, Friend or Foe
IGO	Intergovernmental Organization
IGPS	Global Positioning System
ILO	In-Lieu-Of
IMINT	Imagery Intelligence
IMO	Intermediate Military Objective
IMS	Interagency Management System
INFOSEC	Information Security
IRR	Individual Ready Reserve
INTELSITSUM	Daily Intelligence Summary
INTSUM	Intelligence Summary
IO	Information Operations, International Organization
IPB	Intelligence Preparation of the Battlespace
IPDS	Imagery Processing and Dissemination System
IPG	Initial Planning Guidance
IPIE	Intelligence Preparation of the Information Environment
IPL	Integrated Priority List
IPR	In-Progress Review
IR	Information Requirements
IRINT	Infrared Intelligence
ISB	Intermediate Staging Base
ISR	Intelligence, Surveillance and Reconnaissance
IT	Information Technology
ITV	In-Transit visibility
IW	Irregular Warfare
I&W	Indications and Warnings

J

J-1	Manpower and Personnel Directorate of a Joint Staff
J-2	Intelligence Directorate of a Joint Staff
J-3	Operations Directorate of a Joint Staff
J-4	Logistics Directorate of a Joint Staff
J-5	Plans Directorate of a Joint Staff
J-6	Command, Control, Communications, and Computer System
JAO	Joint Area of Operations
JAOC	Joint Air Operations Center
JATF	Joint Amphibious Task Force
JAWS	Joint Advanced Warfighting School
JC2WC	Joint Command and Control Warfare Center
JCA	Joint Capability Area
JCAT	Joint Crisis Action Team
JCATF	Joint Civil Affairs Task Force
JCCA	Joint Combat Capability Assessment
JCIOC	Joint Counterintelligence Operations Center
J/CIPS	Joint/Combined Interoperability Planning System
JCMOTF	Joint Civil-Military Operations Task Force
JCS	Joint Chiefs of Staff
JDB	Joint Deployment Board
JDS	Joint Deployment System
JERPAT	Joint Execution Readiness Plan Assessment Tool
JFACC	Joint Force Air Component Commander
JFAST	Joint Flow and Analysis System for Transportation
JFC	Joint Force Commander
JFCC-ISR	Joint Functional Component Command for Intelligence, Surveillance and Reconnaissance (USSTRATCOM)
JFLCC	Joint Force Land Component Commander
JFMCC	Joint Force Maritime Component Commander
JFP	Joint Force Provider
JFSOCC	Joint Force Special Operations Component Commander

JIACG	Joint Interagency Coordination Group
JIA	Joint Individual Augmentee
JIB	Joint Information Bureau
JIC	Joint Intelligence Center
JICG	Joint Information Coordination Group
JIOC	Joint Intelligence Operations Center
JIPOE	Joint Intelligence Preparation of the Operational Environment
JIPTL	Joint Integrated Prioritized Target List
JISE	Joint Intelligence Support Element
JLOTS	Joint Logistics Over-the-Shore
JMC	Joint Movement Center
JMD	Joint Manning Document
JMET	Joint Mission Essential Task
JMETL	Joint Mission Essential Task List
JMO	Joint Maritime Operations
JOA	Joint Operations Area
JOC	Joint Operations Center
JOPES	Joint Operation Planning and Execution System
JOPP	Joint Operation Planning Process
JOWPD	Joint Operational War Plans Division
JP	Joint Pub
JPEC	Joint Planning and Execution Community
JPG	Joint Planning Group
JRA	Joint Rear Area
JRSOI	Joint Reception, Staging, Onward Movement, and Integration
JS	Joint Staff
JSAR	Joint Search and Rescue
JSCP	Joint Strategic Capabilities Plan
JSE	Joint Support Elements
JSOA	Joint Special Operations Area
JSOTF	Joint Special Operations Task Force
JSPS	Joint Strategic Planning System
JTB	Joint Transportation Board
JTF	Joint Task Force
JTF HQ	Joint Task Force Headquarters
JTMD	Joint Theater Missile Defense
JTTP	Joint Tactics, Techniques, and Procedures
JULLS	Joint Universal Lessons Learned System
JWFC	Joint Warfighting Center
JWICS	Joint Worldwide Intelligence Communications System

K

KC-130	Stratotanker
Km	Kilometer

L

L-hour	Specific hour on C-day at which a deployment operation commences or is to commence
LAD	Latest Arrival Date
LAN	Local Area Network
LEA	Law Enforcement Agencies
LFA	Lead Federal Agency
LLO	Logical Lines of Operation
LNO	Liaison Officer
LOAC	Law of Armed Conflict
LOC	Lines of Communications
LOE	Line of Effort
LOI	Letter of Instruction
LOO	Line of Operation
LOTS	Logistics Over-the-Shore
LSA	Logistics Supportability Analysis

M

M-day	Unnamed day on which full mobilization of forces commences or is to commence
MAG	Marine Aircraft Group
MAGTF	Marine Air-Ground Task Force
MARFOR	Marine Corps Forces
MASINT	Measurement and Signature Intelligence
MCC	Millennium Challenge Corporation; Movement Control Center
MCIA	Marine Corps Intelligence Activity
MCOO	Modified Combined Operations Overlay
MDCOA	Most Dangerous Course of Action
MEF	Marine Expeditionary Force
MEU (SOC)	Marine Expeditionary Force (Special Operations Capable)
METT-TC	Mission, Enemy, Terrain, Troops, Time Available and Civilian
MLCOA	Most Likely Course of Action
MFR	Memorandum for the Record
MILDEC	Military Deception
MIO	Maritime Intercept Operations
MISO	Military Information Support Operations
MOE	Measure of Effectiveness
MOOTW	Military Operations Other Than War
MOP	Measure of Performance

MOU	Memorandum of Understanding
MSC	Military Sealift Command
MTT	Mobile Training Team

N

NAI	Named Area of Interest
NAVFOR	Navy Forces
NATO	North Atlantic Treaty Organization
NCR	National Capital Region
NDCS	National Drug Control Strategy
NDS	National Defense Strategy
NEC	National Economic Council
NEO	Noncombatant Evacuation Operation
NGA	National Geospatial-Intelligence Agency
NGB	National Guard Bureau
NGO	Nongovernmental Organization
NIMA	National Imagery and Mapping Agency
NISP	National Intelligence Support Plan
NIST	National Intelligence Support Team
NLT	Not Later Than
NMD	National Missile Defense
NMIST	National Military Intelligence Support Team (DIA)
NMS	National Military Strategy
NOPLAN	No Operation Plan Available or Prepared
NORAD	North American Aerospace Defence Command
NRO	National Reconnaissance Office
NSA	National Security Agency
NSC	National Security Council
NSC/DC	National Security Council/Deputies Committee
NSC/PC	National Security Council/Principals Committee
NSC/IPC	National Security Council/Interagency Policy Committee
NSCS	National Security Council System
NSPD	National Security Presidential Directive
NSS	National Security Strategy
NWP	Naval Warfare Publication

O

OA	Operational Area
OB	Order of Battle
OCOKA	Observation and fields of fire, concealment and cover, obstacles, key terrain, and avenues of approach
OCONUS	Outside the Continental United States
OCJCS	Office of the Chairman of the Joint Chiefs of Staff
OE	Operating Environment
OEF	Operation Enduring Freedom
OFDA	Office of Foreign Disaster Assistance
OGA	Other Governmental Agency
OIC	Officer In Charge
OIF	Operation Iraqi Freedom
O/O	On Order
OOB	Order of Battle
OOTW	Operations Other Than War
ONA	Operational Net Assessment
OPCON	Operational Control
OPFOR	Opposing Forces
OPG	Operations Planning Group
OPLAN	Operation Plan
OPORD	Operation Order
OPREP	Operational Report
OPSEC	Operations Security
OPTINT	Optical Intelligence
OSAD (PA)	Office of the Assistant Secretary of Defense (Public Affairs)
OSD	Office of the Secretary of Defense
OSINT	Open-Source Intelligence
OUSD(P)	Office of the Under Secretary of Defense for Policy

P

P&S	Peace and Security
PA	Public Affairs
PAG	Public Affairs Guidance
PAO	Public Affairs Office; Public Affairs Officer
PAT	Public Affairs Team
PB	Peace Building
PC	Promote Cooperation
PD	Presidential Directive
PDD	Presidential Decision Directive
PEL	Priority Effects List
PEO	Peace Enforcement Operations
PERMREP	U.S. Permanent Representative to the United Nations
PGM	Precision-Guided Munitions
PHIBGRU	Amphibious Group
PHIBRON	Amphibious Squadron
PIR	Priority Intelligence Requirements
PKO	Peacekeeping Operations
PLANORD	Planning Order

PM	Bureau of Political-Military Affairs; Provost Marshal
PMESII	Political, military, economic, social, information, infrastructure
PMIS	Psychological Operations Management Information Subsystem
PO	Peace Operations
POC	Point of Contact
POD	Port of Debarkation
POE	Port of Embarkation
POL	Petroleum, Oil, and Lubricants
POLAD	Political Advisor
POLMIL	Political-Military
PPBE	Planning, Programing, Budgeting and Execution
PPD	Presidential Policy Document
PRT	Provincial Reconstruction Team
PSYOP	Psychological Operations
PTDO	Prepare to Deploy Order
PVO	Private Volunteer Organizations

Q

QDR	Quadrennial Defense Review

R

RATE	Refinement, Adaptation, Termination, or Execution
RC	Reserve Component
RDD	Required Delivery Date (at destination)
RFF	Request for Forces
RFI	Request for Information
RMS	Requirements Management System
ROC	Rehearsal of Concept
ROE	Rules of Engagement
ROK	Republic of Korea

S

SA	Situational Awareness
SAO	Security Assistance Organization
SAT	Satellite
SATCOM	Satellite Communications
SC	Strategic Communication
SCG	Security Cooperation Guidance
SCI	Sensitive Compartmented Information
SCIF	Sensitive Compartmented Information Facility
S/CRS	Office of the Coordinator for Reconstruction and Stabilization
SCP	Subordinate Campaign Plans
SDIO	Strategic Defense Initiative Organization
SDOB	Secretary of Defense Operations Book
SECDEF	Secretary of Defense
SF	Special Forces
SGS	Strategic Guidance Statements
SIG	Signal
SIGINT	Signals Intelligence
SINCGARS	Single-channel and Airborne Radio System
SIPRNET	Secret Internet Protocol Router Network
SITREP	Situation Report
SJA	Staff Judge Advocate
SJFHQ(CE)	Standing Joint Force Headquarters (Command Element)
SLOC	Sea Line of Communication
SME	Subject Matter Expert
SO	Special Operations
SOC	Special Operations Command
SOF	Special Operations Forces
SOFA	Status of Forces Agreement
SOP	Standing Operating Procedures
SPECOPS	Special Operations
SPG	Strategic Planning Guidance
SPOD	Seaport of Debarkation
SPOE	Seaport of Embarkation
SSTR	Stability, Security, Transition and Reconstruction
SVC	Service(s)
SVS	Secure Voice System

T

TACAIR	Tactical Air
TACON	Tactical Control
TACP	Tactical Air Control Party
TAD	Temporary Additional Duty (non-unit related personnel)
TADC	Tactical Air Direction Center
TADIL	Tactical Digital Information Link
TAI	Target Area of Interest
TALO	Theater Airlift Liaison Officer
TAOC	Tactical Air Operations Center (USMC)
TAOR	Tactical Area of Responsibility
TBD	To Be Determined
TBM	Tactical Ballistic Missile
TBMCS	Theater Battle Management Core System

TCP Theater Campaign Plan
TCN Third Country National
TDY Temporary Duty
TF Task Force
TNG Training
TOE Table of Organization and Equipment
TPFDD Time-Phased Force and Deployment Data
TPFDL Time-Phased Force and Deployment List
TSG Theater Security Guidance
TTP Tactics, Techniques, and Procedures

U

UAV Unmanned Aerial Vehicle
UCMJ Uniform Code of Military Justice
UCP Unified Command Plan
UJT Universal Joint Task
UJTL Universal Joint Task List
ULN Unit Line Number
UN United Nations
UNAAF United Action Armed Forces
UNITAF Unified Task Force
U.S. United States
USA United States Army; United States of America
USAR United States Army Reserve
USACIDC United States Army Criminal Investigations Command
USAF United States Air Force
USAFR United States Air Force Reserve
USAFCOM United States African Command
USAID United States Agency for International Development
USCENTCOM United States Central Command
USCG United States Coast Guard
USCGR United States Coast Guard Reserve
USDAO United States Defense Attaché's Office
USD(P) Under Secretary of Defense for Policy
USDR United States Defense Representative
USELEMNORAD United States Element North American Aerospace Defense Command
USEUCOM United States European Command
USFJ United States Forces Japan
USFK United States Forces Korea
USFORAZORES United States Forces Azores
USG United States Government
USIA United States Information Agency
USMC United States Marine Corps
USMCR United States Marine Corps Reserve
USMILGP United States Military Group
USMTM United States Military Training Mission
USN United States Navy
USNORTHCOM United States Northern Command
USNR United States Navy Reserve
USPACOM United States Pacific Command
USSOCOM United States Special Operations Command
USSOUTHCOM United States Southern Command
USSS United States Signals Intelligence (SIGINT) System
USSTRATCOM United States Strategic Command
USTRANSCOM United States Transportation Command
UW Unconventional Warfare

V

VCJCS Vice Chairman of the Joint Chiefs of Staff
VOCO Voice Orders of the Commander
VTC Video Teleconferencing

W

WARNORD Warning Order
WMD Weapons of Mass Destruction
WX Weather

Y

YR Year

Z

Z Zulu
ZULU Time Zone Indicator for Universal Time

Index

E

F

G

H

I

J

K

L

M

N

O

Related SMARTbooks (Joint/Interagency)

Joint operation planning consists of planning activities associated with joint military operations by combatant commanders (CCDRs) and their subordinate joint force commanders (JFCs) in response to contingencies and crises. It transforms national strategic objectives into activities of joint forces. The Department of Defense (DOD) conducts interorganizational coordination across a range of operations, with each type of operation involving different communities of interests and structures. This is especially pronounced for domestic and foreign operations, which are governed by different authorities and have considerably different US Government (USG) governing structures and stakeholders.

The Joint Forces Operations & Doctrine SMARTbook (3rd Rev. Ed.)

Guide to Joint, Multinational & Interagency Operations

Designed for use by ALL SERVICES and JOINT FORCES across the tactical, operational and strategic levels of war.

Joint/Interagency Smartbook 1: Joint Strategic & Operational Planning

Joint planning integrates military actions with those of other instruments of national power and our multinational partners in time, space, and purpose to achieve a specified endstate.

Joint/Interagency Smartbook 2: Interagency Planning & Process

The Department of Defense conducts interorganizational coordination across a range of operations, with each type of operation involving different communities of interests and structures.

Joint/Interagency Smartbook 3: Official Visits, Protocol & Liaison

In diplomatic services and governmental fields, protocols specify the proper and generally accepted behavior in matters of state and diplomacy. This is a guide designed for international business, government, and military professionals.

View, download samples and purchase online at: **www.TheLightningPress.com.** Join our SMARTnews mailing list to receive email notification of SMARTupdates, member-only discounts, new titles & revisions to your SMARTbooks!

---

www.TheLightningPress.com